Containers for Developers Handbook

A practical guide to developing and delivering applications using software containers

Francisco Javier Ramírez Urea

‹packt›

BIRMINGHAM—MUMBAI

Containers for Developers Handbook

Group Product Manager: Preet Ahuja

Publishing Product Manager: Suwarna Rajput

Book Project Manager: Neil D'mello

Senior Editor: Shruti Menon

Technical Editor: Nithik Cheruvakodan

Copy Editor: Safis Editing

Language Support Editor: Safis Editing

Proofreader: Safis Editing

Indexer: Manju Arasan

Production Designer: Prashant Ghare

DevRel Marketing Coordinator: Rohan Dobhal

Senior Marketing Coordinator: Linda Pearlson

First published: October 2023

Production reference: 1311023

Published by Packt Publishing Ltd.

Grosvenor House

11 St Paul's Square

Birmingham

B3 1RB, UK

ISBN 978-1-80512-798-7

www.packtpub.com

To my wife, Raquel, my kids, Jorge and Andrea, and my friends and colleagues at SatCen.

– Francisco Javier Ramírez Urea

Contributors

About the author

Francisco Javier Ramírez Urea has been working in the IT industry since 1999 after receiving his bachelor's degree in physical sciences. He worked as a systems administrator for years before specializing in the monitoring of networking devices, operating systems, and applications. In 2015, he started working with software container technologies as an application architecture consultant and has developed his career and evolved with these technologies ever since. He became a Docker Captain in 2018 alongside being a Docker and Kubernetes teacher with various certifications (CKA, CKS, Docker DCA and MTA, and RedHat RHCE/RHCSA). He currently works at the European Union Satellite Centre as a SecDevOps officer.

I want to thank the people who have been close to me and supported me, especially my wife, Raquel, and my kids, Jorge and Andrea. I also want to thank my colleagues at SatCen for their patience and encouragement throughout the long process of writing this book.

About the reviewer

Luis Carlos Sampaio is a systems engineer with over a decade of experience managing servers. Since 2010, he has acquired expertise in Red Hat, Ubuntu, and Debian and has obtained certifications such as RHCE, RHCSA, and CCNP. Luis has extensive experience in diverse private and hybrid cloud architectures, working with platforms such as Amazon Web Services and Azure. His proficiency extends to DevOps, with hands-on experience in tools such as Ansible, Docker, and Kubernetes. Luis is passionate about sharing his technical insights and experiments with the community. In his leisure time, he enjoys visiting new cities.

I'd like to extend my deepest gratitude to my wonderful wife, Yolanda, and my terrific and smart son, Bruno Miguel.

Yolanda, not only are you my best friend, but your unwavering love and support have been the pillars upon which I've leaned throughout this journey. Without you by my side, completing the review of this book would have been an insurmountable challenge.

Table of Contents

3

Sharing Docker Images   81

4

Running Docker Containers   109

5

Creating Multi-Container Applications   149

Part 2: Container Orchestration

6

Fundamentals of Container Orchestration   185

7

Orchestrating with Swarm 199

8

Deploying Applications with the Kubernetes Orchestrator 231

Part 3: Application Deployment

9

Implementing Architecture Patterns   275

10

Leveraging Application Data Management in Kubernetes   309

11

Publishing Applications   337

12

Part 4: Improving Applications' Development Workflow

13

Preface

This book is a practical introduction to creating applications using containers. Readers will learn what containers are and why they are a new application deployment standard. We will start the journey with key concepts about containers, their features and usage, and how they will help us secure and speed up an application's life cycle. You will also learn how to create secure container images for your applications and how to share and run them in different development stages and production.

The book is divided into different parts to help you use the software container technology. First, you will learn how to use a host environment to build and run your applications using multiple components, and then you will learn how to run them when distributed in complex container orchestrators.

This book focuses on the Kubernetes container orchestrator because of its unique features and popularity. You will learn how to use different Kubernetes resources to securely accomplish different application architecture models. You will also find different software options in this book that will help you create and test your applications on your desktop, using fully functional Kubernetes-packaged platforms. We will teach you how to use these Kubernetes desktop environments to prepare your applications, using the best security practices to deliver them on any Kubernetes platform. We will cover important topics such as data management (configurations, sensitive data, and application data), different mechanisms to publish applications, and application observability (monitoring, logging, and traceability).

Finally, we will show you how to automate the building, testing, and delivery of your applications, creating continuous integration and continuous delivery workflows that run within Kubernetes clusters.

Who this book is for

This book is intended for different roles within the software development life cycle:

- Developers learning how to prepare their applications using software containers, running distributed in container orchestrators and within modern microservices architectures

- DevOps personas who need to implement secure software supply chains, using software container technologies

What this book covers

Chapter 1, Modern Infrastructure and Applications with Docker, explains the evolution of software architecture and how microservices fit with container-based applications, due to their special features and characteristics.

Chapter 2, Building Docker Images, teaches you what container images are, explaining the layers model and the use of a Dockerfile to build these images using the best security practices.

Chapter 3, Shipping Docker Images, shows you how to store and share container images for your projects.

Chapter 4, Running Docker Containers, covers how to run containers using different software container clients, explaining how to manage container isolation, security, and resource usage.

Chapter 5, Creating Multi-Container Applications, teaches you how to run applications based on multiple components with containers, using Docker Compose to build, run, and deploy your applications in different environments.

Chapter 6, Fundamentals of Orchestration, introduces the container orchestration concept to define and manage application component logic within distributed container runtimes, running as part of a cluster.

Chapter 7, Orchestrating with Swarm, examines Docker Swarm orchestrators with examples of their features and usage.

Chapter 8, Deploying Applications with the Kubernetes Orchestrator, introduces the Kubernetes orchestrator, showing you its components and features, and explaining how to prepare your applications to run on your own Kubernetes platform on your desktop computer.

Chapter 9, Implementing Architecture Patterns, shows you how different application architecture models can be delivered and secured with Kubernetes, thanks to its unique features.

Chapter 10, Leveraging Application Data Management in Kubernetes, dives deep into different Kubernetes resources used to manage sensitive, temporal, and persistent distributed data.

Chapter 11, Publishing Applications, describes different architecture strategies to securely publish your application frontends on Kubernetes.

Chapter 12, Gaining Application Insights, covers managing applications observability within Kubernetes, using open source tools to monitor metrics and provide logging and traceability.

Chapter 13, Managing the Application Life Cycle, introduces the application software life cycle concept and stages, covering how we can manage them by working with containers. This chapter also delves into automating and improving the life cycle by using continuous integration and continuous deployment models.

To get the most out of this book

In this book, we will use different open source tools to create, run, and share software containers. The book presents different options for the labs, and you can choose the environment in which you feel more comfortable. To follow the labs in this book, you will need the following:

- A common laptop or desktop computer with a modern CPU (Intel Core i5 or i7, or an equivalent AMD CPU) and 16 GB of RAM is recommended. You would probably be able to run the labs with lower resources, but your experience may be impacted.

- A Microsoft Windows or Linux operating system. Although Linux will be referred to in this book, you can use either.

- You are expected to have some user-level knowledge of Linux/Windows.

- Some experience in coding in common programming languages, such as Go, JavaScript, Java, or .NET Core, will be useful, although the examples are not complicated and will be easy to follow.

Software/hardware covered in the book	Operating system requirements
Docker and other software container tools	Windows, macOS, or Linux
Docker Swarm orchestrator	Windows, macOS, or Linux
Kubernetes orchestrator desktop environments, such as Docker Desktop, Rancher Desktop, and Minikube	Windows, macOS, or Linux
Tools for monitoring, logging, and tracing, such as Prometheus, Grafana Loki, and OpenTelemetry	Windows, macOS, or Linux

The Kubernetes features used during the labs will work on any of the aforementioned operating system options. If any special characteristics are expected from the underlying platform, they are described in the lab.

If you are using the digital version of this book, we advise you to type the code yourself or access the code from the book's GitHub repository (a link is available in the next section). Doing so will help you avoid any potential errors related to the copying and pasting of code.

Download the example code files

You can download the example code files for this book from GitHub at `https://github.com/PacktPublishing/Containers-for-Developers-Handbook`. If there's an update to the code, it will be updated in the GitHub repository.

We also have other code bundles from our rich catalog of books and videos available at `https://github.com/PacktPublishing/`. Check them out!

Code in Action

The *Code in Action* videos for this book can be viewed at `https://packt.link/JdOIY`.

Conventions used

There are a number of text conventions used throughout this book.

`Code in text`: Indicates code words in text, database table names, folder names, filenames, file extensions, pathnames, dummy URLs, user input, and Twitter handles. Here is an example: "We can verify the image of our system and review its information by executing `docker image inspect`."

A block of code is set as follows:

```
apiVersion: apps/v1
kind: ReplicaSet
metadata:
  name: replicated-webserver
spec:
  replicas: 3
  selector:
    matchLabels:
      application: webserver
  template:
    metadata:
        application: webserver
      spec:
        containers:
        - name: webserver-container
          image: docker.io/nginx:alpine
```

Any command-line input or output is written as follows:

```
$ docker rm -f webserver webserver2
```

Bold: Indicates a new term, an important word, or words that you see on screen. For instance, words in menus or dialog boxes appear in **bold**. Here is an example: "You can verify this by quickly accessing your Docker Desktop's settings by navigating to **Settings | Resources | WSL Integration**."

> **Tips or important notes**
> Appear like this.

Get in touch

Feedback from our readers is always welcome.

General feedback: If you have questions about any aspect of this book, email us at `customercare@packtpub.com` and mention the book title in the subject of your message.

Errata: Although we have taken every care to ensure the accuracy of our content, mistakes do happen. If you have found a mistake in this book, we would be grateful if you would report this to us. Please visit `www.packtpub.com/support/errata` and fill in the form.

Piracy: If you come across any illegal copies of our works in any form on the internet, we would be grateful if you would provide us with the location address or website name. Please contact us at `copyright@packt.com` with a link to the material.

If you are interested in becoming an author: If there is a topic that you have expertise in and you are interested in either writing or contributing to a book, please visit `authors.packtpub.com`.

Share Your Thoughts

Once you've read *Containers for Developers Handbook*, we'd love to hear your thoughts! Scan the QR code below to go straight to the Amazon review page for this book and share your feedback.

`https://packt.link/r/1805127985`

Your review is important to us and the tech community and will help us make sure we're delivering excellent quality content.

Download a free PDF copy of this book

Thanks for purchasing this book!

Do you like to read on the go but are unable to carry your print books everywhere? Is your eBook purchase not compatible with the device of your choice?

Don't worry, now with every Packt book you get a DRM-free PDF version of that book at no cost.

Read anywhere, any place, on any device. Search, copy, and paste code from your favorite technical books directly into your application.

The perks don't stop there, you can get exclusive access to discounts, newsletters, and great free content in your inbox daily

Follow these simple steps to get the benefits:

1. Scan the QR code or visit the link below

https://packt.link/free-ebook/978-1-80512-798-7

2. Submit your proof of purchase
3. That's it! We'll send your free PDF and other benefits to your email directly

Part 1:
Key Concepts of Containers

This part will explain the key concepts of software containers. We will learn their main features, how they implement the *security by default* concept using well-known operating system features, and how to create and deploy applications based on containers in Linux and Windows environments.

This part has the following chapters:

- *Chapter 1, Modern Infrastructure and Applications with Docker*
- *Chapter 2, Building Docker Images*
- *Chapter 3, Shipping Docker Images*
- *Chapter 4, Running Docker Containers*
- *Chapter 5, Creating Multi-Container Applications*

1

Modern Infrastructure and Applications with Docker

Software engineering and development is always evolving and introducing new technologies in its architectures and workflows. Software containers appeared more than a decade ago, becoming particularly popular over the last five years thanks to Docker, which made the concept mainstream. Currently, every enterprise manages its container-based application infrastructure in production in both the cloud and on-premises distributed infrastructures. This book will teach you how to increase your development productivity using software containers so that you can create, test, share, and run your applications. You will use a container-based workflow and your final application artifact will be a Docker image-based deployment, ready to run in production environments.

This chapter will introduce software containers in the context of the current software development culture, which needs faster software supply chains made of moving, distributed pieces. We will review how containers work and how they fit into modern application architectures based on distributed components with very specific functionalities (microservices). This allows developers to choose the best language for each application component and distribute the total application load. We will learn about the kernel features that make software containers possible and learn how to create, share, and run application components as software containers. At the end of this chapter, we will learn about the different tools that can help us work with software containers and provide specific use cases for your laptop, desktop computer, and servers.

In this chapter, we will cover the following topics:

- Evolution of application architecture, from monoliths to distributed microservice architectures
- Developing microservice-based applications
- How containers fit in the microservices model
- Understanding the main concepts, features, and components of software containers
- Comparing virtualization and containers

- Building, sharing, and running containers
- Explaining Windows containers
- Improving security using software containers

Technical requirements

This book will teach you how to use software containers to improve your application development. We will use open source tools for building, sharing, and running containers, along with a few commercial ones that don't require licensing for non-professional use. Also included in this book are some labs to help you practically understand the content that we'll work through. These labs can be found at `https://github.com/PacktPublishing/Containers-for-Developers-Handbook/tree/main/Chapter1`. The *Code In Action* video for this chapter can be found at `https://packt.link/JdOIY`.

From monoliths to distributed microservice architectures

Application architectures are continuously evolving due to technological improvements. Throughout the history of computation, every time a technical gap is resolved in hardware and software engineering, software architects rethink how applications can be improved to take advantage of the new developments. For example, network speed increases made distributing application components across different servers possible, and nowadays, it's not even a problem to distribute these components across data centers in multiple countries.

To take a quick look at how computers were adopted by enterprises, we must go back in time to the old mainframe days (before the 1990s). This can be considered the base for what we call **unitary architecture** – one big computer with all the processing functionality, accessed by users through terminals. Following this, the **client-server** model became very popular as technology also advanced on the user side. Server technologies improved while clients gained more and more functionality, freeing up the server load for publishing applications. We consider both models as **monoliths** as all application components run on one server; even if the databases are decoupled from the rest of the components, running all important components in a dedicated server is still considered monolithic. Both of these models were very difficult to upgrade when performance started to drop. In these cases, newer hardware with higher specifications was always required. These models also suffer from availability issues, meaning that any maintenance tasks required on either the server or application layer will probably lead to service outages, which affects the normal system uptime.

Exploring monolithic applications

Monolithic applications are those in which all functionalities are provided by just one component, or a set of them so tightly integrated that they cannot be decoupled from one another. This makes them hard to maintain. They weren't designed with reusability or modularity in mind, meaning that every time developers need to fix an issue, add some new functionality, or change an application's behavior, the entire application is affected due to, for example, having to recompile the whole application's code.

Providing high availability to monolithic applications required duplicated hardware, quorum resources, and continuous visibility between application nodes. This may not have changed too much today but we have many other resources for providing high availability. As applications grew in complexity and gained responsibility for many tasks and functionalities, we started to decouple them into a few smaller components (with specific functions such as the web server, database, and more), although core components were kept immutable. Running all application components together on the same server was better than distributing them into smaller pieces because network communication speeds weren't high enough. Local filesystems were usually used for sharing information between application processes. These applications were difficult to scale (more hardware resources were required, usually leading to acquiring newer servers) and difficult to upgrade (testing, staging, and certification environments before production require the same hardware or at least compatible ones). In fact, some applications could run only on specific hardware and operating system versions, and developers needed workstations or servers with the same hardware or operating system to be able to develop fixes or new functionality for these applications.

Now that we know how applications were designed in the early days, let's introduce virtualization in data centers.

Virtual machines

The concept of **virtualization** – providing a set of physical hardware resources for specific purposes – was already present in the mainframe days before the 1990s, but in those days, it was closer to the definition of **time-sharing** at the compute level. The concept we commonly associate with virtualization comes from the introduction of the **hypervisor** and the new technology introduced in the late 1990s that allowed for the creation of complete virtual servers running their own virtualized operating systems. This hypervisor software component was able to virtualize and share host resources in virtualized guest operating systems. In the 1990s, the adoption of Microsoft Windows and the emergence of Linux as a server operating system in the enterprise world established x86 servers as the industry standard, and virtualization helped the growth of both of these in our data centers, improving hardware usage and server upgrades. The virtualization layer simplified virtual hardware upgrades when applications required more memory or CPU and also improved the process of providing services with high availability. Data centers became smaller as newer servers could run dozens of virtual servers, and as physical servers' hardware capabilities increased, the number of virtualized servers per node increased.

In the late 1990s, the servers became services. This means that companies started to think about the services they provided instead of the way they did it. Cloud providers arrived to provide services to small businesses that didn't want to acquire and maintain their own data centers. Thus, a new architecture model was created, which became pretty popular: the **cloud computing infrastructure** model. Amazon launched **Amazon Web Services (AWS)**, providing storage, computation, databases, and other infrastructure resources. And pretty soon after that, Elastic Compute Cloud entered the arena of virtualization, allowing you to run your own servers with a few clicks. Cloud providers also allowed users to use their well-documented **application programming interfaces (APIs)** for automation, and the concept of **Infrastructure as Code (IaC)** was introduced. We were able to create our virtualization instances using programmatic and reusable code. This model also changed the service/hardware relationship and what started as a good idea at first – using cloud platforms for every enterprise service – became a problem for big enterprises, which saw increased costs pretty quickly based on network bandwidth usage and as a result of not sufficiently controlling their use of cloud resources. Controlling cloud service costs soon became a priority for many enterprises, and many open source projects started with the premise of providing cloud-like infrastructures. **Infrastructure elasticity** and **easy provisioning** are the keys to these projects. OpenStack was the first one, distributed in smaller projects, each one focused on different functionalities (storage, networking, compute, provisioning, and so on). The idea of having on-premises cloud infrastructure led software and infrastructure vendors into new alliances with each other, in the end providing new technologies for data centers with the required flexibility and resource distribution. They also provided APIs for quickly deploying and managing provisioned infrastructure, and nowadays, we can provision either cloud infrastructure resources or resources on our data centers using the same code with few changes.

Now that we have a good idea of how server infrastructures work today, let's go back to applications.

Three-tier architecture

Even with these decoupled infrastructures, applications can still be monoliths if we don't prepare them for separation into different components. Elastic infrastructures allow us to distribute resources and it would be nice to have distributed components. Network communications are essential and technological evolution has increased speeds, allowing us to consume network-provided services as if they were local and facilitating the use of distributed components.

Three-tier architecture is a software application architecture where the application is decoupled into three to five logical and physical computing layers. We have the **presentation tier**, or user interface; the **application tier**, or backend, where data is processed; and the **data tier**, where the data for use in the application is stored and managed, such as in a database. This model was used even before virtualization arrived on the scene, but you can imagine the improvement of being able to distribute application components across different virtual servers instead of increasing the number of servers in your data center.

Just to recap before continuing our journey: the evolution of infrastructure and network communications has allowed us to run component-distributed applications, but we just have a few components per application in the three-tier model. Note that in this model, different roles are involved in application maintenance as different software technologies are usually employed. For example, we need database administrators, middleware administrators, and infrastructure administrators for systems and network communications. In this model, although we are still forced to use servers (virtual or physical), application component maintenance, scalability, and availability are significantly improved. We can manage each component in isolation, executing different maintenance tasks and fixes and adding new functionalities decoupled from the application core. In this model, developers can focus on either frontend or backend components. Some coding languages are specialized for each layer – for example, JavaScript was the language of choice for frontend developers (although it evolved for backend services too).

As Linux systems grew in popularity in the late 1990s, applications were distributed into different components, and eventually different applications working together and running in different operating systems became a new requirement. Shared files, provided by network filesystems using either **network-attached storage** (**NAS**) or more complex **storage area network** (**SAN**) storage backends were used at first, but **Simple Object Access Protocol** (**SOAP**) and other queueing message technologies helped applications to distribute data between components and manage their information without filesystem interactions. This helped decouple applications into more and more distributed components running on top of different operating systems.

Microservices architecture

The **microservices architecture** model goes a step further, decoupling applications into smaller pieces with enough functionality to be considered components. This model allows us to manage a completely independent component life cycle, freeing us to choose whatever coding language fits best with the functionality in question. Application components are kept light in terms of functionality and content, which should lead to them using fewer host resources and responding faster to start and stop commands. Faster restarts are key to resilience and help us maintain our applications while up, with fewer outages. Application health should not depend on component-external infrastructure; we should improve components' logic and resilience so that they can start and stop as fast as possible. This means that we can ensure that changes to an application are applied quickly, and in the case of failure, the required processes will be up and running in seconds. This also helps in managing the application components' life cycle as we can upgrade components very fast and prepare circuit breakers to manage stopped dependencies.

Microservices use the **stateless** paradigm; therefore, application components should be stateless. This means that a microservice's state must be abstracted from its logic or execution. This is key to being able to run multiple replicas of an application component, allowing us to run them distributed on different nodes from a pool.

This model also introduced the concept of *run everywhere*, where an application should be able to run its components on either cloud or on-premise infrastructures, or even a mix of both (for example, the presentation layer for components could run on cloud infrastructure while the data resides in our data center).

Microservices architecture provides the following helpful features:

- Applications are decoupled into different smaller pieces that provide different features or functionalities; thus, we can change any of them at any time without impacting the whole application.

- Decoupling applications into smaller pieces lets developers focus on specific functionalities and allows them to use the most appropriate programming language for each component.

- Interaction between application components is usually provided via **Representational State Transfer (REST)** API calls using HTTP. RESTful systems aim for fast performance and reliability and can scale without any problem.

- Developers describe which methods, actions, and data they provide in their microservice, which are then consumed by other developers or users. Software architects must standardize how application components talk with each other and how microservices are consumed.

- Distributing application components across different nodes allows us to group microservices into nodes for the best performance, closer to data sources and with better security. We can create nodes with different features to provide the best fit for our application components.

Now that we've learned what microservices architecture is, let's take a look at its impact on the development process.

Developing distributed applications

Monolith applications, as we saw in the previous section, are applications in which all functionalities run together. Most of these applications were created for specific hardware, operating systems, libraries, binary versions, and so on. To run these applications in production, you need a least one dedicated server with the right hardware, operating system, libraries, and so on, and developers require a similar node architecture and resources even just for fixing possible application issues. Adding to this, the pre-production environments for tasks such as certification and testing will multiply the number of servers significantly. Even if your enterprise had the budget for all these servers, any maintenance task as a result of any upgrade in any operating system-related component in production should always be replicated on all other environments. Automation helps in replicating changes between environments, but this is not easy. You have to replicate environments and maintain them. On the other hand, new node provisioning could have taken months in the old days (preparing the specifications for a new node, drawing up the budget, submitting it to your company's approvals workflow, looking for a hardware provider, and so on). Virtualization helped system administrators provision new nodes for developers faster, and automation (provided by tools such as Chef, Puppet, and, my favorite, Ansible) allowed for the alignment of changes between all environments. Therefore, developers were able to obtain their development environments quickly and ensure they were using an aligned version of system resources, improving the process of application maintenance.

Virtualization also worked very well with the three-tier application architecture. It was easy to run application components for developers in need of a database server to connect to while coding new changes. The problem with virtualization comes from the concept of replicating a complete operating system with server application components when we only need the software part. A lot of hardware resources are consumed for the operating system alone, and restarting these nodes takes some time as they are a complete operating system running on top of a hypervisor, itself running on a physical server with its own operating system.

Anyhow, developers were hampered by outdated operating system releases and packages, making it difficult for them to enable the evolution of their applications. System administrators started to manage hundreds of virtual hosts and even with automation, they weren't able to maintain operating systems and application life cycles in alignment. Provisioning virtual machines on cloud providers using their **Infrastructure-as-a-Service** (**IaaS**) platforms or using their **Platform-as-a-Service** (**PaaS**) environments and scripting the infrastructure using their APIs (IaC) helped but the problem wasn't fully resolved due to the quickly growing number of applications and required changes. The application life cycle changed from one or two updates per year to dozens per day.

Developers started to use cloud-provided services and using scripts and applications quickly became more important than the infrastructure on which they were running, which today seems completely normal and logical. Faster network communications and distributed reliability made it easier to start deploying our applications anywhere, and data centers became smaller. We can say that developers started this movement and it became so popular that we finished decoupling application components from the underlying operating systems.

Software containers are the evolution of process isolation features that were learned throughout the development of computer history. Mainframe computers allowed us to share CPU time and memory resources many years ago. Chroot and jail environments were common ways of sharing operating system resources with users, who were able to use all the binaries and libraries prepared for them by system administrators in BSD operating systems. On Solaris systems, we had **zones** as resource containers, which acted as completely isolated virtual servers within a single operating system instance.

So, why don't we just isolate processes instead of full operating systems? This is the main idea behind containers. Containers use kernel features to provide process isolation at the operating system level, and all processes run on the same host but are isolated from each other. So, every process has its own set of resources sharing the same host kernel.

Linux kernels have featured this design of process grouping since the late 2000s in the form of **control groups** (**cgroups**). This feature allows the Linux kernel to manage, restrict, and audit groups of processes.

Another very important Linux kernel feature that's used with containers is **kernel namespaces**, which allow Linux to run processes wrapped with their process hierarchy, along with their own network interfaces, users, filesystem mounts, and inter-process communication. Using kernel namespaces and control groups, we can completely isolate a process within an operating system. It will run as if it were on its own, using its own operating system and limited CPU and memory (we can even limit its disk I/O).

The **Linux Containers** (**LXC**) project took this idea further and created the first working implementation of it. This project is still available, is still in progress, and was the key to what we now know as **Docker containers**. LXC introduced terms such as **templates** to describe the creation of encapsulated processes using kernel namespaces.

Docker containers took all these concepts and created Docker Inc., an open source project that made it easy to run software containers on our systems. Containers ushered in a great revolution, just as virtualization did more than 20 years ago.

Going back to microservices architecture, the ideal application decoupling would mean running defined and specific application functionalities as completely standalone and isolated processes. This led to the idea of running microservice applications' components within containers, with minimum operating system overhead.

What are containers?

We can define a container as a process with all its requirements isolated using cgroups and namespace kernel features. A **process** is the way we execute a task within the operating system. If we define a **program** as the set of instructions developed using a programming language, included in an executable format on disk, we can say that a process is a program in action.

The execution of a process involves the use of some system resources, such as CPU and memory, and although it runs on its own environment, it can use the same information as other processes sharing the same host system.

Operating systems provide tools for manipulating the behavior of processes during execution, allowing system administrators to prioritize the critical ones. Each process running on a system is uniquely identified by a **Process Identifier** (**PID**). A parent-child relationship between processes is developed when one process executes a new process (or creates a new thread) during its execution. The new process (or sub-process) that's created will have as its parent the previous one, and so on. The operating system stores information about process relations using PIDs and parent PIDs. Processes may inherit a parent hierarchy from the user who runs them, so users own and manage their own processes. Only administrators and privileged users can interact with other users' processes. This behavior also applies to child processes created by our executions.

Each process runs on its own environment and we can manipulate its behavior using operating system features. Processes can access files as needed and use pointers to descriptors during execution to manage these filesystem resources.

The operating system kernel manages all processes, scheduling them on its physical or virtualized CPUs, giving them appropriate CPU time, and providing them with memory or network resources (among others).

These definitions are common to all modern operating systems and are key for understanding software containers, which we will discuss in detail in the next section.

Understanding the main concepts of containers

We have learned that as opposed to virtualization, containers are processes running in isolation and sharing the host operating system kernel. In this section, we will review the components that make containers possible.

Kernel process isolation

We already introduced kernel process namespace isolation as a key feature for running software containers. Operating system kernels provide namespace-based **isolation**. This feature has been present in Linux kernels since 2006 and provides different layers of isolation associated with the properties or attributes a process has when it runs on a host. When we apply these namespaces to processes, they will run their own set of properties and will not see the other processes running alongside them. Hence, kernel resources are partitioned such that each set of processes sees different sets of resources. Resources may exist in multiple spaces and processes may share them.

Containers, as they are host processes, run with their own associated set of kernel namespaces, such as the following:

- **Processes**: The container's main process is the parent of others within the container. All these processes share the same process namespace.

- **Network**: Each container receives a network stack with unique interfaces and IP addresses. Processes (or containers) sharing the same network namespace will get the same IP address. Communications between containers pass through host bridge interfaces.

- **Users**: Users within containers are unique; therefore, each container gets its own set of users, but these users are mapped to real host user identifiers.

- **Inter-process communication (IPC)**: Each container receives its own set of shared memory, semaphores, and message queues so that it doesn't conflict with other processes on the host.

- **Mounts**: Each container mounts a root filesystem; we can also attach remote and host local mounts.

- **Unix time-sharing (UTS)**: Each container is assigned a hostname and the time is synced with the underlying host.

Processes running inside a container sharing the same process kernel namespace will receive PIDs as if they were running alone inside their own kernel. The container's main process is assigned PID 1 and other sub-processes or threads will get subsequent IDs, inheriting the main process hierarchy. The container will die if the main process dies (or is stopped).

The following diagram shows how our system manages container PIDs inside the container's PID namespace (represented by the gray box) and outside, at the host level:

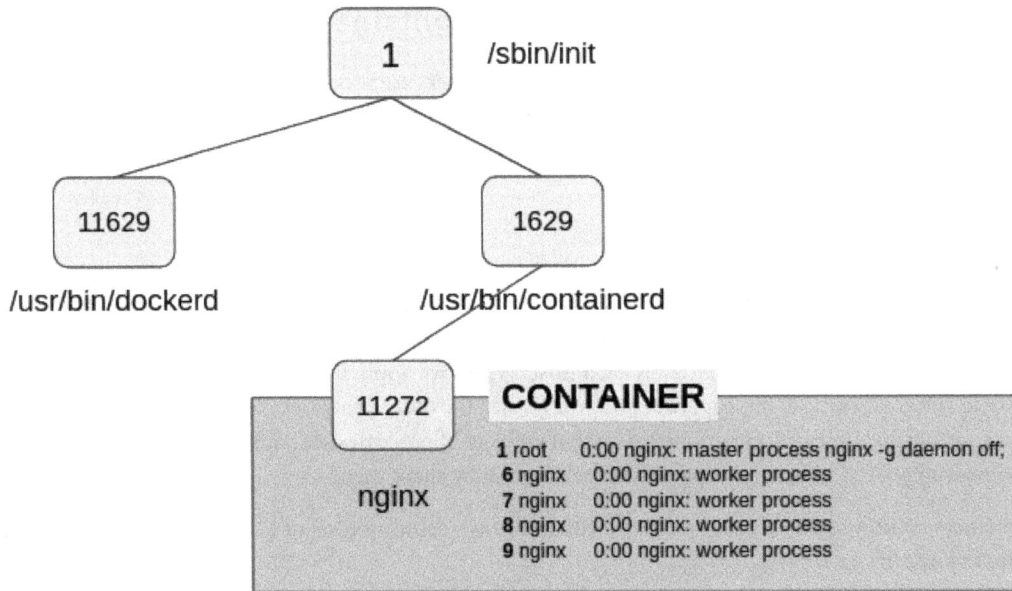

Figure 1.1 – Schema showing a hierarchy of PIDs when you execute
an NGINX web server with four worker processes

In the preceding figure, the main process running inside a container is assigned PID 1, while the other processes are its children. The host runs its own PID 1 process and all other processes run in association with this initial process.

Control groups

A **cgroup** is a feature provided by the Linux kernel that enables us to limit and isolate the host resources associated with processes (such as CPU, memory, and disk I/O). This provides the following features:

- **Resource limits**: Host resources are limited by using a cgroup and thus, the number of resources that a process can use, including CPU or memory

- **Prioritization**: If resource contention is observed, the amount of host resources (CPU, disk, or network) that a process can use compared to processes in another cgroup can be controlled

- **Accounting**: Cgroups monitor and report resource limits usage at the cgroup level

- **Control**: We can manage the status of all processes in a cgroup

The isolation of cgroups will not allow containers to bring down a host by exhausting its resources. An interesting fact is that you can use cgroups without software containers just by mounting a cgroup (cgroup type system), adjusting the CPU limits of this group, and finally adding a set of PIDs to this group. This procedure will apply to either cgroups-V1 or the newer cgroups-V2.

Container runtime

A **container runtime**, or **container engine**, is a piece of software that runs containers on a host. It is responsible for downloading container images from a registry to create containers, monitoring the resources available in the host to run the images, and managing the isolation layers provided by the operating system. The container runtime also reviews the current status of containers and manages their life cycle, starting again when their main process dies (if we declare them to be available whenever this happens).

We generally group container runtimes into **low-level runtimes** and **high-level runtimes**.

Low-level runtimes are those simple runtimes focused only on software container execution. We can consider **runC** and **crun** in this group. Created by Docker and the **Open Container Initiative** (**OCI**), runC is still the de facto standard. Red Hat created crun, which is faster than runC with a lower memory footprint. These low-level runtimes do not require container images to run – we can use a configuration file and a folder with our application and all its required files (which is the content of a Docker image, but without any metadata information). This folder usually contains a file structure resembling a Linux root filesystem, which, as we mentioned before, is everything required by an application (or component) to work. Imagine that we execute the `ldd` command on our binaries and libraries and iterate this process with all its dependencies, and so on. We will get a complete list of all the files strictly required for the process and this would become the smallest image for the application.

High-level container runtimes usually implement the **Container Runtime Interface** (**CRI**) specification of the OCI. This was created to make container orchestration more runtime-agnostic. In this group, we have Docker, CRI-O, and Windows/Hyper-V containers.

The CRI interface defines the rules so that we can integrate our container runtimes into container orchestrators, such as Kubernetes. Container runtimes should have the following characteristics:

- Be capable of starting/stopping pods
- Deal with all containers (start, pause, stop, and delete them)
- Manage container images
- Provide metrics collection and access to container logs

The Docker container runtime became mainstream in 2016, making the execution of containers very easy for users. CRI-O was created explicitly for the Kubernetes orchestrator by Red Hat to allow the execution of containers using any OCI-compliant low-level runtime. High-level runtimes provide tools for interacting with them, and that's why most people choose them.

A middle ground between low-level and high-level container runtimes is provided by Containerd, which is an industry-standard container runtime. It runs on Linux and Windows and can manage the complete container life cycle.

The technology behind runtimes is evolving very fast; we can even improve the interaction between containers and hosts using sandboxes (**gVisor** from Google) and virtualized runtimes (**Kata Containers**). The former increases containers' isolation by not sharing the host's kernel with them. A specific kernel (the small **unikernel** with restricted capabilities) is provided to containers as a proxy to the real kernel. Virtualized runtimes, on the other hand, use virtualization technology to isolate a container within a very small virtual machine. Although both cases add some load to the underlying operating system, security is increased as containers don't interact directly with the host's kernel.

Container runtimes only review the main process execution. If any other process running inside a container dies and the main process isn't affected, the container will continue running.

Kernel capabilities

Starting with Linux kernel release 2.2, the operating system divides process privileges into distinct units, known as **capabilities**. These capabilities can be enabled or disabled by operating system and system administrators.

Previously, we learned that containers run processes in isolation using the host's kernel. However, it is important to know that only a restricted set of these kernel capabilities are allowed inside containers unless they are explicitly declared. Therefore, containers improve their processes' security at the host level because those processes can't do anything they want. The capabilities that are currently available inside a container running on top of the Docker container runtime are SETPCAP, MKNOD, AUDIT_ WRITE, CHOWN, NET_RAW, DAC_OVERRIDE, FOWNER, FSETID, KILL, SETGID, SETUID, NET_BIND_SERVICE, SYS_CHROOT, and SETFCAP.

This set of capabilities allows, for example, processes inside a container to attach and listen on ports below 1024 (the NET_BIND_SERVICE capability) or use ICMP (the NET_RAW capability).

If our process inside a container requires us to, for example, create a new network interface (perhaps to run a containerized OpenVPN server), the NET_ADMIN capability should be included.

> **Important note**
> Container runtimes allow containers to run with full privileges using special parameters. The processes within these containers will run with all kernel capabilities and it could be very dangerous. You should avoid using privileged containers – it is best to take some time to verify which capabilities are needed by an application to work correctly.

Container orchestrators

Now that we know that we need a runtime to execute containers, we must also understand that this will work in a standalone environment, without hardware high availability. This means that server maintenance, operating system upgrades, and any other problem at the software, operating system, or hardware levels may affect your application.

High availability requires resource duplicity and thus more servers and/or hardware. These resources will allow containers to run on multiple hosts, each one with a container runtime. However, maintaining application availability in this situation isn't easy. We need to ensure that containers will be able to run on any of these nodes; in the *Overlay filesystems* section, we'll learn that synchronizing container-related resources within nodes involves more than just copying a few files. **Container orchestrators** manage node resources and provide them to containers. They schedule containers as needed, take care of their status, provide resources for persistence, and manage internal and external communications (in *Chapter 6, Fundamentals of Orchestration*, we will learn how some orchestrators delegate some of these features to different modules to optimize their work).

The most famous and widely used container orchestrator today is **Kubernetes**. It has a lot of great features to help manage clustered containers, although the learning curve can be tough. Also, **Docker Swarm** is quite simple and allows you to quickly execute your applications with high availability (or resilience). We will cover both in detail in *Chapter 7, Orchestrating with Swarm*, and *Chapter 8, Deploying Applications with the Kubernetes Orchestrator*. There were other opponents in this race but they stayed by the wayside while Kubernetes took the lead.

HashiCorp's **Nomad** and Apache's **Mesos** are still being used for very special projects but are out of scope for most enterprises and users. Kubernetes and Docker Swarm are community projects and some vendors even include them within their enterprise-ready solutions. Red Hat's **OpenShift**, SUSE's **Rancher**, Mirantis' **Kubernetes Engine** (old Docker Enterprise platform), and VMware's **Tanzu**, among others, all provide on-premises and some cloud-prepared custom Kubernetes platforms. But those who made Kubernetes the most-used platform were the well-known cloud providers – Google, Amazon, Azure, and Alibaba, among others, serve their own container orchestration tools, such as Amazon's **Elastic Container Service** or **Fargate**, Google's **Cloud Run**, and Microsoft's **Azure Container Instances**, and they also package and manage their own Kubernetes infrastructures for us to use (Google's GKE, Amazon's EKS, Microsoft's AKS, and so on). They provide **Kubernetes-as-a-Service** platforms where you only need an account to start deploying your applications. They also serve you storage, advanced networking tools, resources for publishing your applications, and even *follow-the-sun* or worldwide distributed architectures.

There are many Kubernetes implementations. The most popular is probably OpenShift or its open source project, OKD. There are others based on a binary that launches and creates all of the Kubernetes components using automated procedures, such as Rancher RKE (or its government-prepared release, RKE2), and those featuring only the strictly necessary Kubernetes components, such as K3S or K0S, to provide the lightest platform for IoT and more modest hardware. And finally, we have some Kubernetes distributions for desktop computers, offering all the features of Kubernetes ready to develop and test applications with. In this group, we have Docker Desktop, Rancher Desktop, Minikube, and **Kubernetes in Docker (KinD)**. We will learn how to use them in this book to develop, package, and prepare applications for production.

We shouldn't forget solutions for running orchestrated applications based on multiple containers on standalone servers or desktop computers, such as **Docker Compose**. Docker has prepared a simple Python-based orchestrator for quick application development, managing the container dependencies for us. It is very convenient for testing all of our components together on a laptop with minimum overhead, instead of running a full Kubernetes or Swarm cluster. We will cover this tool, seeing as it has evolved a lot and is now part of the common Docker client command line, in *Chapter 5, Creating Multi-Container Applications*.

Container images

Earlier in this chapter, we mentioned that containers run thanks to **container images**, which are used as templates for executing processes in isolation and attached to a filesystem; therefore, a container image contains all the files (binaries, libraries, configurations, and so on) required by its processes. These files can be a subset of some operating system or just a few binaries with configurations built by yourself.

Virtual machine templates are immutable, as are container templates. This immutability means that they don't change between executions. This feature is key because it ensures that we get the same results every time we use an image for creating a container. Container behavior can be changed using configurations or command-line arguments through the container runtime. This ensures that images created by developers will work in production as expected, and moving applications to production (or even creating upgrades between different releases) will be smooth and fast, reducing the time to market.

Container images are a collection of files distributed in layers. We shouldn't add anything more than the files required by the application. As images are immutable, all these layers will be presented to containerized processes as read-only sets of files. But we don't duplicate files between layers. Only files modified on one layer will be stored in the next layer above – this way, each layer retains the changes from the original base layer (referenced as the base image).

The following diagram shows how we create a container image using multiple layers:

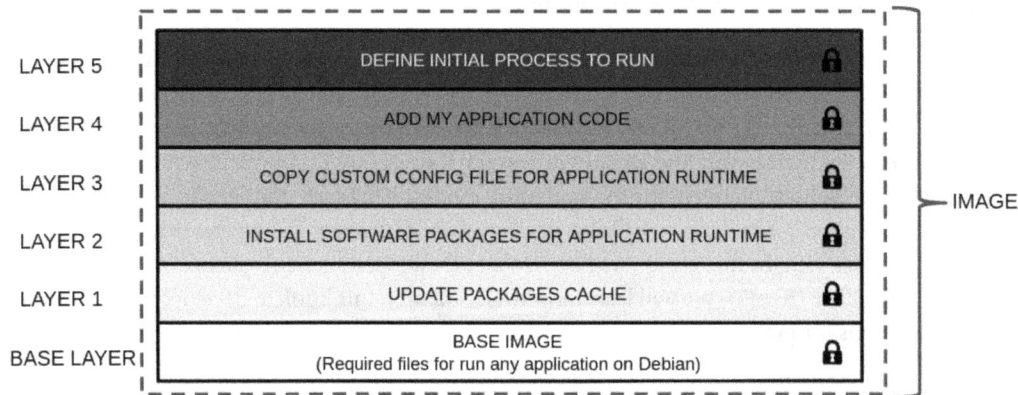

LAYER 5	DEFINE INITIAL PROCESS TO RUN	🔒	
LAYER 4	ADD MY APPLICATION CODE	🔒	
LAYER 3	COPY CUSTOM CONFIG FILE FOR APPLICATION RUNTIME	🔒	IMAGE
LAYER 2	INSTALL SOFTWARE PACKAGES FOR APPLICATION RUNTIME	🔒	
LAYER 1	UPDATE PACKAGES CACHE	🔒	
BASE LAYER	BASE IMAGE (Required files for run any application on Debian)	🔒	

Figure 1.2 – Schema of stacked layers representing a container image

A base layer is always included, although it could be empty. The layers above this base layer may include new binaries or just include new meta-information (which does not create a layer but a meta-information modification).

To easily share these templates between computers or even environments, these file layers are packaged into .tar files, which are finally what we call images. These packages contain all layered files, along with meta-information that describes the content, specifies the process to be executed, identifies the ports that will be exposed to communicate with other containerized processes, specifies the user who will own it, indicates the directories that will be kept out of container life cycle, and so on.

We use different methods to create these images, but we aim to make the process reproducible, and thus we use Dockerfiles as recipes. In *Chapter 2, Building Container Images*, we will learn about the image creation workflow while utilizing best practices and diving deep into command-line options.

These container images are stored on registries. This application software is intended to store file layers and meta-information in a centralized location, making it easy to share common layers between different images. This means that two images using a common Debian base image (a subset of files from the complete operating system) will share these base files, thus optimizing disk space usage. This can also be employed on containers' underlying host local filesystems, saving a lot of space.

Another result of the use of these layers is that containers using the same template image to execute their processes will use the same set of files, and only those files that get modified will be stored.

All these behaviors related to the optimized use of files shared between different images and containers are provided by operating systems thanks to overlay filesystems.

Overlay filesystems

An **overlay filesystem** is a union mount filesystem (a way of combining multiple directories into one that appears to contain their whole combined content) that combines multiple underlying mount points. This results in a structure with a single directory that contains all underlying files and sub-directories from all sources.

Overlay filesystems merge content from directories, combining the file objects (if any) yielded by different processes, with the *upper* filesystem taking precedence. This is the magic behind container-image layers' reusability and disk space saving.

Now that we understand how images are packaged and how they share content, let's go back to learning a bit more about containers. As you may have learned in this section, containers are processes that run in isolation on top of a host operating system thanks to a container runtime. Although the kernel host is shared by multiple containers, features such as kernel namespaces and cgroups provide special containment layers that allow us to isolate them. Container processes need some files to work, which are included in the container space as immutable templates. As you may think, these processes will probably need to modify or create some new files found on container image layers, and a new read-write layer will be used to store these changes. The container runtime presents this new layer to the container to enable changes – we usually refer to this as the **container layer**.

The following schema outlines the read-write layers coming from the container image template with the newly added container layer, where the container's running processes store their file modifications:

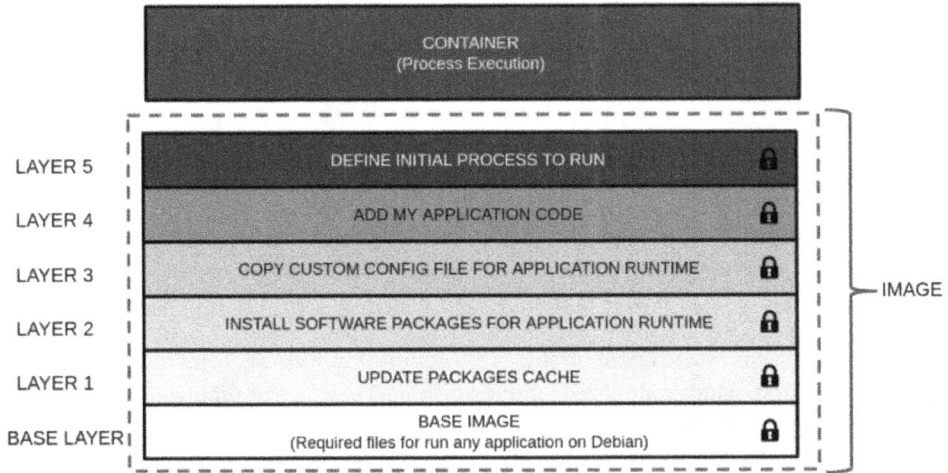

Figure 1.3 – Container image layers will always be read-only; the
container adds a new layer with read-write capabilities

The changes made by container processes are always *ephemeral* as the container layer will be lost whenever we remove the container, while image layers are immutable and will remain unchanged. With this behavior in mind, it is easy to understand that we can run multiple containers using the same container image.

The following figure represents this situation where three different running containers were created from the same image:

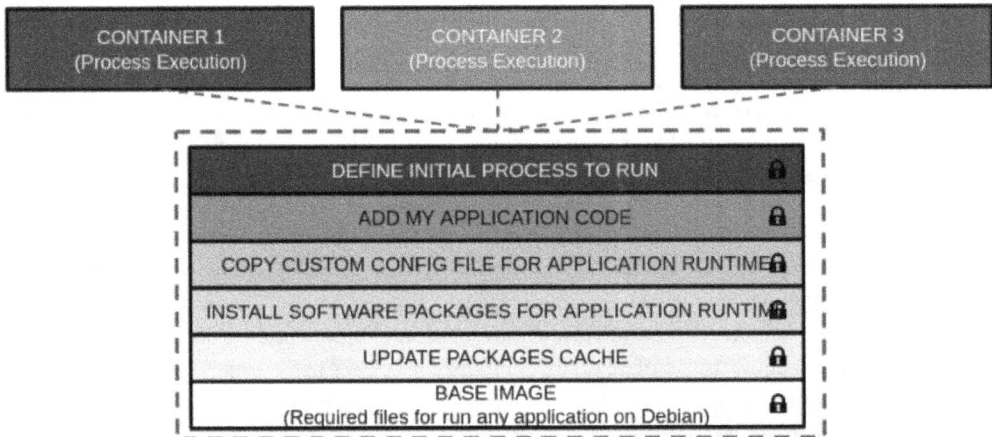

Figure 1.4 – Three different containers run using the same container image

As you may have noticed, this behavior leaves a very small footprint on our operating systems in terms of disk space. Container layers are very small (or at least they should be, and you as a developer will learn which files shouldn't be left inside the container life cycle).

Container runtimes manage how these overlay folders will be included inside containers and the magic behind that. The mechanism for this is based on specific operating system drivers that implement **copy-on-write** filesystems. Layers are arranged one on top of the other and only files modified within them are merged on the upper layer. This process is managed at speed by operating system drivers, but some small overhead is always expected, so keep in mind that all files that are modified continuously by your application (logs, for example) should never be part of the container.

> **Important note**
>
> *Copy-on-write* uses small layered filesystems or folders. Files from any layer are accessible to read access, but *write* requires searching for the file within the underlying layers and copying this file to the upper layer to store the changes. Therefore, the I/O overhead from reading files is very small and we can keep multiple layers for better file distribution between containers. In contrast, writing requires more resources and it would be better to leave big files and those subject to many or continuous modifications out of the container layer.

It is also important to notice that containers are not ephemeral at all. As mentioned previously, changes in the container layer are retained until the container is removed from the operating system; so, if you create a 10 GB file in the container layer, it will reside on your host's disk. Container orchestrators manage this behavior, but be careful where you store your persistent files. Administrators should do container housekeeping and disk maintenance to avoid disk-pressure problems.

Developers should keep this in mind and prepare their applications using containers to be logically ephemeral and store persistent data outside the container's layers. We will learn about options for persistence in *Chapter 10*, *Leveraging Application Data Management in Kubernetes*.

This thinking leads us to the next section, where we will discuss the intrinsic dynamism of container environments.

Understanding dynamism in container-based applications

We have seen how containers run using immutable storage (container images) and how the container runtime adds a new layer for managing changed files. Although we mentioned in the previous section that containers are not ephemeral in terms of disk usage, we have to include this feature in our application's design. Containers will start and stop whenever you upgrade your application's components. Whenever you change the base image, a completely new container will be created (remember the layers ecosystem described in the previous section). This will become even worse if you want to distribute these application components across a cluster – even using the same image will result in different containers being created on different hosts. Thus, this **dynamism** is inherited in these platforms.

In the context of networking communications inside containers, we know that processes running inside a container share its network namespace, and thus they all get the same network stack and IP address. But every time a new container is created, the container runtime will provide a new IP address. Thanks to container orchestration and the **Domain Name System** (**DNS**) included, we can communicate with our containers. As IP addresses are dynamically managed by the container runtime's internal **IP Address Management** (**IPAM**) using defined pools, every time a container dies (whether the main process is stopped, killed manually, or ended by an error), it will free its IP address and IPAM will assign it to a new container that might be part of a completely different application. Hence, we can trust the IP address assignment although we shouldn't use container IP addresses in our application configurations (or even worse, write them in our code, which is a bad practice in every scenario). IP addresses will be dynamically managed by the IPAM container runtime component by default. We will learn about better mechanisms we can use to reference our application's containers, such as service names, in *Chapter 4, Running Docker Containers*.

Applications use fully qualified domain names (or short names if we are using internal domain communications, as we will learn when we use Docker Compose to run multi-container applications, and also when applications run in more complicated container orchestrations).

Because IP addresses are dynamic, special resources should be used to assign sets of IP addresses (or unique IP addresses, if we have just one process replica) to service names. In the same way, publishing application components requires some resource mappings, using **network address translation** (**NAT**) for communicating between users and external services and those running inside containers, distributed across a cluster in different servers or even different infrastructures (such as cloud-provided container orchestrators, for example).

Since we're reviewing the main concepts related to containers in this chapter, we can't miss out on the tools that are used for creating, executing, and sharing containers.

Tools for managing containers

As we learned previously, the container runtime will manage most of the actions we can achieve with containers. Most of these runtimes run as **daemons** and provide an interface for interacting with them. Among these tools, Docker stands out as it provides *all the tools in a box*. Docker acts as a client-server application and in newer releases, both the client and server components are packaged separately, but in any case, both are needed by users. At first, when Docker Engine was the most popular and reliable container engine, Kubernetes adopted it as its runtime. But this marriage did not last long, and Docker Engine was deprecated in Kubernetes release 1.22. This happened because Docker manages its own integration of Containerd, which is not standardized nor directly usable by the Kubernetes CRI. Despite this fact, Docker is still the most widely used option for developing container-based applications and the de facto standard for building images.

We mentioned Docker Desktop and Rancher Desktop earlier in this section. Both act as container runtime clients that use either the `docker` or `nerdctl` command lines. We can use such clients because in both cases, `dockerd` or `containerd` act as container runtimes.

Developers and the wider community pushed Docker to provide a solution for users who prefer to run containers without having to run a privileged system daemon, which is dockerd's default behavior. It took some time but finally, a few years ago, Docker published its rootless runtime with user privileges. During this development phase, another container executor arrived, called Podman, created by Red Hat to solve the same problem. This solution can run without root privileges and aims to avoid the use of a daemonized container runtime. The host user can run containers without any system privilege by default; only a few tweaks are required by administrators if the containers are to be run in a security-hardened environment. This made Podman a very secure option for running containers in production (without orchestration). Docker also included rootless containers by the end of 2019, making both options secure by default.

As you learned at the beginning of this section, containers are processes that run on top of an operating system, isolated using its kernel features. It is quite evident why containers are so popular in microservice environments (one container runs a process, which is ultimately a microservice), although we can still build microservice-based applications without containers. It is also possible to use containers to run whole application components together, although this isn't an ideal situation.

> **Important note**
>
> In this chapter, we'll largely focus on software containers in the context of Linux operating systems. This is because they were only introduced in Windows systems much later. However, we will also briefly discuss them in the context of Windows.

We shouldn't compare containers with virtual nodes. As discussed earlier in this section, containers are mainly based on cgroups and kernel namespaces while virtual nodes are based on hypervisor software. This software provides sandboxing capabilities and specific virtualized hardware resources to guest hosts. We still need to prepare operating systems to run these virtual guest hosts. Each guest node will receive a piece of virtualized hardware and we must manage servers' interactions as if they were physical.

We'll compare these models side by side in the following section.

Comparing virtualization and containers

The following schema represents a couple of virtual guest nodes running on top of a physical host:

VIRTUAL NODE 1   VIRTUAL NODE 2

Application 1 **Application 2**	**Application 3**
Guest Operating System 1	Guest Operating System 2
Guest Virtual Hardware 1	Guest Virtual Hardware 2

Hypervisor (KVM, Hyper-V, VirtualBox, VMWare)
Host Operating System
Hardware (CPU, Memory, Disk I/O, Network)

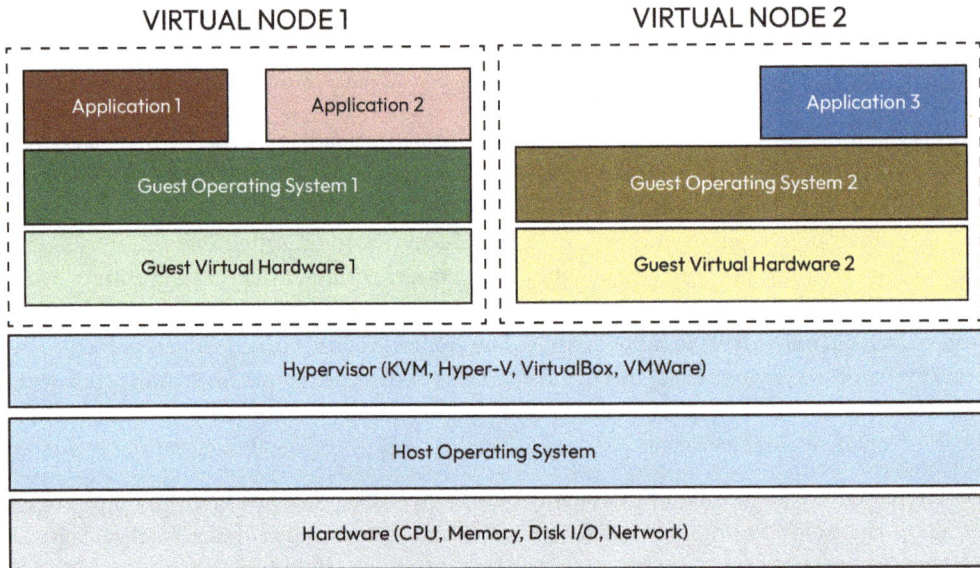

Figure 1.5 – Applications running on top of virtual guest nodes, running on top of a physical server

A physical server running its own operating system executes a hypervisor software layer to provide virtualization capabilities. A specific amount of hardware resources is virtualized and provisioned to these new virtual guest nodes. We should install new operating systems for these new hosts and after that, we will be able to run applications. Physical host resources are partitioned for guest hosts and both nodes are completely isolated from each other. Each virtual machine executes its own kernel and its operating system runs on top of the host. There is complete isolation between guests' operating systems because the underlying host's hypervisor software keeps them separated.

In this model, we require a lot of resources, even if we just need to run a couple of processes per virtual host. Starting and stopping virtual hosts will take time. Lots of non-required software and processes will probably run on our guest host and it will require some tuning to remove them.

As we have learned, the microservices model is based on the idea of applications running decoupled in different processes with complete functionality. Thus, running a complete operating system within just a couple of processes doesn't seem like a good idea.

Although automation will help us, we need to maintain and configure those guest operating systems in terms of running the required processes and managing users, access rights, and network communications, among other things. System administrators maintain these hosts as if they were physical. Developers require their own copies to develop, test, and certify application components. Scaling up these virtual servers can be a problem because in most cases, increasing resources require a complete reboot to apply the changes.

Modern virtualization software provides API-based management, which enhances their usage and virtual node maintenance, but it is not enough for microservice environments. Elastic environments, where components should be able to scale up or down on demand, will not fit well in virtual machines.

Now, let's review the following schema, which represents a set of containers running on physical and virtual hosts:

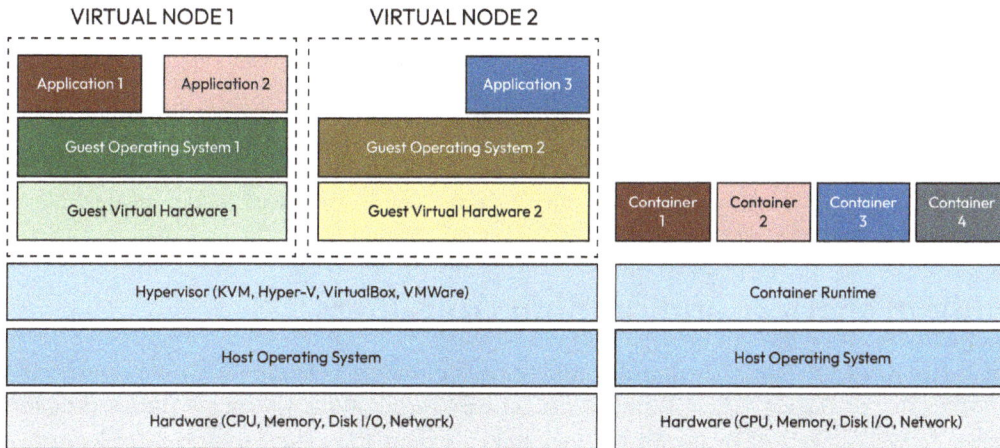

Figure 1.6 – A set of containers running on top of physical and virtual hosts

All containers in this schema share the same host kernel as they are just processes running on top of an operating system. In this case, we don't care whether they run on a virtual or a physical host; we expect the same behavior. Instead of hypervisor software, we have a **container runtime** for running containers. Only a template filesystem and a set of defined resources are required for each container. To clarify, a complete operating system filesystem is not required – we just need the specific files required by our process to work. For example, if a process runs on a Linux kernel and is going to use some network capabilities, then the `/etc/hosts` and `/etc/nsswitch.conf` files would probably be required (along with some network libraries and their dependencies). The **attack surface** will be completely different than having a whole operating system full of binaries, libraries, and running services, regardless of whether the application uses them or not.

Containers are designed to run just one main process (and its threads or sub-processes) and this makes them lightweight. They can start and stop as fast as their main process does.

All the resources consumed by a container are related to the given process, which is great in terms of the allocation of hardware resources. We can calculate our application's resource consumption by observing the load of all its microservices.

We define **images** as templates for running containers. These images contain all the files required by the container to work plus some meta-information providing its features, capabilities, and which commands or binaries will be used to start the process. Using images, we can ensure that all the containers created with one template will run the same. This eliminates infrastructure friction and helps developers prepare their applications to run in production. The configuration (and of course security information such as credentials) is the only thing that differs between the development, testing, certification, and production environments.

Software containers also improve application security because they run by default with limited privileges and allow only a set of system calls. They run anywhere; all we need is a container runtime to be able to create, share, and run containers.

Now that we know what containers are and the most important concepts involved, let's try to understand how they fit into development processes.

Building, sharing, and running containers

Build, ship, and run: you might have heard or read this quote years ago. Docker Inc. used it to promote the ease of using containers. When creating container-based applications, we can use Docker to build container images, share these images within environments, move the content from our development workstations to testing and staging environments, execute them as containers, and finally use these packages in production. Only a few changes are required throughout, mainly at the application's configuration level. This workflow ensures application usage and immutability between the development, testing, and staging stages. Depending on the container runtime and container orchestrator chosen for each stage, Docker could be present throughout (Docker Engine and Docker Swarm). Either way, most people still use the Docker command line to create container images due to its great, always-evolving features that allow us, for example, to build images for different processor architectures using our desktop computers.

Adding **continuous integration** (**CI**) and **continuous deployment** (**CD**) (or **continuous delivery**, depending on the source) to the equation simplifies developers' lives so they can focus on their application's architecture and code.

They can code on their workstations and push their code to a source code repository, and this event will trigger a CI/CD automation to build applications artifacts, compiling their code and providing the artifacts in the form of binaries or libraries. This automation can also include these artifacts inside container images. These become the new application artifacts and are stored in image registries (the backends that store container images). Different executions can be chained to test this newly compiled component together with other components in the integration phase, achieve verification via some tests in the testing phase, and so on, passing through different stages until it gets to production. All these chained workflows are based on containers, configuration, and the images used for execution. In this workflow, developers never explicitly create a release image; they only build and test development ones, but the same Dockerfile recipe is used on their workstations and in the CI/CD phases executed on servers. Reproducibility is key.

Developers can run multiple containers on their developer workstations as if they were using the real environment. They can test their code along with other components in their environment, allowing them to evaluate and discover problems faster and fix them even before moving their components to the CI/CD pipelines. When their code is ready, they can push it to their code repository and trigger the automation. Developers can build their development images, test them locally (be it a standalone component, multiple components, or even a full application), prepare their release code, then push it, and the CI/CD orchestrator will build the release image for them.

In these contexts, images are shared between environments via the use of image registries. *Shipping images from server to server is easy as the host's container runtime will download the images from the given registries – but only those layers not already present on the servers will be downloaded, hence the layer distribution within container images is key.

The following schema outlines this simplified workflow:

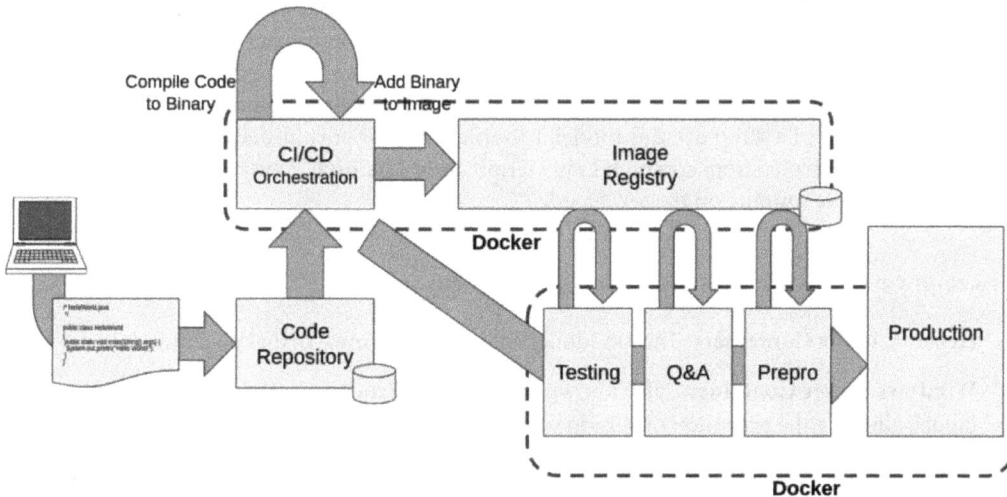

Figure 1.7 – Simplified schema representing a CI/CD workflow example
using software containers to deliver applications to production

Servers running these different stages can be either standalone servers, pools of nodes from orchestrated clusters, or even more complex dedicated infrastructures, including in some cases cloud-provided hosts or whole clusters. Using container images ensures the artifact's content and infrastructure-specific configurations will run in the custom application environment in each case.

With this in mind, we can imagine how we could build a full development chain using containers. We talked about Linux kernel namespaces already, so let's continue by understanding how these isolation mechanisms work on Microsoft Windows.

Explaining Windows containers

During this chapter, we have focused on software containers within Linux operating systems. Software containers started on Linux systems, but due to their importance and advances in technology in terms of host resource usage, Microsoft introduced them in the Microsoft Windows Server 2016 operating system. Before this, Windows users and administrators were only capable of using software containers for Linux through virtualization. Thus, there was the Docker Toolbox solution, of which Docker Desktop formed a part, and installing this software on our Windows-based computer allowed us to have a terminal with the Docker command line, a fancy GUI, and a Hyper-V Linux virtual machine where containers would run. This made it easy for entry-level users to use software containers on their Windows desktops, but Microsoft eventually brought in a game-changer here, creating a new encapsulation model.

> **Important note**
>
> Container runtimes are client-server applications, so we can serve the runtime to local (by default) and remote clients. When we use a remote runtime, we can use our clients to execute commands on this runtime using different clients, such as `docker` or `nerdctl`, depending on the server side. Earlier in this chapter, we mentioned that desktop solutions such as Docker Desktop or Rancher Desktop use this model, running a container runtime server where the common clients, executed from common Linux terminals or Microsoft PowerShell, can manage software containers running on the server side.

Microsoft provided two different software container models:

- **Hyper-V Linux Containers**: The old model, which uses a Linux virtual machine
- **Windows Server Containers**, also known as **Windows Process Containers**: This is the new model, allowing the execution of Windows operating-system-based applications

From the user's perspective, the management and execution of containers running on Windows are the same, no matter which of the preceding models is in use, but only one model can be used per server, thus applying to all containers on that server. The differences here come from the isolation used in each model.

Process isolation on Windows works in the same way it does on Linux. Multiple processes run on a host, accessing the host's kernel, and the host provides isolation using namespaces and resources control (along with other specific methods, depending on the underlying operating system). As we already know, processes get their own filesystem, network, processes identifiers, and so on, but in this case, they also get their own Windows registry and object namespace.

Due to the very nature of the Microsoft Windows operating system, some system services and **dynamic linked libraries (DLLs)** are required within the containers and cannot be shared from the host. Thus, process containers need to contain a copy of these resources, which makes Windows images quite a lot bigger than Linux-based container images. You may also encounter some compatibility issues within image releases, depending on which base operating system (files tree) was used to generate it.

The following schema represents both models side by side so that we can observe the main stack differences:

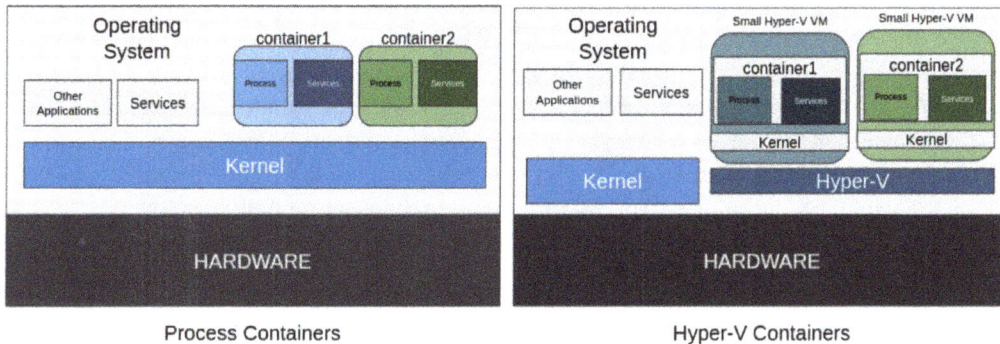

Figure 1.8 – A comparison of Microsoft Windows software container models

We will use Windows Server containers when our application requires strong integration with the Microsoft operating system, for example, for integrating **Group Managed Service Accounts** (**gMSA**) or encapsulating applications that don't run under Linux hosts.

From my experience, Windows Server containers became very popular when they initially arrived, but as Microsoft improved the support of their applications for Linux operating systems, the fact that developers could create their applications in .NET Core for either Microsoft Windows or Linux, and the lack of many cloud providers offering this technology, made them almost disappear from the scene.

It is also important to mention that orchestration technology evolution helped developers move to Linux-only containers. Windows Server containers were supported only on top of Docker Swarm until 2019 when Kubernetes announced their support. Due to the large increase of Kubernetes' adoption in the developer community and even in enterprise environments, Windows Server container usage reduced to very specific and niche use cases.

Nowadays, Kubernetes supports Microsoft Windows Server hosts running as worker roles, allowing process container execution. We will learn about Kubernetes and host roles in *Chapter 8, Deploying Applications with the Kubernetes Orchestrator*. Despite this fact, you will probably not find many Kubernetes clusters running Windows Server container workloads.

We mentioned that containers improve application security. The next section will show you the improvements at the host and container levels that make containers *safer by default*.

Improving security using software containers

In this section, we are going to introduce some of the features found on container platforms that help improve application security.

If we keep in mind how containers run, we know that we first need a **host** with a container runtime. So, having a host with just the software required is the first security measure. We should use dedicated hosts in production for running container workloads. We do not need to concern ourselves with this while developing, but system administrators should prepare production nodes with minimal attack surfaces. We should never share these hosts for use in serving other technologies or services. This feature is so important that we can even find dedicated operating systems, such as Red Hat's CoreOS, SuSE's RancherOS, VMware's PhotonOS, TalOS, or Flatcar Linux, just to mention the most popular ones. These are minimal operating systems that just include a container runtime. You can even create your own by using Moby's LinuxKit project. Some vendors' customized Kubernetes platforms, such as Red Hat's OpenShift, create their clusters using CoreOS, improving the whole environment's security.

We will never connect to any cluster host to execute containers. Container runtimes work in client-server mode. Rather, we expose this engine service and simply using a client on our laptop or desktop computers will be more than enough to execute containers on the host.

Locally, clients connect to container runtimes using **sockets** (/var/run/docker.sock for dockerd, for example). Adding read-write access to this socket to specific users will allow them to use the daemon to build, pull, and push images or execute containers. Configuring the container runtime in this way may be worse if the host has a master role in an orchestrated environment. It is crucial to understand this feature and know which users will be able to run containers on each host. System administrators should keep their container runtimes' sockets safe from untrusted users and only allow authorized access. These sockets are local and, depending on which runtime we are using, TCP or even SSH (in dockerd, for example) can be used to secure remote access. Always ensure **Transport Layer Security (TLS)** is used to secure socket access.

It is important to note that container runtimes do not provide any **role-based access control** (**RBAC**). We will need to add this layer later with other tools. Docker Swarm does not provide RBAC, but Kubernetes does. RBAC is key for managing user privileges and multiple application isolation.

We should say here that, currently, desktop environments (Docker Desktop and Rancher Desktop) also work with this model, in which you don't connect to the host running the container runtime. A virtualized environment is deployed on your system (using Qemu if on Linux, or Hyper-V or the newer Windows Subsystem for Linux on Windows hosts) and our client, using a terminal, will connect to this virtual container runtime (or the Kubernetes API when deploying workloads on Kubernetes, as we will learn in *Chapter 8, Deploying Applications with the Kubernetes Orchestrator*).

Here, we have to reiterate that container runtimes add only a subset of kernel capabilities by default to container processes. But this may not be enough in some cases. To improve containers' security behavior, container runtimes also include a default **Secure Computing Mode (Seccomp)** profile. Seccomp is a Linux security facility that filters the system calls allowed inside containers. Specific profiles can be included and used by runtimes to add some required system calls. You, as the developer, need to notice when your application requires extra capabilities or uncommon system calls. The special features described in this section are used on host monitoring tools, for example, or if we need to add a new kernel module using system administration containers.

Container runtimes usually run as daemons; thus, they will quite probably run as root users. This means that any container can contain the host's files inside (we will learn how we can mount volumes and host paths within containers in *Chapter 4*, *Running Docker Containers*) or include the host's namespaces (container processes may access host's PIDs, networks, IPCs, and so on). To avoid the undesired effects of running container runtime privileges, system administrators should apply special security measures using **Linux Security Modules** (**LSM**), such as SELinux or AppArmor, among others.

SELinux should be integrated into container runtimes and container orchestration. These integrations can be used to ensure, for example, that only certain paths are allowed inside containers. If your application requires access to the host's files, non-default SELinux labels should be included to modify the default runtime behavior. Container runtimes' software installation packages include these settings, among others, to ensure that common applications will run without problems. However, those with special requirements, such as those that are prepared to read hosts' logs, will require further security configurations.

So far in this chapter, we have provided a quick overview of the key concepts related to containers. In the following section, we'll put this into practice.

Labs

In this first chapter, we covered a lot of content, learning what containers are and how they fit into the modern microservices architecture.

In this lab, we will install a fully functional development environment for container-based applications. We will use Docker Desktop because it includes a container runtime, its client, and a minimal but fully functional Kubernetes orchestration solution.

We could use Docker Engine in Linux directly (the container runtime only, following the instructions at `https://docs.docker.com/`) for most labs but we will need to install a new tool for the Kubernetes labs, which requires a minimal Kubernetes cluster installation. Thus, even for just using the command line, we will use the Docker Desktop environment.

> **Important note**
> We will use a Kubernetes desktop environment to minimize CPU and memory requirements. There are even lighter Kubernetes cluster alternatives such as KinD or K3S, but these may require some customization. Of course, you can also use any cloud provider's Kubernetes environment if you feel more comfortable doing so.

Installing Docker Desktop

This lab will guide you through the installation of **Docker Desktop** on your laptop or workstation and how to execute a test to verify that it works correctly.

Docker Desktop can be installed on Microsoft Windows 10, most of the common Linux flavors, and macOS (the arm64 and amd64 architectures are both supported). This lab will show you how to install this software on Windows 10, but I will use Windows and Linux interchangeably in other labs as they mostly work the same – we will review any differences between the platforms when required.

We will follow the simple steps documented at `https://docs.docker.com/get-docker/`. Docker Desktop can be deployed on Windows using **Hyper-V** or the newer **Windows Subsystem for Linux** 2 (**WSL 2**). This second option uses less compute and memory resources and is nicely integrated into Microsoft Windows, making it the preferred installation method, but note that WSL2 is required on your host before installing Docker Desktop. Please follow the instructions from Microsoft at `https://learn.microsoft.com/en-us/windows/wsl/install` before installing Docker Desktop. You can install any Linux distribution because the integration will be automatically included.

We will use the **Ubuntu** WSL distribution. It is available from the **Microsoft Store** and is simple to install:

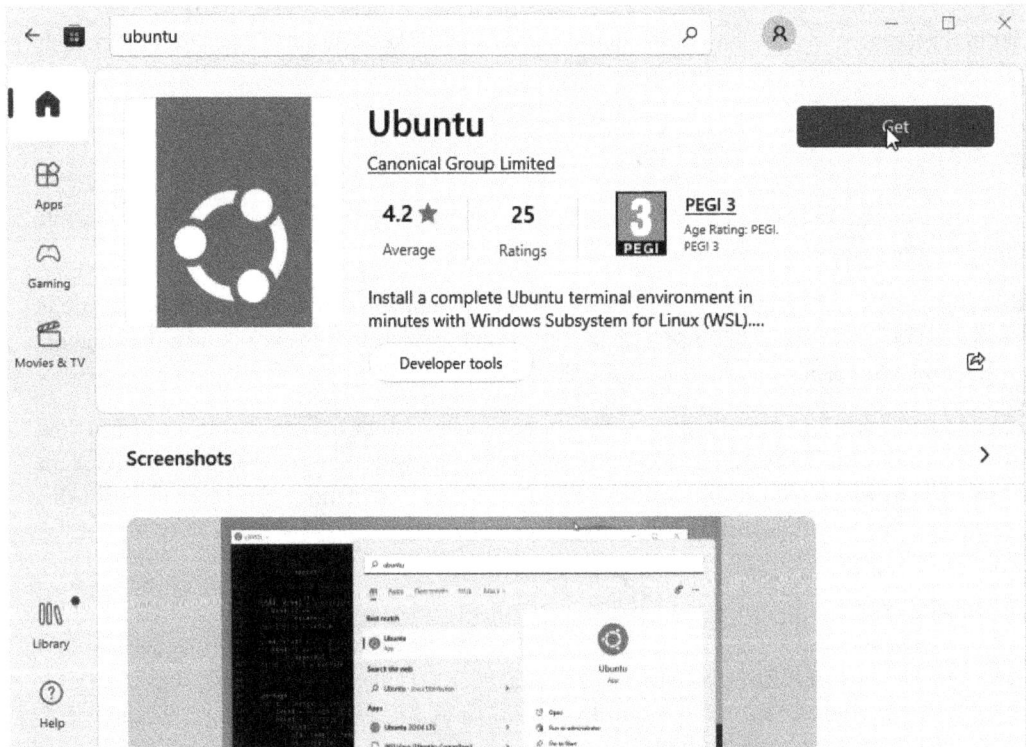

Figure 1.9 – Ubuntu in the Microsoft Store

During the installation, you will be prompted for **username** and **password** details for this Windows subsystem installation:

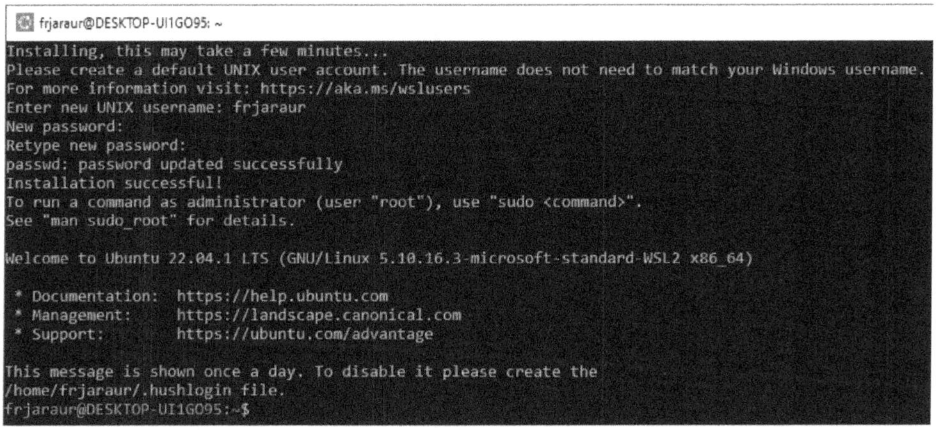

Figure 1.10 – After installing Ubuntu, you will have a fully functional Linux Terminal

You can close this Ubuntu Terminal as the Docker Desktop integration will require you to open a new one once it has been configured.

> **Important note**
>
> You may need to execute some additional steps at `https://docs.microsoft.com/windows/wsl/wsl2-kernel` to update WSL2 if your operating system hasn't been updated.

Now, let's continue with the Docker Desktop installation:

1. Download the installer from `https://docs.docker.com/get-docker/`:

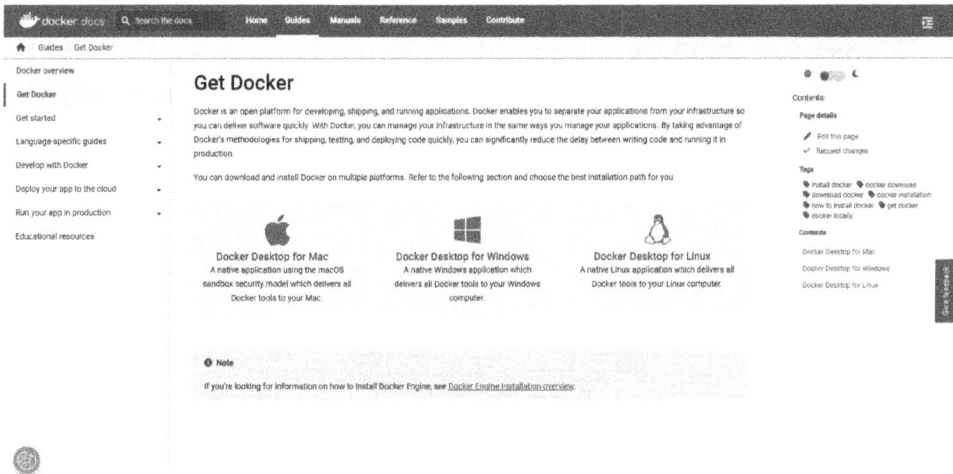

Figure 1.11 – Docker Desktop download section

2. Once downloaded, execute the `Docker Desktop Installer.exe` binary. You will be asked to choose between Hyper-V or WSL2 backend virtualization; we will choose WSL2:

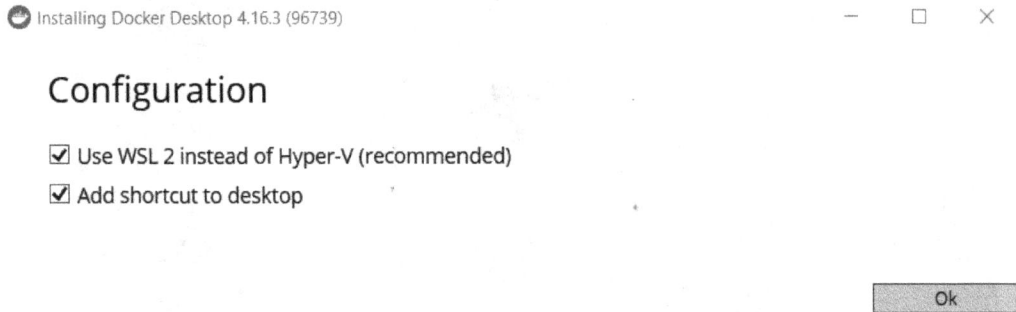

Installing Docker Desktop 4.16.3 (96739)   — □ ×

Configuration

☑ Use WSL 2 instead of Hyper-V (recommended)
☑ Add shortcut to desktop

Ok

Figure 1.12 – Choosing the WSL2 integration for better performance

3. After clicking **Ok**, the installation process will begin decompressing the required files (libraries, binaries, default configurations, and so on). This could take some time (1 to 3 minutes), depending on your host's disk speed and compute resources:

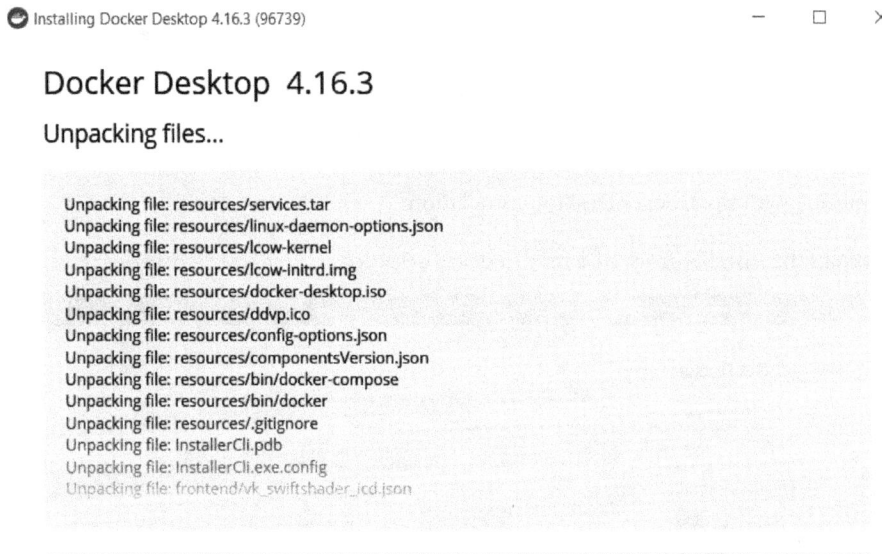

Installing Docker Desktop 4.16.3 (96739)   — □ ×

Docker Desktop 4.16.3

Unpacking files...

```
Unpacking file: resources/services.tar
Unpacking file: resources/linux-daemon-options.json
Unpacking file: resources/lcow-kernel
Unpacking file: resources/lcow-initrd.img
Unpacking file: resources/docker-desktop.iso
Unpacking file: resources/ddvp.ico
Unpacking file: resources/config-options.json
Unpacking file: resources/componentsVersion.json
Unpacking file: resources/bin/docker-compose
Unpacking file: resources/bin/docker
Unpacking file: resources/.gitignore
Unpacking file: InstallerCli.pdb
Unpacking file: InstallerCli.exe.config
Unpacking file: frontend/vk_swiftshader_icd.json
```

Figure 1.13 – The installation process will take a while as the application
files are decompressed and installed on your system

4. To finish the installation, we will be asked to log out and log in again because our user was added to new system groups (Docker) to enable access to the remote Docker daemon via operating system pipes (similar to Unix sockets):

Installing Docker Desktop 4.16.3 (96739)   —   □   ×

Docker Desktop 4.16.3

Installation succeeded

You must log out of Windows to complete installation.

Close and log out

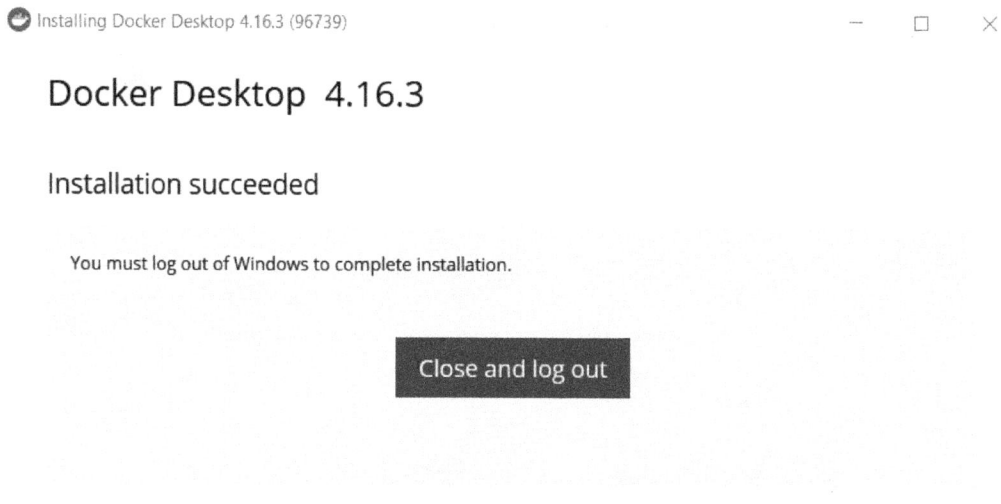

Figure 1.14 – Docker Desktop has been successfully installed and we must log out

5. Once we log in, we can execute Docker Desktop using the newly added application icon. We can enable Docker Desktop execution on start, which could be very useful, but it may slow down your computer if you are short on resources. I recommend starting Docker Desktop only when you are going to use it.

    Once we've accepted the Docker Subscription license terms, Docker Desktop will start. This may take a minute:

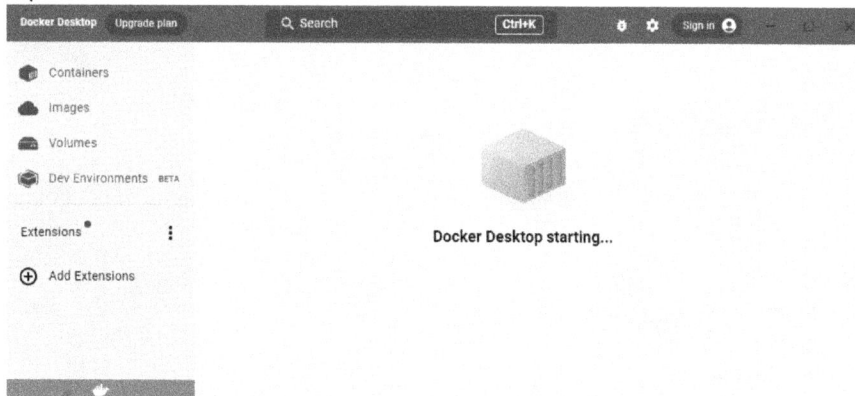

Figure 1.15 – Docker Desktop is starting

You can skip the quick guide that will appear when Docker Desktop is running because we will learn more about this in the following chapters as we deep dive into building container images and container execution.

6.   We will get the following screen, showing us that Docker Desktop is ready:

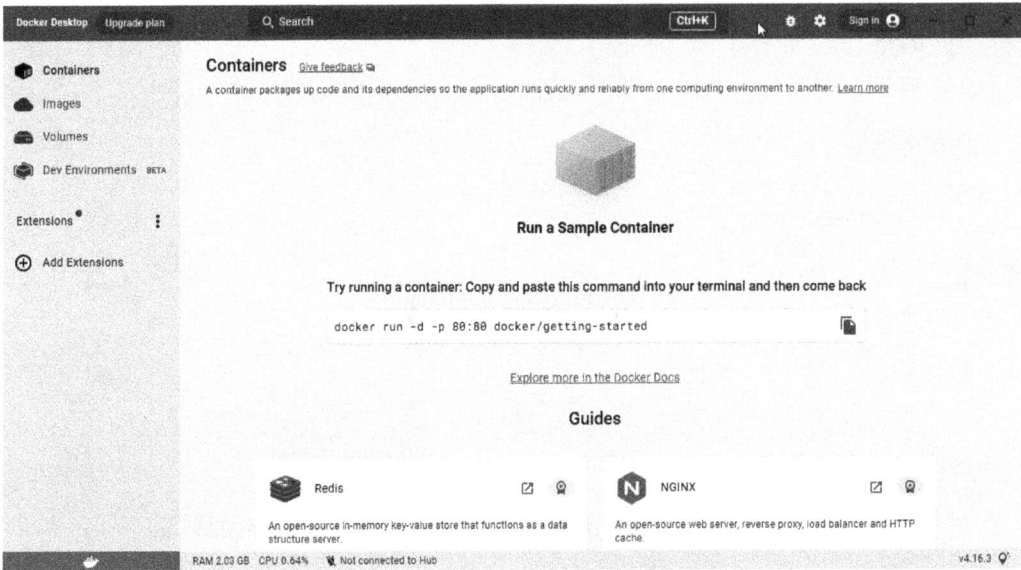

Figure 1.16 – Docker Desktop main screen

7.   We need to enable WSL2 integration with our favorite Linux distribution:

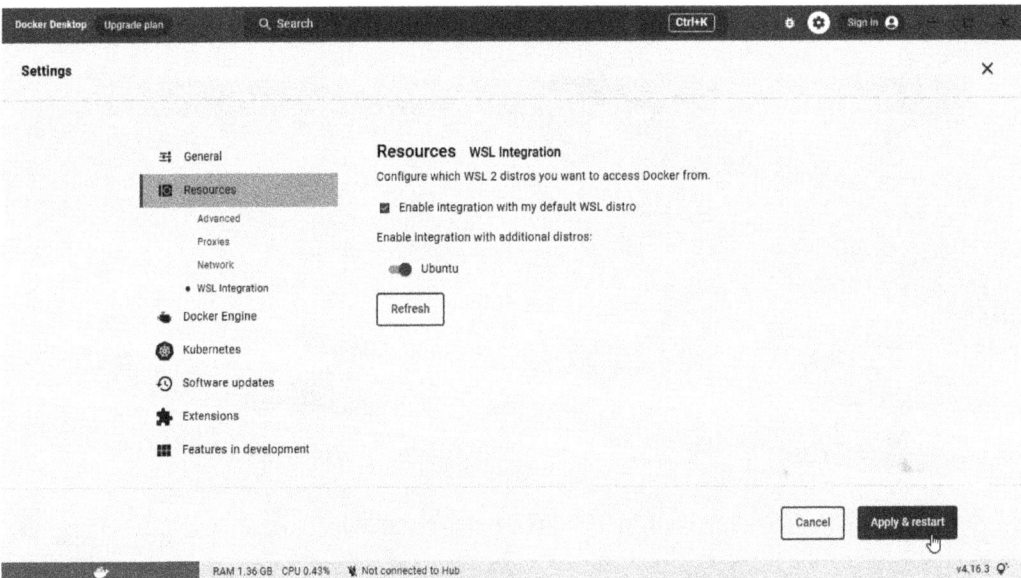

Figure 1.17 – Enabling our previously installed Ubuntu using WSL2

8. After this step, we are finally ready to work with Docker Desktop. Let's open a terminal using our Ubuntu distribution, execute `docker`, and, after that, `docker info`:

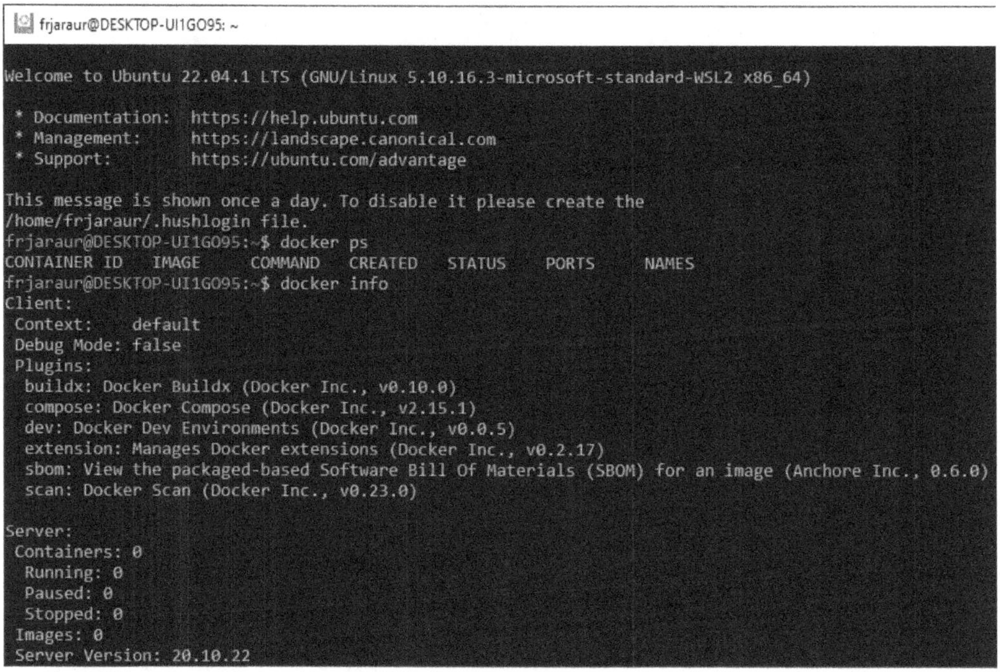

```
frjaraur@DESKTOP-UI1GO95: ~

Welcome to Ubuntu 22.04.1 LTS (GNU/Linux 5.10.16.3-microsoft-standard-WSL2 x86_64)

 * Documentation:  https://help.ubuntu.com
 * Management:     https://landscape.canonical.com
 * Support:        https://ubuntu.com/advantage

This message is shown once a day. To disable it please create the
/home/frjaraur/.hushlogin file.
frjaraur@DESKTOP-UI1GO95:~$ docker ps
CONTAINER ID   IMAGE      COMMAND    CREATED    STATUS    PORTS      NAMES
frjaraur@DESKTOP-UI1GO95:~$ docker info
Client:
 Context:    default
 Debug Mode: false
 Plugins:
  buildx: Docker Buildx (Docker Inc., v0.10.0)
  compose: Docker Compose (Docker Inc., v2.15.1)
  dev: Docker Dev Environments (Docker Inc., v0.0.5)
  extension: Manages Docker extensions (Docker Inc., v0.2.17)
  sbom: View the packaged-based Software Bill Of Materials (SBOM) for an image (Anchore Inc., 0.6.0)
  scan: Docker Scan (Docker Inc., v0.23.0)

Server:
 Containers: 0
  Running: 0
  Paused: 0
  Stopped: 0
 Images: 0
 Server Version: 20.10.22
```

Figure 1.18 – Executing some Docker commands just to verify container runtime integration

As you can see, we have a fully functional Docker client command line associated with the Docker Desktop WSL2 server.

9. We will end this lab by executing an **Alpine container** (a small Linux distribution), reviewing its process tree and the list of its root filesystem.

We can execute `docker run-ti alpine` to download the Alpine image and execute a container using it:

```
Select frjaraur@DESKTOP-UI1GO95: ~
frjaraur@DESKTOP-UI1GO95:~$ docker run -ti alpine
Unable to find image 'alpine:latest' locally
latest: Pulling from library/alpine
63b65145d645: Pull complete
Digest: sha256:69665d02cb32192e52e07644d76bc6f25abeb5410edc1c7a81a10ba3f0efb90a
Status: Downloaded newer image for alpine:latest
/ # ps -ef
PID   USER     TIME  COMMAND
    1 root      0:00 /bin/sh
    7 root      0:00 ps -ef
/ # ls -lrt /
total 56
drwxr-xr-x   12 root     root          4096 Feb 10 16:45 var
drwxr-xr-x    7 root     root          4096 Feb 10 16:45 usr
drwxrwxrwt    2 root     root          4096 Feb 10 16:45 tmp
drwxr-xr-x    2 root     root          4096 Feb 10 16:45 srv
drwxr-xr-x    2 root     root          4096 Feb 10 16:45 run
drwxr-xr-x    2 root     root          4096 Feb 10 16:45 opt
drwxr-xr-x    2 root     root          4096 Feb 10 16:45 mnt
drwxr-xr-x    5 root     root          4096 Feb 10 16:45 media
drwxr-xr-x    2 root     root          4096 Feb 10 16:45 home
drwxr-xr-x    2 root     root          4096 Feb 10 16:45 sbin
drwxr-xr-x    7 root     root          4096 Feb 10 16:45 lib
drwxr-xr-x    2 root     root          4096 Feb 10 16:45 bin
drwxr-xr-x    1 root     root          4096 Feb 12 17:06 etc
dr-xr-xr-x   11 root     root             0 Feb 12 17:06 sys
dr-xr-xr-x  515 root     root             0 Feb 12 17:06 proc
drwxr-xr-x    5 root     root           360 Feb 12 17:06 dev
drwx------    1 root     root          4096 Feb 12 17:07 root
/ # exit
frjaraur@DESKTOP-UI1GO95:~$
```

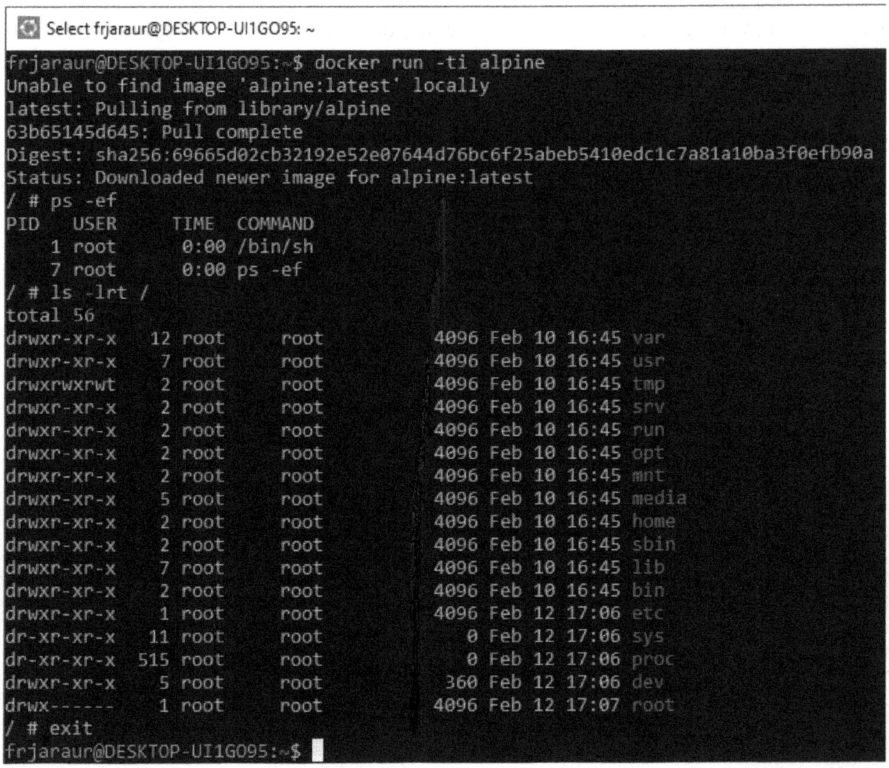

Figure 1.19 – Creating a container and executing some commands inside before exiting

10. This container execution left changes in Docker Desktop; we can review the current images present in our container runtime:

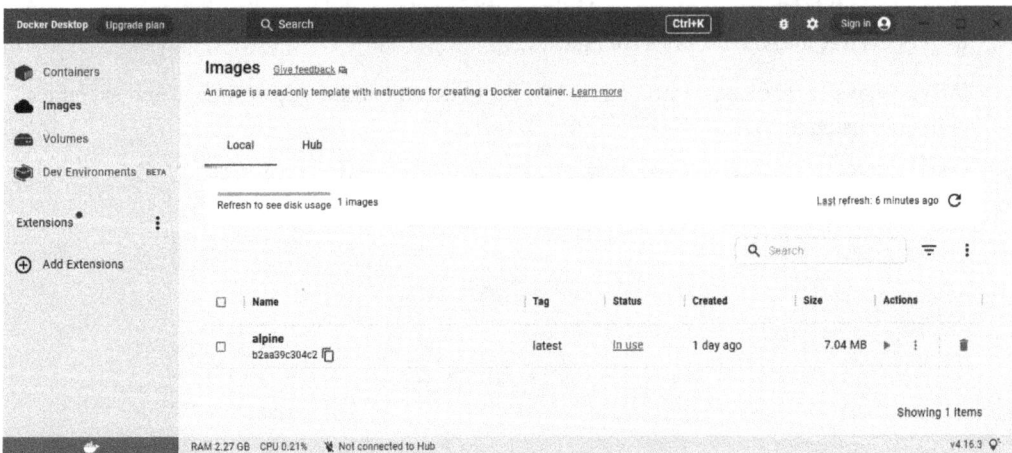

Figure 1.20 – Docker Desktop – the Images view

11. We can also review the container, which is already dead because we exited by simply executing `exit` inside its shell:

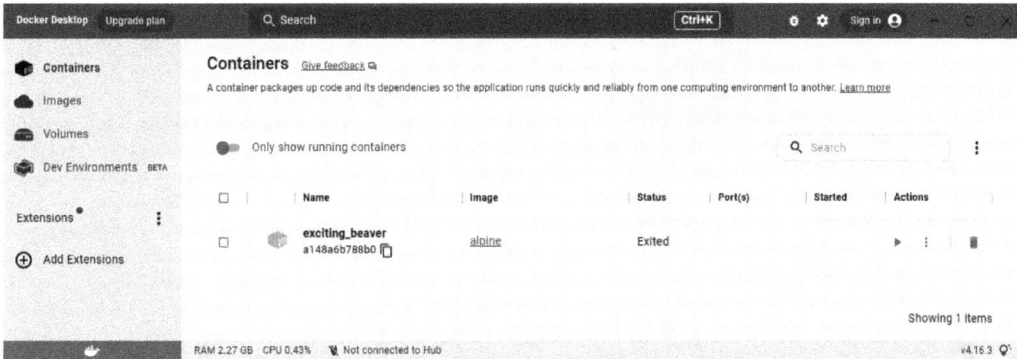

Figure 1.21 – Docker Desktop – the Containers view

Now, Docker Desktop works and we are ready to work through the following labs using our WSL2 Ubuntu Linux distribution.

Summary

In this chapter, we learned the basics around containers and how they fit into modern microservices applications. The content presented in this chapter has helped you understand how to implement containers in distributed architectures, using already-present host operating system isolation features and container runtimes, which are the pieces of software required for building, sharing, and executing containers.

Software containers assist application development by providing resilience, high availability, scalability, and portability thanks to their very nature, and will help you create and manage the application life cycle.

In the next chapter, we will deep dive into the process of creating container images.

2

Building Docker Images

Applications that have components running as software containers are quite a new development and a great way of avoiding problems with underlying infrastructure. As we learned in the previous chapter, containers are processes that are executed on hosts using their kernels, isolated using features present in these kernels (in some cases, for years), and encapsulated in their own filesystems.

In this chapter, we will use container images, which are template-like objects, to create containers. Building these images is the first step to creating your own container-based applications. We will learn different procedures to create container images. These images will be our new application's artifacts, and as such, we need to build them securely and be ready to run them on our laptops or computers, staging and production servers, or even cloud-provisioned infrastructures.

In this chapter, we will cover the following topics:

- Understanding how copy-on-write filesystems work

- Building container images

- Understanding common Dockerfile keys

- The command line for creating images

- Advanced image creation techniques

- Best practices for container image creation

Technical requirements

In this chapter, we will teach you how to build container images and use them in your code-compiling workflow. We will use open source tools, as well as a few commercial ones that can run without licensing for non-professional use, to build images and verify their security. We have included some labs in this chapter to help you understand the content presented. These labs have been published at the following GitHub repository: `https://github.com/PacktPublishing/Containers-for-Developers-Handbook/tree/main/Chapter2`. Here, you will find some extended explanations that have been omitted from this book's content to make the chapters easier to follow. The *Code In Action* video for this chapter can be found at `https://packt.link/JdOIY`.

Understanding how copy-on-write filesystems work

Building a container image is the first step that's required when you develop an application using containers. In this chapter, we will learn about different methods to build images. But first, it will be interesting to deep dive into how images can be created in terms of filesystems.

Containers are processes that run isolated thanks to kernel features. They run on top of a host system with its own filesystem as if they were running completely independently within their own sub-system. Files included in this filesystem are grouped in different layers, one layer on top of another. Files that have to be modified from a lower layer are copied to the layer where the modification is going to be made, and these changes are then committed. New files are only created on the upper layer. This is the basis of **copy-on-write (CoW)** filesystems.

As we can expect with this model, the container runtime will manage all these changes. Every file modification requires host resources to copy the file between layers and, thus, makes this mechanism a problem to create files continuously. Before creating a new file in the upper layer, all layers must be read to verify that the file isn't present yet to copy its content to the upper layer.

All these layers are presented in **read-only** mode to a container every time we create a container using a specific container image as a template, and a new layer is added on top of other layers in **read-write** mode. This new layer is the layer that will contain all the file changes since the container started. However, this behavior will occur in all containers running on your system. All containers based on the same container images share these read-only layers, which is very important in terms of disk usage. Only the **container layer** differs every time a new container is executed.

> **Important note**
>
> All data that should persist across different container executions must be declared and used outside the containers' life cycle – for instance, by using **volumes**, as we will learn in *Chapter 4, Running Docker Containers*. We can declare volumes during the container-image-building process, which indicates that the content exists outside of the image's layers.

As we can see, using these templates speeds up container creation and reduces the size of all containers in our systems. If we compare this with virtual machines, it works like virtual machine templates or snapshots. Only changes are stored at the host level, although it is important to mention here that containers use very little space.

However, performance is always affected when using CoW filesystems, which you should be aware of. Never store logs in a container layer as they may be lost if you remove the container, and it is very important to remember that due to the searching-copying-writing process for any file, your application performance may also be impacted. Therefore, we will never use a container layer to store logs, where processes are continuously writing files or monitoring data. You should write these files on remote backends or use the container volumes feature. This performance decrease applies when you write a lot of small files (thousands), the opposite (a few enormous files), or lots of directories with quite a deep tree structure. You, as a developer, must avoid any of these cases in your applications, and you should prepare your containers to avoid them.

Now that we know the behavior of these CoW filesystems, applied to both container image creation and their execution, let's learn how to build images.

Creating container images

In this section, we will review the different methods to build container images, along with their pros and cons and use cases, so that you can choose the right one, depending on your needs.

There are three ways to create container images:

- Using a base image within a Dockerfile, which is a recipe file that contains different automated steps that are executed to create an image

- Interactively and manually executing commands and storing the resulting filesystem

- From an empty filesystem, using a Dockerfile recipe file and copying only the binaries and libraries required for our application

It is easy to see that the last method is the best in terms of security, but this can be difficult to implement if your code has many dependencies and is very integrated with operating system files. Let's explore these methods, starting with the most common.

Using Dockerfiles to create container images

Before we describe this method, let's learn what a **Dockerfile** is.

A Dockerfile is an **Open Container Initiative** (**OSI**)-compliant file that works as a recipe, containing a step-by-step procedure to create a container image. It contains a set of key-value pairs that describe different executions and meta-information regarding the image's behavior. We can use variables to expand arguments that are passed when building images, and it is perfect for automation. If a Dockerfile is well written, we can ensure its reproducibility.

The following is an example of a Dockerfile:

```
FROM debian:stable-slim
RUN apt-get update -qq && apt-get install -qq package1 package2
COPY . /myapp
RUN make /myapp
CMD python /myapp/app.py
EXPOSE 5000
```

As mentioned before, this file describes all the steps required to assemble an image. Let's provide a quick overview of the steps taken in the presented Dockerfile.

The first line, FROM debian:stable-slim, indicates that this container image will be taken as a base image; hence, all its layers will be used. The container runtime will download (*pull*) all these layers if they are not present in our host. If any of them are already in our host, they will be used. This layer could have come from any other image already in our host. Container image layers are reused.

The second line, RUN apt-get update -qq && apt-get install -qq package1 package2, executes all the content included as values. First, apt-get update -qq will be executed, and if it's successful, apt-get install -qq package1 package2 will be executed. This full step creates just one layer, on top of the previous one. This layer will automatically be enabled for any other image using the same execution, using the same debian:stable-slim base image.

The third line, COPY . /myapp, will copy all the files available in the current directory to a directory named /myapp, in a new layer. As mentioned in the second line, this also creates a reusable layer for any new image that contains the same entry.

The fourth line, RUN make /myapp, executes the make /myapp command and creates a new line. Remember that this is an example. We added a make sentence to build our source code. In this step, for example, we run a compiler, previously installed in the image, and build our binary artifact. All executing layers (those that include a RUN key) should exit correctly. If this doesn't happen, the image build process will break and be stopped. If this happens, all previous layers will remain in your system. The container runtime creates a layers cache, and all following executions will reuse them by default. This behavior can be avoided by recreating all previous images during the build process.

The two final steps don't add layers. The CMD key declares which command line will be executed (remember that a container runs a main process), and EXPOSE adds the meta-information regarding which port should be exposed (listening). This way, we explicitly declare in which port our application will listen to any kind of communication.

> **Important note**
>
> You should declare all relevant meta-information in your Dockerfiles, such as the *ports exposed*, the *volumes* for persistent data, the *username* (or *userid*) defined for your main process, and the command line that should run on startup. This information may be required by your container's orchestrator administrators because it is very important to avoid security issues. They will probably force some security policies in the production platform that disallow your application's execution. Ask them whether some security policies are applied to ensure you added the required information. Anyway, if you follow the security practices described in this book, you probably won't have any problems in production.

As you can see, this is a pretty reproducible process. This recipe will create the same image every time if we don't change anything. This helps developers focus on their code. However, creating reproducible images is not that easy. If you take a closer look at the used FROM value, we use `debian:stable-slim`, which means that the default image **registry** will be used, `docker.io`. For now, you just have to know that a registry is a store for all container image layers. The value of the FROM key indicates that a `debian` image, with a specific tag of `stable-slim`, will be used, and thus, if Docker changes this image, all your image builds will also change. Tags are the way we identify images, but they are not uniquely identified. Each image and layer within images are uniquely identified by **digest hashes**, and these are the real relevant values that you should closely monitor. To get these values, we have to either pull the image or review the information in the defined registry. The easier method is to pull the image, which happens when you execute your build process, but in this example, we used a mocked Dockerfile, so it won't work as-is.

So, let's pull the image from the official Docker images registry, at `https://hub.docker.com`, or by using the `docker.io` command-line tool:

```
$ docker image pull debian:stable-slim
stable-slim: Pulling from library/debian
de661c304c1d: Pull complete
Digest: sha256:f711bda490b4e5803ee7f634483c4e6fa7dae54102654f2c-
231ca58eb233a2f1
Status: Downloaded newer image for debian:stable-slim
docker.io/library/debian:stable-slim
```

Here, we execute `docker image pull debian:stable-slim` to download this image from `docker.io`. All its layers will be downloaded. The Docker Hub website provides lots of useful information, such as all the tags associated with an image and the vulnerabilities detected in the contained files.

The digest shown in the previous code snippet will identify this image uniquely. We can verify the image of our system and review its information by executing `docker image inspect`, using its **image ID**:

```
$ docker image ls --no-trunc
REPOSITORY     TAG            IMAGE
ID                                                                    CR
EATED          SIZE
debian         stable-slim    sha256:4ea5047878b3bb91d62ac9a99cd-
cf9e53f4958b01000d85f541004ba587c1cb1   9 days ago    80.5MB
$ docker image inspect 4ea5047878b3bb91d62ac9a99cdcf9e53f4958b01000d-
85f541004ba587c1cb1 |grep -A1 -i repodigest
        "RepoDigests": [
            "debian@sha256:f711bda490b4e5803ee7f634483c4e6fa-
7dae54102654f2c231ca58eb233a2f1"
```

All containers' related objects are identified by object IDs, and as such, we can use them to refer to each object. In this example, we used the image ID to inspect the object.

Important note

We can use `--digests` when listing local images to retrieve all their digests – for example, with the image used in this section:

```
$ docker image ls --digests
REPOSITORY    TAG          DIGEST                     IMAGE
ID        CREATED      SIZE
debian        stable-slim   sha256:f711bda490b4e5803ee7f634483c4e-
6fa7dae54102654f2c231ca58eb233a2f1    4ea5047878b3   9 days
ago    80.5MB
```

It is important to note that image IDs are different from their digests. The ID represents the current compilation or identifier generated on your system, while the digest represents the compendium of all the layers and essentially identifies the image anywhere – on your laptop, on your servers, or even in the registry where it is remotely stored. The image digest is associated with the image content manifest (`https://docs.docker.com/registry/spec/manifest-v2-2/`) and is used in V2 registries (the current version for most modern registry implementations). Since your local builds are not in a registry format, the digest will be displayed as `none`. Pushing your images to a V2 registry will change this.

Let's review this process, as well as the image IDs and digests, by looking at a quick and simple example. We will build a couple of images using the following two lines of a Dockerfile:

```
FROM debian:stable-slim
RUN apt-get update -qq && apt-get install -qq curl
```

Using the current `debian:stable-slim` image, we will update its content and install the `curl` package.

We will build two images, `one` and `two`, as shown in the following screenshot:

```
frjaraur@sirius:~$ docker build -t one .
[+] Building 0.3s (6/6) FINISHED
 => [internal] load build definition from Dockerfile
 => => transferring dockerfile: 37B
 => [internal] load .dockerignore
 => => transferring context: 2B
 => [internal] load metadata for docker.io/library/debian:stable-slim
 => [1/2] FROM docker.io/library/debian:stable-slim
 => CACHED [2/2] RUN apt-get update -qq && apt-get install -qq curl
 => exporting to image
 => => exporting layers
 => => writing image sha256:331ed31f881616f4e635e6d53e2b6b5107013af8f96748e2cd1493c703872d75
 => => naming to docker.io/library/one

Use 'docker scan' to run Snyk tests against images to find vulnerabilities and learn how to fix them
frjaraur@sirius:~$ docker build -t two .
[+] Building 0.3s (6/6) FINISHED
 => [internal] load build definition from Dockerfile
 => => transferring dockerfile: 37B
 => [internal] load .dockerignore
 => => transferring context: 2B
 => [internal] load metadata for docker.io/library/debian:stable-slim
 => [1/2] FROM docker.io/library/debian:stable-slim
 => CACHED [2/2] RUN apt-get update -qq && apt-get install -qq curl
 => exporting to image
 => => exporting layers
 => => writing image sha256:331ed31f881616f4e635e6d53e2b6b5107013af8f96748e2cd1493c703872d75
 => => naming to docker.io/library/two

Use 'docker scan' to run Snyk tests against images to find vulnerabilities and learn how to fix them
```

Figure 2.1 – The execution of two consecutive container image builds.
No changes are expected; hence, the images are equal

The first build process will create a layers cache, and thus, the second build will reuse them and the process will be faster as the layers are the same. No installation process will be triggered. We have used the current directory as the **build context**. Container runtimes such as Docker are executed in a client-server model, and as such, we talk with the Docker daemon using our Docker command line. The building process sends all files in the *current context* (*path*) to the daemon so that it can use them to create the image's filesystem. This is critical because if we choose the wrong context, a lot of files will be sent to the daemon, and this will impact the building process. We should correctly specify which directory contains our code, and this will be used during the build process. In this context folder, we should avoid binaries, libraries, documentation, and so on. It is important to note that we can use Git repositories (in URL format) as the build context, which makes it very interesting for CI/CD integrations.

To avoid sending irrelevant files to the daemon during the build process, we can use the `.dockerignore` file. In this file, we will add the list of files and folders that should be excluded, even if they are present in our build context.

Let's review the information we have from these images on our system. If we execute `docker image ls -digest`, we will obtain their image IDs and their digests:

```
frjaraur@sirius:~$ docker image ls --digests
REPOSITORY   TAG          DIGEST                                                                        IMAGE ID       CREATED          SIZE
two          latest       <none>                                                                        331ed31f8816   19 minutes ago   105MB
one          latest       <none>                                                                        331ed31f8816   19 minutes ago   105MB
debian       stable-slim  sha256:f711bda490b4e5803ee7f634483c4e6fa7dae54102654f2c231ca58eb233a2f1       4ea5047878b3   9 days ago       80.5MB
```

Figure 2.2 – A list of the created container images, showing their completely equal IDs

The first thing we can see is that both images, `one` and `two`, have the same image ID. This is because we reused their layers. In the second build process, using the same Dockerfile, the container runtime reuses all previous equal image layers (those coming from the same execution), and the image was created very fast. They are the same image with two different tags.

We can also see that only the base image shows its digest. As mentioned previously, it is the only one that comes from a V2 registry. If we upload one of our images to *Docker Hub* (or any other V2-compatible registry), its digest will be created.

> **Important note**
>
> To be able to upload images to Docker Hub, you need a working account. Create your account by going to `https://hub.docker.com/signup`. The process is pretty simple, and you will have a Docker Hub registry account within a minute.

Let's see how uploading the image works and how it will have its immutable and unique reference digest.

Before initiating the process, we will just log in to Docker Hub using our account name. We will be prompted for our password, after which we should receive a `Login Succeeded` message:

```
$ docker login --username <YOUR_USERNAME>
Password:
Login Succeeded
Logging in with your password grants your terminal complete access to
your account.
For better security, log in with a limited-privilege personal access
token. Learn more at https://docs.docker.com/go/access-tokens/
```

Now that we are logged in, we need to retag our image. Image tags are the human-readable format we use to reference images. In the build processes used for this example, we used `one` and `two` as tags through the command line by writing `docker build -t <TAG>`. However, we saw that both were the same image; hence, we can say that tags are names for an image ID, which may cause you some confusion. *Can we trust image tags?* The short answer is, *no, we can't.* They don't represent a unique image state. We can have different tags for an image ID and change these images, but if you still use those tags, you will be using completely different images. In our example, anyone can change our `debian:stable-slim` image. If we rebuild some of our images, based on this tag, we will create a new image with completely different content. What if the new image contains some code exploitation because a malicious attacker included it in that base image? This should not happen in very controlled image registries such as Docker Hub, but this problem does exist.

Let's retag and upload our image by using `docker tag` and then `docker push`:

```
frjaraur@sirius:~$ docker image ls --digests
REPOSITORY    TAG          DIGEST                                                                   IMAGE ID      CREATED        SIZE
frjaraur/one  180223       <none>                                                                   331ed31f8816  21 minutes ago 105MB
one           latest       <none>                                                                   331ed31f8816  21 minutes ago 105MB
two           latest       <none>                                                                   331ed31f8816  21 minutes ago 105MB
debian        stable-slim  sha256:f711bda490b4e5803ee7f634483c4e6fa7dae54102654f2c231ca58eb233a2f1  4ea5047878b3  9 days ago     80.5MB
frjaraur@sirius:~$ docker push  frjaraur/one:180223
The push refers to repository [docker.io/frjaraur/one]
b1fe51447de8: Pushed
b9ebbed0d983: Mounted from library/debian
180223: digest: sha256:a2f09396d5fcd805df2f546cf89793c9d75ab24a6bd69d23c525ffbd37309d7a size: 740
frjaraur@sirius:~$ docker image ls --digests
REPOSITORY    TAG          DIGEST                                                                   IMAGE ID      CREATED        SIZE
frjaraur/one  180223       sha256:a2f09396d5fcd805df2f546cf89793c9d75ab24a6bd69d23c525ffbd37309d7a  331ed31f8816  22 minutes ago 105MB
one           latest       <none>                                                                   331ed31f8816  22 minutes ago 105MB
two           latest       <none>                                                                   331ed31f8816  22 minutes ago 105MB
debian        stable-slim  sha256:f711bda490b4e5803ee7f634483c4e6fa7dae54102654f2c231ca58eb233a2f1  4ea5047878b3  9 days ago     80.5MB
```

Figure 2.3 – Tagging and pushing an image to obtain its digest

Note that we need to push the image. Just re-tagging does not work. Now, we have a unique image, and anyone can use our tag to reference it. If we update our `one` image, by adding some new content or changing the command line to be executed, this digest will change. And even if we still use the same `frjaraur/one` tag, a new build process using our image will create new content.

> **Important note**
>
> As a developer, you should be aware of any changes introduced in the images you use as a reference for creating your images. You might be wondering which method is correct to manage these changes. The short answer would be always using image digests (following the example tags and digest, we will use `FROM debian:stable-slim@sha256:f711b-da490b4e5803ee7f634483c4e6fa7dae54102654f2c231ca58eb233a2f1`). This method can be very complex, but it is the most secure. Another method would be using your own registry, isolated from the internet, where you store your images. With your own managed private registry, you may be comfortable using image tags. You will be the only one able to update your base images; hence, you manage the complete image life cycle.

As we mentioned at the beginning of this example, we built two images using the same Dockerfile, and we realized that both images have the same image ID; hence, they are exactly the same. Let's change this a bit and use the `docker build --no-cache` option, which avoids reusing previously created layers:

```
frjaraur@sirius:~$ docker build -t two --no-cache .
[+] Building 11.3s (6/6) FINISHED
 => [internal] load build definition from Dockerfile
 => => transferring dockerfile: 370
 => [internal] load .dockerignore
 => => transferring context: 20
 => [internal] load metadata for docker.io/library/debian:stable-slim
 => CACHED [1/2] FROM docker.io/library/debian:stable-slim
 => [2/2] RUN apt-get update -qq && apt-get install -qq curl
 => exporting to image
 => => exporting layers
 => => writing image sha256:8e64333d60334472ff1bf6901d249dd3595d37801bf2d42d2b2f51042086b9d96
 => => naming to docker.io/library/two

Use 'docker scan' to run Snyk tests against images to find vulnerabilities and learn how to fix them
frjaraur@sirius:~$ docker image ls --digests
REPOSITORY      TAG          DIGEST                                                                     IMAGE ID      CREATED         SIZE
two             latest       <none>                                                                     8e64333d6033  3 seconds ago   105MB
one             latest       <none>                                                                     331ed31f8816  31 minutes ago  105MB
frjaraur/one    180223       sha256:a2f09396d5fcd805df2f546cf89793c9d75ab24a6bd69d23c525ffbd37309d7a  331ed31f8816  31 minutes ago  105MB
debian          stable-slim  sha256:f711bda490b4e5803ee7f634483c4e6fa7dae54102654f2c231ca58eb233a2f1  4ea5047878b3  9 days ago      80.5MB
```

Figure 2.4 – Executing the image-building process without a cache

We can see that a completely new image was built, even though we are using the same Dockerfile. This is due to the time between executions. Layers change between build executions because we made modifications at two different points in time. Of course, we can also include new changes due to package updates, but in this case, it is even simpler.

What we can learn from this is that reusing layers helps us maintain image sizes and build times (we didn't notice this in this example because we used a simple two-line Dockerfile, but when you are compiling or downloading a bunch of modules for your code, it can take a lot of time), but when we need to refresh the image content, disabling the cache is a must. This is very useful when we create base image files for our projects – for example, our own .NET Core and Python projects. We will use these base images, uploaded into our registry, and we will be sure of their content. When a new release arrives, we can rebuild these images and all their dependent images (our applications' images). This process should be part of our automated CI/CD pipelines.

Now that we understand how to build images using Dockerfiles, we will move on to a new method that can be helpful in very specific cases.

Creating container images interactively

We haven't mentioned it before, but it is important to comment here that the Dockerfile RUN lines create intermediate containers to execute the commands, written as values after the RUN key. Hence, the docker build command launches a series of chained containers that create the different layers that are finally part of an image. Before executing a new container, this process stores the modified files (container layer) in the system, using the container runtime's commit feature.

These containers run one after another, using the layer created by the previous one. The interactive process we are about to describe follows this workflow in a simplified way. We will run a container, using an image as a base, and manually run and copy all the commands and content required by our application. Changes will be created on the fly, and we will commit the created container layer when we have finished. This method may be interesting when we need to install software that asks for different configurations interactively and we can't automate the process.

This method lacks reproducibility and shouldn't be used if we can find a way to automate the image creation process. No one will have any clue of how you installed the content inside the image (the shell history will contain the steps if you didn't remove it, but interactive commands will not be there). Let's introduce a command that will help us understand how images were built – `docker image history`. This command shows all the steps taken to create an image, including the meta-information added in the process, in reverse order.

Let's take a look at this output using one of the images from the previous section, *Using Dockerfiles to create container images*:

```
frjaraur@sirius:~$ docker image history one  --no-trunc
IMAGE                                                                       CREATED       CRE
ATED BY
     SIZE        COMMENT
sha256:331ed31f881616f4e635e6d53e2b6b5107013af8f96748e2cd1493c703872d75     16 hours ago  RUN
  /bin/sh -c apt-get update -qq && apt-get install -qq curl # buildkit
     24.6MB      buildkit.dockerfile.v0
<missing>                                                                   10 days ago   /bi
n/sh -c #(nop)  CMD ["bash"]
     0B
<missing>                                                                   10 days ago   /bi
n/sh -c #(nop) ADD file:bd3bc6e983b68e9cb252c4172dfce4a5e247e8d2b9823cdb20a14f946870be04 in
/     80.5MB
```

Figure 2.5 – Reviewing all the steps that were used to create a container image

Image history must be read in reverse order, starting from the latest line. We will start with an ADD key, which represents the initial FROM key from our Dockerfile. This is because the FROM key is interpreted as copying all the base image content on top of the base layer.

We used `--no-trunc` to be able to read the full command line from the output. We can easily see that this image was created using the `/bin/sh -c apt-get update -q && apt-get install -qq curl` command. The `docker image history` command will show us the steps that were executed to build any image created from a Dockerfile, but it won't work for interactively created ones.

Let's see a simple example of installing a Postfix mail server using the *Debian* image:

```
frjaraur@sirius:~$ docker run -ti debian:stable-slim@sha256:f711bda490b4e5803ee7f634483c4e6fa7dae54102654f2c231ca58eb233a2f1
root@a648df9c17c7:/# apt-get update -qq
root@a648df9c17c7:/# apt-get install -qq postfix
debconf: delaying package configuration, since apt-utils is not installed
Selecting previously unselected package libpython3.9-minimal:amd64.
(Reading database ... 6661 files and directories currently installed.)
Preparing to unpack .../libpython3.9-minimal_3.9.2-1_amd64.deb ...
Unpacking libpython3.9-minimal:amd64 (3.9.2-1) ...
Selecting previously unselected package libexpat1:amd64.
Preparing to unpack .../libexpat1_2.2.10-2+deb11u5_amd64.deb ...
Unpacking libexpat1:amd64 (2.2.10-2+deb11u5) ...
Selecting previously unselected package python3.9-minimal.
Preparing to unpack .../python3.9-minimal_3.9.2-1_amd64.deb ...
Unpacking python3.9-minimal (3.9.2-1) ...
Setting up libpython3.9-minimal:amd64 (3.9.2-1) ...
Setting up libexpat1:amd64 (2.2.10-2+deb11u5) ...
Setting up python3.9-minimal (3.9.2-1) ...
Selecting previously unselected package python3-minimal.
(Reading database ... 6955 files and directories currently installed.)
Preparing to unpack .../0-python3-minimal_3.9.2-3_amd64.deb ...
Unpacking python3-minimal (3.9.2-3) ...
Selecting previously unselected package media-types.
Preparing to unpack .../1-media-types_4.0.0_all.deb ...
Unpacking media-types (4.0.0) ...
Selecting previously unselected package libmpdec3:amd64.
Preparing to unpack .../2-libmpdec3_2.5.1-1_amd64.deb ...
Unpacking libmpdec3:amd64 (2.5.1-1) ...
Selecting previously unselected package libncursesw6:amd64.
Preparing to unpack .../3-libncursesw6_6.2+20201114-2_amd64.deb ...
Unpacking libncursesw6:amd64 (6.2+20201114-2) ...
Selecting previously unselected package readline-common.
```

Figure 2.6 – The manual execution of a Postfix mail package

Once the installation process has finished, we will be prompted to configure various aspects of the server. This configuration is completely interactive:

```
/share/perl/5.32.1 /usr/lib/x86_64-linux-gnu/perl5/5.32 /usr/share/perl5 /usr/lib/x86_64-linux-gnu/perl-base /usr/lib/x86_64-linux-gnu/perl/5.32 /usr/share/perl/5.32 /usr/local/l
/site_perl) at /usr/share/perl5/Debconf/FrontEnd/Readline.pm line 7.)
debconf: falling back to frontend: Teletype
Postfix Configuration
---------------------

Please select the mail server configuration type that best meets your needs.

 No configuration:
  Should be chosen to leave the current configuration unchanged.
 Internet site:
  Mail is sent and received directly using SMTP.
 Internet with smarthost:
  Mail is received directly using SMTP or by running a utility such
  as fetchmail. Outgoing mail is sent using a smarthost.
 Satellite system:
  All mail is sent to another machine, called a 'smarthost', for delivery.
 Local only:
  The only delivered mail is the mail for local users. There is no network.

  1. No configuration  2. Internet Site  3. Internet with smarthost  4. Satellite system  5. Local only
General type of mail configuration: 3

The "mail name" is the domain name used to "qualify" _ALL_ mail addresses without a domain name. This includes mail to and from <root>: please do not make your machine send out
mail from root@example.org unless root@example.org has told you to.

This name will also be used by other programs. It should be the single, fully qualified domain name (FQDN).

Thus, if a mail address on the local host is foo@example.org, the correct value for this option would be example.org.

System mail name: whatever@example.com

Please specify a domain, host, host:port, [address] or [address]:port. Use the form [destination] to turn off MX lookups. Leave this blank for no relay host.

Do not specify more than one host.

The relayhost parameter specifies the default host to send mail to when no entry is matched in the optional transport(5) table. When no relay host is given, mail is routed
directly to the destination.

SMTP relay host (blank for none):
```

Figure 2.7 – The Postfix installation is interactive because it asks users for specific configurations

The installation process will ask you for some configurations interactively and after that, the Postfix server will be ready to work. We can exit the container process by executing `exit`, and we will commit the container layer as a new image. We use `docker container ls -l` to only list the last container executed, and then we execute `docker commit` (or `docker container commit` – both commands will work as they both refer to containers) to save the current container layer as a new image:

```
frjaraur@sirius:~$ docker container ls -l
CONTAINER ID   IMAGE               COMMAND   CREATED         STATUS               PORTS     NAMES
a648df9c17c7   debian:stable-slim  "bash"    7 minutes ago   Exited (0) 12 seconds ago      recursing_goldwasser
frjaraur@sirius:~$ docker commit a648df9c17c7 postfix:test
sha256:a8768bd1ec8f7ef1c0916f7b7b604af3d2d3027c642751eb69849e822df0ca41
frjaraur@sirius:~$ docker image ls
REPOSITORY     TAG           IMAGE ID       CREATED         SIZE
postfix        test          a8768bd1ec8f   18 seconds ago  169MB
four           latest        3de85feddb15   4 hours ago     105MB
three          latest        55f07527310e   16 hours ago    105MB
frjaraur/two   180223        8e64333d6033   16 hours ago    105MB
two            latest        8e64333d6033   16 hours ago    105MB
frjaraur/one   180223        331ed31f8816   17 hours ago    105MB
one            latest        331ed31f8816   17 hours ago    105MB
alpine         latest        b2aa39c304c2   8 days ago      7.05MB
debian         stable-slim   4ea5047878b3   10 days ago     80.5MB
```

Figure 2.8 – Committing the container layer to create an image

However, as we previously mentioned about this method, we can't know the steps taken to create the image.

Let's try using the `docker image history` command:

```
frjaraur@sirius:~$ docker image history postfix:test
IMAGE          CREATED       CREATED BY                                   SIZE      COMMENT
a8768bd1ec8f   2 hours ago   bash                                         88.2MB
4ea5047878b3   10 days ago   /bin/sh -c #(nop)  CMD ["bash"]              0B
<missing>      10 days ago   /bin/sh -c #(nop) ADD file:bd3bc6e983b68e9cb_  80.5MB
frjaraur@sirius:~$ docker image history postfix:test --no-trunc
IMAGE                                                                        CREATED      CREATED BY
                                                                     SIZE    COMMENT
sha256:a8768bd1ec8f7ef1c0916f7b7b604af3d2d3027c642751eb69849e822df0ca41      2 hours ago  bash
                                                                     88.2MB
sha256:4ea5047878b3bb91d62ac9a99cdcf9e53f4958b01000d85f541004ba587c1cb1      10 days ago  /bin/sh -c #(nop)  CMD ["bash"]
                                                                     0B
<missing>                                                                    10 days ago  /bin/sh -c #(nop) ADD file:bd3bc6e983
b68e9cb252c4172dfce4a5e247e8d2b9823cdb20a14f946870be04 in /          80.5MB
```

Figure 2.9 – History does not show any commands when an interactive process was followed

All we can see in the output is that we used `bash` to do something. We will have the commands in its `.bash_history` file, but this is not how things should be done. If you must use this method in specific cases, such as when your application's installation requires some interactive steps, remember to document all the changes you made in the file to let other developers understand your process.

This method is not recommended because it is not reproducible, and we can't add any meta-information to the container image. In the next section, we will describe possibly the best method to remedy this, but it requires a lot of knowledge about your application binary files, libraries, and hidden dependencies.

Creating images from scratch

In this method, as its name already indicates, we will create an empty layer and all files will be introduced, using a packaged set of files. You may have noticed that all the image history we have seen so far involved using the ADD key as the first step. This is how a container runtime starts the building process – by copying the content of the base image.

Using this method, you can ensure that only explicit required files will be included in the container image. It works very well with coding languages such as Go because you can include all their dependencies in binaries; hence, adding your compiled artifacts will probably be enough for your application to work correctly. This method also uses Dockerfile files, but in this case, we will start with a simple FROM scratch line. This creates an empty layer for our files. Let's take a look at a simple example Dockerfile:

```
FROM scratch
ADD hello /
CMD ["/hello"]
```

This is a simple Dockerfile in which we just add files and meta-information. It will contain our binary file, on top of an empty structure, and the meta-information required to build a complete container image. As you can imagine, this method creates the most secure images because the attack surface is completely reduced to our own application. Developers can create images from scratch, packaging all the files required for their applications. This can be very tricky and lots of effort is required to include all dependencies. As mentioned earlier in this section, it works very well with applications running static binaries, which include all their dependencies.

This method can also be used to create images based on exotic or highly customized operating systems for which we don't have base images. In these cases, you should remove all non-required files and all references to the underlying hardware. This can be very difficult, and that's why it is usually recommended to use official container images. We will learn a bit more about the different types of images and how to ensure their origin, immutability, and ownership in *Chapter 3, Shipping Docker Images*.

Now that we know how to make container images using different methods, we should review the most important keys we will use in Dockerfiles.

Understanding common Dockerfile keys

In this section, we will take a look at the most important keys and their best practices. For full reference, it is better to review the documentation provided by Docker Inc. (https://docs.docker.com/engine/reference/builder/).

Container runtimes can create container images by reading a series of instructions written in a Dockerfile. Following this recipe-like file, a container runtime will assemble a container image.

FROM

All Dockerfiles always start with a FROM key. This key is used to set the base image and initialize the build process. We can use any valid container image as a valid value for the FROM key, and a scratch keyword is reserved to build images based on an empty layer.

A Dockerfile can include multiple image build processes, although usually, we will use different files for each process.

We can refer to images using their names and tags, and we can include their digests to ensure image uniqueness. If no tag is used, latest will be used automatically. Try to avoid this bad practice and always use the appropriate tag, or, even better, add its digest if you use public image registries. It is also possible to define a reference for each building process using the AS key. This way, we can share content between container images built with a unique Dockerfile. **Multi-stage building** is a practice in which we copy content from an image into others. We will explore a use case in the *Advanced image build processes* section later in this chapter.

As mentioned previously, a Dockerfile can include multiple build definitions, and we will name them using the AS key, which allows us to execute only specific targets, and the --target command.

To modify the behavior of the building process, we will use the ARG and ENV keys. We can use the --build-arg option to include additional arguments in the build process, and the container runtime will evaluate these values whenever the ARG key is found. The following line shows an example of how arguments can be passed to the build command:

```
$ docker image build -build-arg myvariable=myvalue -tag
mynewimage:mytag context-directory -file myDockerfile
```

Note here that we used a specific context and non-default Dockerfile by adding the -file argument. We also added myvalue to the myvariable variable, and we should have included the ARG key in the myDockerfile file to expand this value.

ARG

ARG is the only key that can be used before FROM to use build arguments – for example, to choose a specific base image. As a developer, you may want to have two different images for production and development, with some small changes, such as enabling debugging flags. We will use only one Dockerfile, but two build processes will be triggered, depending on the arguments passed. The following simple example may help you understand this use case:

```
ARG CODE_VERSION=dev
FROM base:${CODE_VERSION}
CMD /code/run-app
```

We will use -build-arg CODE_VERSION=prod whenever we need to build a production image, using a specific base image, base:prod, which may contain fewer files and binaries.

It is also usual to add the ENV key with ARG. The ENV key is used to add or modify environment variables for the containers that are used during the build process – for example, to add some path to LD_LIBRARY or change the PATH variable. ARG can then be used to modify environment variables at runtime.

To include meta-information in our final container image, we can use the LABEL key. Labels will help us identify a framework that's been used, a release version, the creator and maintainer of the content, and so on, or even a short description of its usage.

> **Important note**
>
> The OCI defines some conventional labels that may be used, and it would be interesting to use them instead of creating your own as many applications integrate this standard. You can review these labels at https://github.com/opencontainers/image-spec/blob/main/annotations.md. You will find labels such as org.opencontainers.image.authors, org.opencontainers.image.vendor, and org.opencontainers.artifact.description, all of which are standard and integrated into many container-related tools.

WORKDIR

All command executions defined in a Dockerfile will run relative to a working directory. We can change this by using the WORKDIR key. Once defined in our Dockerfile, all subsequent defined steps will use this environment – for example, to copy files inside the image layers.

COPY and ADD

Adding files to image layers is always needed. We will include our code or binaries, libraries, some static files, and so on. However, we shouldn't add certificates, tokens, passwords, and so on. In general, any content that requires some security or may change frequently must be included during runtime, and not in the image layers.

We can use the COPY and ADD keys to add files to image layers. The COPY instruction copies files and directories into specified image paths. If relative paths are used for the source, files must be included in the build context directory. If relative paths are used for the destination, the WORKDIR key will be used as the reference path. We can also copy files from other images declared in the same Dockerfile by using --from=<IMAGE_TARGET_NAME>. It is important to note that file ownership can be changed using the --chown=<USERNAME or USERID>:<GROUPNAME or GROUPID> command; if omitted, the user from the current container execution step will be used.

ADD works like COPY, but in this case, you can use remote URLs as a source, as well as TAR and gzip packaged files. If you use a compressed and packaged file, it will be unpackaged and uncompressed automatically for you in the specified destination.

Each file that's passed is verified against the checksums of image files, but the modification time isn't recorded, so you must be aware of the changes you make to your files before executing the building process. It is better to add a separate COPY line for those files you are often editing (for example, your application's code), or simply disable caching if you are not sure whether your file changes were correctly copied.

To avoid copying some files inside our project folders, we can use the .dockerignore file. This file contains a list of files that shouldn't be included in the Docker build context; hence, they will not be copied into the image layers.

RUN

The RUN key is used to execute the command line inside containers that were created during the build process. This action is fundamental to creating container images. All commands passed as a value to this key will be executed, and the resulting container layer will be committed as a new image layer; hence, all the RUN keys create a layer. Only the COPY, ADD, and RUN keys create layers; none of the other keys increase image size because they modify the resulting image behavior and add meta-information. You will probably see the RUN values use multiple lines, starting with && and ending with \. This simple trick will avoid the creation of new layers for each command executed. This way, you can concatenate multiple executions in one line and separate them into multiple lines for easy reading. Lines will be treated as if they were just one line, and thus, only one layer will be created. You should take care here because you may lose layer reusability, and this method can also mask errors during building processes. If you are having issues with one long line that contains a lot of commands, decouple them into multiple executions to isolate the error and, once solved, concatenate the lines again to create just one line.

A simple example would look like this:

```
RUN apt-get update -qq \
&& apt-get install --no-install-recommends --no-install-suggests -qq \
curl \
ca-certificates \
&& apt-get clean
```

These five lines will be interpreted like three different executions in the same container, so they will just create one layer for the final image.

It is important to understand that the build process does not store process states. This means that if we run a process and we expect it to be running upon the next RUN line, it won't because the container runtime only stores files from the container layer. This also applies to services or daemons. The build process will not work if you expect to have some processes already running and you apply some data or files to them. Each execution ends when the RUN line is processed.

USER

By default, the container runtime will execute all commands inside containers with `userid`, which is defined in the base image, and `root` if we are creating an image from scratch. You will find that most official Docker container images will run as `root`. Docker Inc. and other vendors prepare their images to allow you to install and manage additional software and binaries. You should ensure that your images run with the principle of *less privilege*, and thus, you must declare which user will run a container's main process. Dockerfile's USER key will help us define this user and even switch them multiple times in the same Dockerfile. Switching users will ensure that each Dockerfile line runs with the appropriate user, and containers created with this image will also run with the right user.

It is mandatory to avoid using containers with privileged users. This will essentially protect your applications and the underlying infrastructure. If you need to use `root` or any other privileged users, you should declare this situation explicitly. You can use a label, for example, to indicate that your image requires a privileged account to run.

> **Important note**
>
> If you are developing an application that requires a root user for its execution, you can use user namespace mappings. This feature lets us map a container's root user with a normal user in our host. If you need to set up this feature, you can follow the instructions provided at `https://docs.docker.com/engine/security/userns-remap/`.

ENTRYPOINT

Now, let's introduce how to declare which processes will run inside our container. The following keys add the meta-information required in an image to define which binary or script will run.

We will use the `ENTRYPOINT` key to define the main process the container will run. If this key isn't defined, the `/bin/sh` shell will be used for Linux containers and `cmd.exe` for Microsoft Windows containers. This key can come already modified in our base images, with a custom value, but we can also override it in our Dockerfile declaration to modify our container's behavior.

You can also use the `CMD` key, which allows you to specify which arguments should be passed to the shell, Windows command, or any other defined `ENTRYPOINT`. As such, we can think of the main process execution as the concatenation or sum of the `ENTRYPOINT` and `CMD` keys. For example, if we use the default `/bin/sh` shell's `ENTRYPOINT`, and we define our CMD key as `ping 8.8.8.8`, the final command that executes inside our container will be `/bin/sh -c ping 8.8.8.8`; in other words, a shell is expanded to execute our `ping` command. We can modify any of them during container creation, but remember that the user defined with the USER key will be the process's owner.

As mentioned previously, we can change image behavior by changing these very important keys. `ENTRYPOINT` and `CMD` are managed by the container runtime as arrays, although we can define them in our Dockerfile as strings, which are also commonly used to manually execute a container.

The container runtime concatenates both arrays to build the final command line. Due to this behavior, setting ENTRYPOINT as a string will force CMD to be ignored, but we can use CMD as a string while ENTRYPOINT is an array, and CMD will be treated as an array of 0 size.

Both values can be overridden on container execution, but usually, we will just customize the container arguments by using CMD; as such, this key can be used in the Dockerfile as a default value. As a developer, you should always provide as much information about your application's behavior as possible to make it usable, and LABEL, USER, and CMD must be present in your Dockerfiles.

EXPOSE

We should also add the EXPOSE key to this list, which defines what ports will be used by your application. You can define as many ports as required using ranges and the transport protocol that will be used, be it TCP or UDP. With this information, you will ensure that anyone using your application will know which ports your processes will be listening to.

The following scheme shows a simple Dockerfile stack in practice, including the container layer on top:

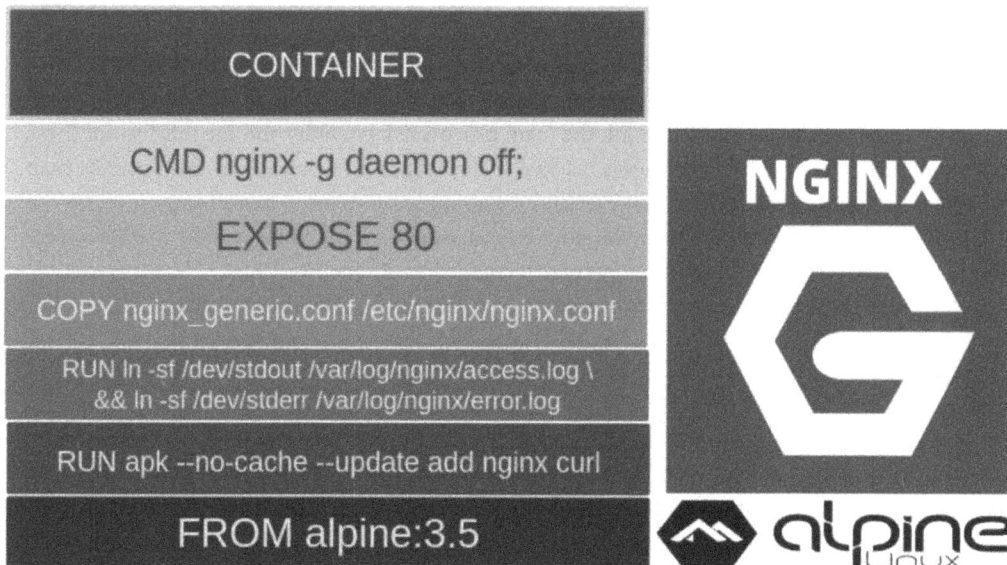

Figure 2.10 – The schema of container image layers created by using a Dockerfile.
The container layer is on top to keep track of changes created by processes

This figure represents the order obtained by using the docker image history command. For this example, we performed the following steps:

1. We used a simple alpine:3.5 base image. We updated the package sources and installed nginx and curl.

2. Next, we prepared NGINX logs to stream their output to `/dev/stdout` and `/dev/stderr`. This will ensure that we can read the application logs through the container runtime because these descriptors will be used by the container's main process.

3. We copied our custom NGINX configuration file, overwriting the default one.

4. We exposed port `80`, indicating that our main process will listen on this port.

5. Finally, we defined the default command line. In this case, `/bin/sh -c "nginx -g daemon off;"` will be executed every time we run a container using this image.

HEALTHCHECK

To ensure that our main process runs correctly within our container, we should add a health probe that will indicate whether this process is healthy or not. Let's imagine we run a web server application and it gets stuck. Processes will continue running but functionality will be completely lost. To remedy this, we can use the `HEALTHCHECK` key to define a command line that will check our main application's health. We can use a script or binary with arguments, such as `curl` for web servers, or a database client if we run a database server. What is very important for health checks is that the command exits correctly (`exit 0`) if the application is healthy. If our check process exits with any other signal, the container will die as a result of the application being set as unhealthy. The `HEALTHCHECK` key will allow us to manage how the checks must be executed, to keep the application up and running. We can modify the number of checks that will mark the main process as unhealthy and the interval for these checks. When the defined number of tries is reached with a negative response (any exit different than 0), the container runtime is informed that even if the main process seems to be running correctly, the service is not working, and the container should die. This usually means a new healthy one is created, but for this process to work, we should configure that container with the `restart: always` option. We will deep dive into container execution in *Chapter 3, Running Docker Containers*.

VOLUME

To end this section, we will review the `VOLUME` key. As the container image build process is based on the execution of multiple containers and storing their layers, this key is used to avoid certain directories from a container's life cycle. It is good practice to include this key to indicate which folders in your image you prepared for persistent storage. You can use this key after all the `RUN` keys to avoid losing an application's folders during the build process.

We have provided clear and simple examples of these keys to help you understand their usage at the end of this chapter, in the *Labs* section.

In the next section, we will present you with some of the most important command-line options that are commonly used to build container images.

The command line for creating images

In this section, we will take a closer look at Docker and other tools that you will commonly use to create container images for your projects.

We will start by reviewing the `docker` command line, which is the most popular tool for developers and users due to its simplicity and friendly environment.

Docker uses a common schema for all its arguments and options. We will use `docker <OBJECT> <ACTION> <OPTIONS>`. As a Docker container runtime identifies its objects by their IDs, it is common to omit the `<OBJECT>` primitive, but you should make sure that you use the right object. It is improbable that you will commit an error, but it is good practice to remember to include the object as part of the command.

Let's start with the basics – that is, learning which command will create an image.

Actions for creating images

We use the `build` action to create images using a Dockerfile. By default, it will search for a file in your current directory, but we can use any name and path to store our build manifests. We must always declare the build context, and usually, we will use the `-tag` option to define a name and tag for our image. Here is an example of its common usage:

```
frjaraur@sirius:~$ docker image build --tag example1:0.0 --file Dockerfile2 context2
[+] Building 0.3s (9/9) FINISHED
 => [internal] load build definition from Dockerfile2                                    0.1s
 => => transferring dockerfile: 39B                                                      0.0s
 => [internal] load .dockerignore                                                        0.1s
 => => transferring context: 2B                                                          0.0s
 => [internal] load metadata for docker.io/library/alpine:latest                         0.0s
 => [1/4] FROM docker.io/library/alpine:latest                                           0.0s
 => [internal] load build context                                                        0.1s
 => => transferring context: 33B                                                         0.0s
 => CACHED [2/4] RUN apk --update --no-cache add ca-certificates    curl    && rm -rf /var/cache/apk/*   0.0s
 => CACHED [3/4] COPY my-root-ca.crt /usr/local/share/ca-certificates                    0.0s
 => CACHED [4/4] RUN update-ca-certificates                                              0.0s
 => exporting to image                                                                   0.0s
 => => exporting layers                                                                  0.0s
 => => writing image sha256:f7bba7eac35eb058dc971defc2ff258cf4e36a556534949b71b18e2af5985b10   0.0s
 => => naming to docker.io/library/example1:0.0                                          0.0s
```

Figure 2.11 – Executing a simple image build process

In this example, `context2` is the name of the folder that contains all the files that should be sent to the container runtime, some of which should be copied to the final image.

Here are the most common options that you will probably add to `docker image build`:

- `--build-arg` is the way we can provide arguments for the build process. It is commonly used with the `ARG` Dockerfile key to modify image creation – for example, we can use `build` arguments to add some **certificate authority (CA)** certificates to the commands.

> **Important note**
>
> When you are behind a proxy server, it is very common to pass the well-known Linux `HTTPS_PROXY`, `HTTP_PROXY`, and `NO_PROXY` variables as arguments using `--build-arg`:
>
> ```
> docker build --build-arg HTTP_PROXY=$http_proxy \
> --build-arg HTTPS_PROXY=$http_proxy --build-arg NO_PROXY="$no_proxy" \
> --build-arg http_proxy=$http_proxy --build-arg https_proxy=$http_proxy \
> --build-arg no_proxy="$no_proxy" -t myimage:tag mycontext
> ```

- `--force-rm` will clean all intermediate containers. By default, all containers created during the building process will remain in your host unless your process ends successfully, hence occupying disk space. It is good practice to clean intermediate containers if you know that your build will create big layers – for example, when your application is compiled in containers and many dependencies are created, after which the process breaks.

- `--label` will let you add further labels to your container image. Adding all the required information, such as special library versions, the author, a short description, and anything that will let other developers understand your content, will be greatly appreciated.

- `--no-cache` will let us decide whether previously created and locally stored layers will be used. Using this argument, your build process will create fresh new layers, even if they already exist in your host. Be aware that without caching, all processes will be executed and store the intermediate container data locally; hence, the build will take more time. You will gain a faster build process by reusing the layers already included in your underlying host as much as possible. This can be very important when you are compiling your applications inside your image build, where a few minor changes will restart processes completely if no caching is used.

- `--target` is used to identify a build definition inside a Dockerfile. This can represent a specific compilation or a stage in a multi-stage build. We can use targets, for example, to maintain a unique Dockerfile with different build definitions, such as `small`, `complete`, and `debug`, each one requiring different steps and base images. We can trigger the build process for one specific definition to build the smallest release for a production environment. This can also be managed with arguments, with different base images chosen depending on variables.

- `--cpuquota`, `--cpu-shares`, and `--memory` will help us manage the resources available per build process. This is especially interesting if you are running out of resources on your desktop computer.

Now that we have learned about the command line to build images, let's look at managing images.

Managing container images

Container images will reside in your host in different directories, decoupling the data files from the meta-information. The location of your files will depend on the container runtime you are using, or in the case of **Podman**, they will probably be in your home directory. This runtime runs in rootless mode and without any daemon, so it is ideal for user containers. Irrespective of this, you will never directly access container image files.

One of the most commonly used actions within Docker (and any other container runtime client) is `list` (or `ls`), which is used to list the objects available in our host (or remote runtime). By default, images can be represented by their names (or repositories – we will learn how to store and manage images in these repositories in *Chapter 3, Shipping Docker Images*), IDs, tags, creation time, and size. In this context, size is the amount of space the image occupies in our host. The smaller the images, the better, and that's why you, as a developer, should be aware of the content of your images. Include only strictly necessary files, and think about your layer strategy if you are working with projects in which you share dependencies. Use the `.dockerignore` file to avoid non-required files as this can help you save a lot of space:

```
$ docker image list
REPOSITORY      TAG           IMAGE ID       CREATED             SIZE
example1        0.0           f7bba7eac35e   22 hours ago        9.51MB
postfix         test          a8768bd1ec8f   2 days ago          169MB
four            latest        3de85feddb15   2 days ago          105MB
three           latest        55f07527310e   2 days ago          105MB
frjaraur/two    180223        8e64333d6033   2 days ago          105MB
frjaraur/one    180223        331ed31f8816   2 days ago          105MB
one             latest        331ed31f8816   2 days ago          105MB
<none>          <none>        7ed6e7202eca   About a minute ago  72.8MB
alpine          latest        b2aa39c304c2   10 days ago         7.05MB
debian          stable-slim   4ea5047878b3   12 days ago         80.5MB
```

The preceding code snippet shows that we have multiple names (repositories) with the same content; we know this because they have the same ID. Images with the same ID are equal; they just differ in their tags. Therefore, we can add more than one tag to an image. We will use `docker tag <ORIGINAL> <NEWTAG>` to tag images. This is necessary to be able to upload images to registries as they are stored in their own repositories. Tags will help you identify images in our registry, but although tags are unique in each repository, we can have a lot to refer to the same image, and you should ensure that you are using the right image.

Developers may choose to tag their images following the application's life cycle, and you will probably encounter many images tagged using the `release.minor.fixes` model. This is good practice, and adding some key labels to identify the author, the project, and so on will improve your work.

You probably also noticed an image without any tag or name. This is a *dangling* container image that has been unused by others, and it is untagged because another one was created using the same repository and tag. It is not referenced by any image and now just occupies space. These dangling images should be removed, and we can use `docker image prune` to delete all of them.

To delete individual images, we can use `docker image rm <IMAGE>`. It is important to understand that images cannot be removed if there are references to them in containers or other images. We can force the removal by using `-force`, but it will only work if containers are stopped (or dead). It is also worth noting that multiple image tags can be deleted by using their ID, instead of their image repository names.

To review all the information included in the container image object, we can use `docker image inspect <IMAGE>`. Very useful information will be presented, including the image digest (if the image has a reference from a registry), the architecture for which the image was built, its labels, its layers, and the configuration that will be used to start the containers, such as environment variables and the main command to be executed.

It is worth introducing some formatting and filtering options we can use with some commands:

- `--filter` will allow us to use defined labels to filter objects from a list. This will work for any list provided by the container runtime – for example, if we labeled our images with the `environment` key, we could use it to obtain only specific images:

```
$ docker image list --filter label=environment
REPOSITORY        TAG             IMAGE ID          CREATED         SIZE
frjaraur/two      180223          8e64333d6033      2 days ago      105MB
frjaraur/one      180223          331ed31f8816      2 days ago      105MB
$ docker image list --filter label=environment=production
REPOSITORY        TAG             IMAGE ID          CREATED         SIZE
frjaraur/one      180223          331ed31f8816      2 days ago      105MB
```

- `--format` works with *Go templates* to manipulate output for listing (and logs from containers). The container runtime and clients work with *JSON* streams; hence, using these templates will help us interpret objects' data. For example, we can use `table` to obtain a table-like output, with the keys we need to review:

```
$ docker image list \
  --format "table {{.Repository}}:{{.Tag}}\t{{.Size}}"
REPOSITORY:TAG          SIZE
example1:0.0            9.51MB
postfix:test            169MB
frjaraur/two:180223     105MB
two:latest              105MB
one:latest              105MB
frjaraur/one:180223     105MB
alpine:latest           7.05MB
debian:stable-slim      80.5MB
```

We can get all the available keys and values by using `docker image ls --format "{{json .}}"`.

To obtain all the labels from a specific image, we can use `docker image inspect <IMAGE> --format "{{ index .Config.Labels }}"`.

In the next section, we will learn about the options available at the command line to share images between hosts or users.

Actions for sharing images

You may be thinking, all these examples were built on a host, so we need to be able to share our images with other developers, or even move them to the servers that are prepared to manage the application's life cycle (such as testing, staging, certification, or production). We can dump our container images and import them to new locations, but using image registries is a better option because these stores will be shared with the containers' orchestrators, and the container runtimes will automate the pull process for us.

Container image registries are considered the best way to share images. We will use `docker image pull` and `docker image push` to pull and push images, respectively. For this to work, you're usually required to log in to your registry. To be able to access your registry, you will require a username and a password. Docker Hub (`docker.io`) is probably the most recognized container registry. It works as a cloud service, providing an image store, scanning, and automations to build images. There are other options; all cloud providers offer registry services, and many code repositories also provide an image store (as they are considered code artifacts). We can deploy some of these solutions on-premises, but we can find also solutions such as Harbor, from VMware, which was prepared specifically for data centers. You may notice that your container runtime also stores images, and in fact, it can be considered a registry – a local registry. The `podman` command line, which supports all actions described in this chapter and can be used instead of the Docker client, will build your images as `localhost/IMAGE_NAME:TAG`, where `IMAGE_NAME` is the name of the repository. We will learn how image registries work in *Chapter 3*, *Shipping Docker Images*; for now, we will just review the most commonly used options to share images.

When someone asks us for an image, we can use `docker image save` to dump a container image to a file. This will completely package all its layers and meta-information. By default, standard output will be used to stream all data, but we can use the `-output` option to specify a file. You can copy this file to another workstation or server and execute `docker image load` to import all image layers and metadata. By default, the command will use standard input, but we can add the `-input` option to specify a file instead:

```
frjaraur@sirius:~$ docker image save debian:stable-slim --output /tmp/debian_stable-slim.tar
frjaraur@sirius:~$ ls -lart /tmp/debian_stable-slim.tar
-rw------- 1 frjaraur frjaraur 84009472 Feb 23 21:54 /tmp/debian_stable-slim.tar
frjaraur@sirius:~$ du -sh /tmp/debian_stable-slim.tar
81M     /tmp/debian_stable-slim.tar
frjaraur@sirius:~$ docker image inspect debian:stable-slim --format='{{ .Size }}'
80514821
frjaraur@sirius:~$ tar -tvf /tmp/debian_stable-slim.tar
-rw-r--r-- 0/0       1463 2023-02-09 04:22 4ea5047878b3bb91d62ac9a99cdcf9e53f4958b01000d85f541004ba587c1cb1.json
drwxr-xr-x 0/0          0 2023-02-09 04:22 a2e3227ecf8ad23add179f1e2ef6e9a64fe26566bd0b66e2b810a97d061aaa76/
-rw-r--r-- 0/0          3 2023-02-09 04:22 a2e3227ecf8ad23add179f1e2ef6e9a64fe26566bd0b66e2b810a97d061aaa76/VERSION
-rw-r--r-- 0/0       1138 2023-02-09 04:22 a2e3227ecf8ad23add179f1e2ef6e9a64fe26566bd0b66e2b810a97d061aaa76/json
-rw-r--r-- 0/0   84000256 2023-02-09 04:22 a2e3227ecf8ad23add179f1e2ef6e9a64fe26566bd0b66e2b810a97d061aaa76/layer.tar
-rw-r--r-- 0/0        207 1970-01-01 01:00 manifest.json
-rw-r--r-- 0/0         94 1970-01-01 01:00 repositories
frjaraur@sirius:~$
```

Figure 2.12 – Saving images to files for sharing is easy

We can verify that the image size is retained, and if we list the files included in the package file, we will obtain the layers and metadata files.

The Docker client can be used with docker image load to integrate this image into our local registry, but we can also use docker image import to only upload image layers. This is interesting as it can be used as the base image for builds from scratch, but be aware that without the metadata manifest JSON file, you would not be able to execute a container. You will need to add its exposed ports, user, main process, arguments, and so on.

As you can imagine, docker image save and docker image load work in small environments, but they don't when you need to distribute files on a dozen servers. Images are hard to sync if you don't maintain good tag maintenance; hence, try to use representative tags and label your images to help others understand their content.

Before reviewing some best practices and recommendations, we will learn about some topics that will help us optimize our workflow so that we can build new images.

Advanced image creation techniques

In this section, we will review some options and techniques available to speed up the building process and optimize image sizes.

In *Chapter 1, Modern Infrastructure and Applications with Docker*, we learned that images are a package of layers. These layers are distributed one over another, containing all the files, and the merging of all these layers gives us a distribution of files optimized for disk space reduction, using CoW filesystems. When a file from a lower layer has to be modified, it is copied to the top layer if it doesn't exist there yet. All unmodified files are used in read-only mode. With that said, it is easy to understand that managing the CoW process correctly will help speed up image creation times.

Whenever we add new RUN commands at the end of our Dockerfile, all previous layers will be used (unless we specify --no-cache); hence, the container runtime just needs to create new layers according to these new changes. However, whenever we add a new line to copy a new file in the middle of the Dockerfile, or even when a file has been modified, the layers included after this change are invalidated. This occurs with COPY, ADD, and RUN because these Dockerfile keys add new layers, but WORKDIR and ENV can also modify the building process behavior and, hence, the subsequent layers. Once a layer changes, the container runtime has to rebuild all downstream layers, even if we didn't modify any line in our Dockerfile after the aforementioned change.

Here are some recommendations that may help your building process:

- Multi-stage builds are key to minimizing and securing container images. We will define different targets in our Dockerfile to use them as stages to compile our code and dependencies, and we will add only the required files to the final image. With this technique, we can ensure that no compilers will be included in the final image. This is a simple example:

```
FROM alpine:3.17.2 as git # First stage, install git on small
Alpine
RUN apk add git
FROM git as fetcher # Second stage, fetching repository
WORKDIR /src
RUN git clone https://gitlab.com/myrepo/mycode.git .
FROM nginx: 1.22.1-alpine as webserver
COPY --from=fetcher /src/html/ /usr/share/nginx/html
```

This is a very simple Dockerfile; the final image contains only the docs directory, retrieved from our Git code repository. We will see a better example in this chapter's *Labs* section.

- Ordering layers is key to speeding up and maintaining application changes. Try to find the best logical order to declare your Dockerfile's recipe. If we have some time-intensive tasks, such as installing a lot of software packages, it is preferable to make these changes at the beginning of the build process. Conversely, the files that we change more often, probably our application's code, should be close to the end of the Dockerfile.

- This also works with the COPY key; if your application has a lot of dependencies, copying all your code and requirements at once can be problematic. It is better to split your files into different COPY sentences and copy your module requirements declaration files, then update these dependencies, and after that, copy the code for building. This ensures that all our code changes will not cause the dependencies to be downloaded again in the container-building process.

- We have to remind you again that you should only keep the necessary files inside container images. Avoid any unnecessary files. This will increase the building time and the final image size, and sometimes, it may be relevant to decide where to store them. Also, using .dockerignore will help you avoid sending unnecessary files to the container runtime, even if they will not be kept in the final image. Avoid copying full directories using COPY . /src if you are unsure of the content, any previous artifact builds, whether you are going to re-build them during image creation, or the logs, for example.

- Avoid non-required dependencies when you install packages. Depending on your base operating system distribution, you will have different arguments or options to only install specific packages, avoiding, for example, the recommended, but not required, associated packages. You will probably need to update the packages list before installing; do this once at the beginning if you don't add or modify any package repository. It is also recommended to clean a package cache when you are not going to install any other package. We can use RUN --mount type=cache,target=DIRECTORY_PATH <INSTALL_EXPRESSION> to install packages. This option will keep the content of the defined directory between different build processes, which will speed up installing new software.

- Sensitive information shouldn't be included inside container images. It is possible to include some files with passwords, certificates, tokens, and so on in your Dockerfile using the COPY or ADD keys, or even as arguments for your docker build command, and remove them before finishing. Although these don't look like bad solutions at first, they are not good enough because unconsciously, you can leave sensible data behind. A multi-stage build can help us if secrets are used to download binaries or libraries, and we can easily copy them to a final stage without adding any sensible data to its layers. However, there is a better solution – using buildx. This Docker tool includes the option to mount secrets only during specific RUN steps, without storing them in any layer, as if they were a file from a volume. Here is a simple example of its usage:

```
FROM python: 3.9.16-alpine3.17
COPY mycript.sh .
RUN --mount=type=secret,id=mysecret ./myscript.sh
```

To pass a value to the mysecret key, we can use an environment variable – for example, we can execute the build process with the following command line:

```
$ SECRETVALUE="mysecretpass" docker image buildx build --secret
id= SECRETVALUE <CONTEXT>
```

> **Important note**
>
> buildx even allows us to mount files with data, such as user credentials, tokens, certificates, and so on, for use as secrets inside containers running within the build process, by using docker image buildx build –secret id=mysecret,src=<FULLPATH_ TO_SECRETFILE>. By default, these files will be included inside containers in /run/ secrets/<SECRETID>, but we can add target to the Dockerfile's mount definition with a full path to the destination file we want to create.

- It is good practice to keep layers as small as possible. We will try to use RUN, COPY, and ADD, executing as many changes as possible, although this may impact layer reusability. We will combine multiple RUN executions into one line. Fewer Dockerfile lines mean smaller caching, which is good, but you can't reuse layers too often for new images. Any small variation between your Dockerfiles will invalidate caching from one Dockerfile to another.

> **Important note**
>
> We can use the heredocs format to combine multiple lines. This improves Dockerfile readability. For example, we can write the following:
>
> ```
> RUN <<EOF
> set -e
> apt-get update -qq
> apt-get install mypackage1 mypackage2
> EOF
> ```

- Docker client installation also provides the unique features of buildx to help us reduce building times and size. We can configure garbage collections to remove unused layers, based on time, and enable remote caching locations. This feature improves CI/CD pipelines that use distributed caches for projects that must compile a lot of dependencies or low-level languages, such as *C* or *Rust*.

- Multiple-processor architectures, such as riscv64 or arm64, can be built by using docker buildx build -platform, with one unique Dockerfile. In the past, we usually had different Dockerfiles, one for each architecture. Machines to use these different processors were also required, and the building process was executed on each one. This new feature allows you to prepare images for different platforms on your laptop with Docker Desktop. We will prepare a container image for arm64 in this chapter's *Labs* section.

- We can considerably reduce the final image size by using -squash when the image contains many layers. Squashing container images is an experimental feature that's available in the Docker container runtime. This means that we need to enable **experimental features** in our Docker Engine's docker.json file, and once configured, we will be able to use the docker image build -squash command. Reducing the number of layers to one will reduce its size, but you will lose the advantage of sharing layers. It's important to mention here that you shouldn't expect miracles. Squashing images depends on the number of layers used; hence, the final size may be pretty much the same as when fewer layers are used.

Before starting with the labs, we will review the content learned in this chapter by providing an overview of the best practices to build your container images.

Best practices for container image creation

In this section, we are going to recommend a list of the best practices you can follow to create your applications, thus improving your applications' security, reusability, and building processes:

- Only include the files that are strictly necessary for your application. Don't install packages, binaries, libraries, and any file your application doesn't need, and keep image content as small as possible, exposing a minimal attack surface.

- Use the `.dockerignore` file to avoid passing unnecessary files from your build context to container runtimes.

- Prepare debugging versions of your images, including some binaries or tools that may help you resolve an issue, but never use these images in production.

- Prepare the logic of your Dockerfiles to accommodate your changes; hence, include your code close to the end of the file, and think about how many modules or dependencies may need to be changed to execute the updates in the proper section.

- Use layer caching whenever it is possible to speed up the build process and remember that using many layers will allow reusability but affect performance when files need runtime changes.

- Never use `root` in your applications unless it is strictly required. If you do, you should understand its risks and manage them. You can use the `USER` key multiple times to change the execution user during builds, but always finish your Dockerfile with a non-root user.

- Never include sensitive information, such as certificates, passwords, and tokens, in your final container images. This information should be provided at runtime. Use Docker's `buildx` to include secrets only during the build process.

- Declare all your application requirements, such as your process user, the exposed ports, and the command line to be executed, in your Dockerfile. This will help other developers use your applications.

- Use labels to add information about your application's life cycle, maintainer, special libraries that are required, and so on. This information will be great for other developers to help them understand how they can integrate their code into your images or evolve your Dockerfiles.

- Image size matters, especially if you are running your containerized applications in a distributed environment. Container runtimes must download images if a container must be created on a host. Depending on the number of changes you make to your images, this can be a challenge, and resilience in the face of application issues may be affected if your platform defines an *always-pull* policy. We have covered some techniques to reduce image size; use them, but remember that a layer's reusability may be affected.

With this list, you can prepare your own container image creation workflow. Some of this advice can be tricky and requires some practice, but I can assure you that it is worth it, and you will deliver quality images for your applications.

Now that we have seen the different methods to build images, the command line we will commonly use, and some advanced techniques and advice to create good and secure images, it's time to put all this into practice with some labs in the next section.

Labs

The following labs will provide examples to help you put the concepts and procedures you've learned in this chapter into practice. We will use Docker Desktop or any other container runtime. We will use different tools such as **Podman** and **nerdctl** to show you some of the possibilities you have at hand, although some of the features that are required for specific labs may be only available with a specific tool (or one tool has a more friendly interface). In these cases, we will ask you to use a specific command-line interface.

The first step for all labs would be to download the most updated version of this book's GitHub repository at `https://github.com/PacktPublishing/Docker-for-Developers-Handbook.git`. To do this, simply execute `git clone https://github.com/PacktPublishing/Docker-for-Developers-Handbook.git` to download all its content. If you have already downloaded it before, ensure you have the newest version by executing `git pull` inside its directory.

We will start this section with a simple lab about using caching to speed up the building process. All commands presented in these labs will be executed inside the `Docker-for-Developers-Handbook/Chapter2` directory.

> **Important note**
>
> To show you the different tools to work with containers, we will use `nerdctl` in these labs, but you can use `podman` or `docker` (standalone or within Docker Desktop). Each tool has features and particularities, but most of the work within containers will execute similarly. We will explicitly notify you if some command shown requires a specific tool. Follow the specific instructions in this book's GitHub code repository to install each tool. We will use **Rancher Desktop**, which runs `containerd` as the container runtime and integrates the `nerdctl` command line inside WSL 2, but all labs can be executed with the Docker command line as well, with `docker` replacing `nerdctl`.

Caching layers

In this first lab, we will review the importance of caching to speed up the building process. We are going to use `nerdctl`, but `docker` or `podman` will work, as well as `buildah` (`https://buildah.io`), which is another open source tool prepared specifically to enhance the build process.

We will build a simple *Node.js* application that I prepared for quick demos a few years ago. Its only purpose is to show some information regarding the container in which it runs, the request headers, and its version. It will be interesting to better understand the load balancing processes within container orchestrators later on in this book, but we will focus on the build process for now:

1. First, we will move inside the `Chapter2/colors/nodejs` folder and execute a simple build, using `ch2lab1:first` as the image name and tag. We will use the following Dockerfile in this process:

```
FROM docker.io/node:18.14.2-alpine3.16
ENV APPDIR /APP
WORKDIR ${APPDIR}
COPY package.json package.json
RUN apk add --no-cache --update curl \
&& rm -rf /var/cache/apk \
&& npm install
COPY app.js app.js
COPY index.html index.html
CMD ["node","app.js","3000"]
EXPOSE 3000
```

Note that here, we have separated the content copy into three lines, although we could have used just one with all the content – for example, by using `COPY . . .`

> **Important note**
>
> As you may have noticed, this Dockerfile does not include any USER directive, but its application runs without any privileges because it is very simple and doesn't use any Linux capability or privileged port. Anyway, it is good practice to include the USER directive, and you can add it to your local repository. Everything described in the following steps will work.

2. We will add `time` to the `build` command to measure the time the build process takes:

```
$ time nerdctl build -t ch2lab1:one \
  --label nodejs=18.14.2 \
  --label=base=alpine3.16 \
  nodejs  --progress plain
#1 [internal] load .dockerignore
#1 transferring context: 2B done
#1 DONE 0.0s

#2 [internal] load build definition from Dockerfile
#2 transferring dockerfile: 311B done
#2 DONE 0.0s
```

```
#3 [internal] load metadata for docker.io/library/node:18.14.2-
alpine3.16
#3 DONE 1.1s


#4 [internal] load build context
#4 transferring context: 90B done
#4 DONE 0.0s
```

After these lines, our Dockerfile starts to be processed by the container runtime:

```
#5 [1/6] FROM docker.io/library/node:18.14.2-alpine3.16@
sha256:84b677af19caffafe781722d4bf42142ad765ac4233960e18bc-
526ce036306fe
#5 resolve docker.io/library/node:18.14.2-alpine3.16@
sha256:84b677af19caffafe781722d4bf42142ad765ac4233960e18bc-
526ce036306fe 0.0s done
#5 DONE 0.1s
#5 [1/6] FROM docker.io/library/node:18.14.2-alpine3.16@
sha256:84b677af19caffafe781722d4bf42142ad765ac4233960e18bc-
526ce036306fe
#5 sha256:aef46d6998490e32dcd27364100923d0c33b-
16165d2ee39c307b6d5b74e7a184 0B / 2.35MB 0.2s
```

Once the required layers have been loaded, our tasks to execute commands start. In our example, many packages must be installed:

```
#8 [4/6] RUN apk add --no-cache --update curl && rm -rf /var/
cache/apk && npm install
#0 0.115 fetch https://dl-cdn.alpinelinux.org/alpine/v3.16/main/
x86_64/APKINDEX.tar.gz
#8 0.273 fetch https://dl-cdn.alpinelinux.org/alpine/v3.16/
community/x86_64/APKINDEX.tar.gz
#8 0.503 (1/5) Installing ca-certificates (20220614-r0)
...
#8 0.601 (5/5) Installing curl (7.83.1-r6)
#8 0.618 Executing busybox-1.35.0-r17.trigger
#8 0.620 Executing ca-certificates-20220614-r0.trigger
#8 0.637 OK: 10 MiB in 21 packages
#8 3.247
#8 3.247 added 3 packages, and audited 4 packages in 2s
#8 3.247
#8 3.247 found 0 vulnerabilities
#8 3.248 npm notice
#8 3.248 npm notice New patch version of npm available! 9.5.0 ->
9.5.1
#8 3.248 npm notice Changelog: <https://github.com/npm/cli/
releases/tag/v9.5.1>
#8 3.248 npm notice Run `npm install -g npm@9.5.1` to update!
#8 3.248 npm notice
```

```
#8 DONE 3.3s
```

Once all execution lines are concluded, a tar file is created with the layer where changes were made:

```
#11 sending tarball
#11 sending tarball 0.6s done
#11 DONE 0.8s
unpacking docker.io/library/ch2lab1:one (sha256:7f63598f21445e-
5c6a051c9eca9c89367152dd59a4f1af366dc3291ae3e01930)…
```

Finally, our image is created, as we can observe in the latest line:

```
Loaded image: docker.io/library/ch2lab1:one
```

Note that we also obtained the time spent for the process because we included the `time` command before `nerdctl build`:

```
real    0m12.588s
user    0m0.009s
sys     0m0.000s
```

You can review the full output in the GitHub code repository for this book, but what's important here is the time it took to build this image: 12.588 seconds. This isn't bad, but as I recall, this project has few dependencies and Node.js is a *just-in-time* code language. Imagine this process if we needed to download code dependencies and compile them to obtain some binaries. It could take minutes or even longer.

3. Let's execute a new build after making some small changes to our code. We will just modify the version variable, which is *line 30* in the `nodejs/app.js` file. Change `var APP_VERSION="1.0";` to any other value, such as the following:

```
var APP_VERSION="1.1";.
```

Execute the first step again with a new tag, and note the CACHED lines in the output:

```
$ time nerdctl build -t ch2lab1:two \
--label nodejs=18.14.2 \
--label=base=alpine3.16  nodejs  \
--progress plain
#1 [internal] load .dockerignore
#1 transferring context: 2B done
#1 DONE 0.0s
```

Lines containing CACHED indicate that the layers were already created; we use these instead of executing the actual line to create a layer:

```
#7 [3/6] COPY package.json package.json
#7 CACHED
```

We changed the content of the app.js file; hence, a new layer must be created:

```
#9 [5/6] COPY app.js app.js
#9 DONE 0.0s


#10 [6/6] COPY index.html index.html
#10 DONE 0.0s
```

All successive lines will also create new layers because we *broke the cache*. A new line of changes was created:

```
#11 sending tarball
#11 sending tarball 0.6s done
#11 DONE 0.7s
unpacking docker.io/library/ch2lab1:two (sha256:bfffba0cd2d7c-
c82f686195b0b996731d0d5a49e4f689a3d39c7b0e6c57dcf0e)...
```

Finally, we obtained our new image:

```
Loaded image: docker.io/library/ch2lab1:two
real    0m1.272s
user    0m0.007s
sys     0m0.000s
```

All steps before copying our app.js file (*Step 9* in the preceding snippet) used the cached layers. Starting from *Step 9*, everything has to be recreated. However, because we used the appropriate logic in our Dockerfile, everything worked as expected. If you copy all the content of your code folder at once, any change will trigger a new layer; hence, if we change the content of index.html or our simple code in app.js, all the packages will be downloaded again.

4. Let's repeat this process by changing the copy process in our Dockerfile:

```
FROM docker.io/node:18.14.2-alpine3.16
ENV APPDIR /APP
WORKDIR ${APPDIR}
COPY . .
RUN apk add --no-cache --update curl \
&& rm -rf /var/cache/apk \
&& npm install
CMD ["node","app.js","3000"]
EXPOSE 3000
```

We execute the build process again. We expect it to last less than 12 seconds because the base image is already in our host:

```
$ time nerdctl build -t ch2lab1:three \
--label nodejs=18.14.2 \
--label=base=alpine3.16  nodejs  \
--progress plain
```

```
...
...
#6 [2/4] WORKDIR /APP
#6 CACHED

...
...
```

The same thing happens here. Changes were made in the Dockerfile; hence, a new build process changed the layers from this COPY step, and no cache can be used:

```
#7 [3/4] COPY . .
#7 DONE 0.0s

#8 [4/4] RUN apk add --no-cache --update curl && rm -rf /var/
cache/apk && npm install
...
...
#8 DONE 2.8s


...
...
#9 sending tarball 0.6s done
#9 DONE 0.8s
unpacking docker.io/library/ch2lab1:three (sha256:b38074f0ee-
5a9e6c4ee7f68e90d8a25575dc7df9560b0b66906b29f3feb8741c)...
Loaded image: docker.io/library/ch2lab1:three
real    0m4.634s
user    0m0.004s
sys     0m0.003s
```

It took 4.634 seconds, which is not bad, but remember that this is an example.

5. Once again, change APP_VERSION to a new value variable to see what happens if we build again. Change it from var APP_VERSION="1.1"; to var APP_VERSION="1.2";, and execute it again:

```
$ time nerdctl build -t ch2lab1:four \
--label nodejs=18.14.2 \
--label=base=alpine3.16 nodejs \
--progress plain
#1 [internal] load build definition from Dockerfile
...
...
#6 [2/4] WORKDIR /APP
#6 CACHED
```

The previous layers were cached, but a minimal change broke all the processes, and the layers must be recreated:

```
#7 [3/4] COPY . .
#7 DONE 0.0s
#8 [4/4] RUN apk add --no-cache --update curl && rm -rf /var/
cache/apk && npm install
#0 0.084 fetch https://dl-cdn.alpinelinux.org/alpine/v3.16/main/
x86_64/APKINDEX.tar.gz
#8 0.172 fetch https://dl-cdn.alpinelinux.org/alpine/v3.16/
community/x86_64/APKINDEX.tar.gz
#8 0.307 (1/5) Installing ca-certificates (20220614-r0)
...
#8 0.376 OK: 10 MiB in 21 packages
...
#8 3.433 added 3 packages, and audited 4 packages in 3s
...
#8 DONE 3.5s
...
...
#9 DONE 0.8s
unpacking docker.io/library/ch2lab1:four
(sha256:75ba902c55459593f792c816b8da55a673ffce3633f-
1504800c90ec9fd214d26)...
Loaded image: docker.io/library/ch2lab1:four
real     0m5.210s
user     0m0.007s
sys      0m0.000s
```

As you can see, it takes the same time as the previous execution because the container runtime can't identify and isolate the small changes and reuse the layers that were created previously.

In this lab, we reviewed how caching layers works and how to avoid build problems by choosing the right logic for our application's Dockerfile.

In the next lab, we will execute a multi-stage build process using an empty layer for the final image.

Executing a multi-stage build process

This is a very interesting use case since our code is in the Go language and we will be including static dependencies:

1. Move to the Chapter2/colors folder and use the go sub-folder this time. The multi-stage Dockerfile looks like this:

    ```
    FROM golang:1.20-alpine3.17 AS builder
    WORKDIR /src
    ```

```
COPY ./src/* .
RUN mkdir bin && go build -o bin/webserver /src/webserver.go

FROM scratch
WORKDIR /app
COPY --from=builder /src/bin/webserver .
CMD ["/app/webserver"]
USER 1000
EXPOSE 3000
```

2. We will use a golang:1.20-alpine3.17 image to compile our code. The compiled binary is copied from the *builder* image to our final image:

```
$ nerdctl build -t ch2lab1:go.1 \
--label golang=1.20 --label=base=alpine3.17  go  \
--progress plain
#1 [internal] load .dockerignore
#1 transferring context: 2B done
...
...
```

3. The first FROM key is reached and the image build process starts:

```
#6 [builder 1/4] FROM docker.io/library/golang:1.20-alpine3.17@
sha256:48f336ef8366b9d6246293e3047259d0f614ee167db1869bdb-
c343d6e09aed8a
...
...
#6 DONE 3.2s

#6 [builder 1/4] FROM docker.io/library/golang:1.20-alpine3.17@
sha256:48f336ef8366b9d6246293e3047259d0f614ee167db1869bdb-
c343d6e09aed8a
#6 extracting sha256:752c438cb1864d6b2151010a811031b48f0c-
3511c7aa49f540322590991c949d
...
...
#6 DONE 4.8s
#7 [builder 2/4] WORKDIR /src
#7 DONE 0.2s
#8 [builder 3/4] COPY ./src/* .
#8 DONE 0.0s
#9 [builder 4/4] RUN mkdir bin && go build -o bin/webserver /
src/webserver.go
#9 DONE 3.3s
```

4. The second FROM key is reached and a new image build process starts – in this case, just copying the content from the previous one:

```
#10 [stage-1 2/2] COPY --from=builder /src/bin/webserver .
#10 DONE 0.0s
#11 exporting to oci image format
...
...
#11 sending tarball 0.1s done
#11 DONE 0.3s
unpacking docker.io/library/ch2lab1:go.1 (sha256:527a2d2f-
49c7ea0083f0ddba1560e0fc725eb26ade22c3990bb05260f1558b0b)...
Loaded image: docker.io/library/ch2lab1:go.1
```

5. The final image is really small because it only contains our application code:

```
$ nerdctl image ls
REPOSITORY     TAG        IMAGE
ID         CREATED          PLATFORM       SIZE      BLOB
SIZE
ch2lab1        one       7f63598f2144    2 hours
ago           linux/amd64     186.6 MiB    51.7 MiB
ch2lab1        go.1      527a2d2f49c7    4 minutes
ago           linux/amd64     6.3 MiB     3.6 MiB
```

In this output, you can compare the different sizes we obtained (sizes may change because some updates may be expected in the code in this book's GitHub repository).

Creating images from scratch using binaries can be very tricky, but they are the best way of delivering our applications.

This lab showed you how you can create a container image from scratch by using static build binaries, which are the best application images you can create.

For the next lab, we will use Docker's buildx features, and therefore, we will use the docker command line.

Building images for different architectures

If you followed the lab with **Rancher Desktop**, **WSL**, and the nerdctl command line, please exit **Rancher Desktop** and launch **Docker Desktop** (or your own Docker engine implementation).

> **Important note**
> Podman and nerdctl also provide multiplatform support on new releases, and a multi-architecture build is commonly available; hence, any of these tools will be right for this lab.

Note that when you change from one container runtime to another, the list of images is completely different. Each container runtime manages its own environment as expected.

We will continue this lab inside the Chapter2/colors folder. We are going to build the image for multiple architectures – that is, amd64 and arm64:

1. We will use buildx with the --platform argument and arm64. But first, we will ensure that we can build images for other architectures by executing the docker buildx ls command:

```
$ docker buildx ls
NAME/NODE       DRIVER/ENDPOINT STATUS  BUILDKIT PLATFORMS
default *       docker
  default       default           running 20.10.22 linux/amd64,
linux/arm64, linux/riscv64, linux/ppc64le, linux/s390x,
linux/386, linux/arm/v7, linux/arm/v6
```

2. Now, we are ready to execute the arm64 architecture build:

```
$ docker buildx build -t ch2lab1:six \
  --label nodejs=18.14.2 \
  --label=base=alpine3.16 \
  nodejs --progress plain \
  --platform arm64 \
  --load –no-cache
#1 [internal] load build definition from Dockerfile
#1 transferring dockerfile: 32B done
#1 DONE 0.0s
...

...
```

Note that the aarch64 architecture image is downloaded during the process:

```
#8 0.348 fetch https://dl-cdn.alpinelinux.org/alpine/v3.17/main/
aarch64/APKINDEX.tar.gz
#8 0.753 fetch https://dl-cdn.alpinelinux.org/alpine/v3.17/
community/aarch64/APKINDEX.tar.gz
#8 1.204 (1/5) Installing ca-certificates (20220614-r4)
...

...
#8 1.341 Executing busybox-1.35.0-r29.trigger
#8 1.366 Executing ca-certificates-20220614-r4.trigger
...

...
#11 writing image sha256:2588e9451f156ca179694c5c5623bf-
1c80b9a36455e5f162dae6b111d8ee00fd done
#11 naming to docker.io/library/ch2lab1:six done
#11 DONE 0.1s
```

As you can see, the `arm64` Alpine image was used, even though we used the same Dockerfile from previous labs.

3. We can verify this image architecture by using `docker inspect`:

```
$ docker image inspect ch2lab1:six \
--format='{{.Architecture}}'
arm64
```

The final image is prepared for the `arm64` architectures and can be used in some QNAP NAS platforms.

In this build process, we also used `--load` and `--no-cache`. The first argument is used to load the image that was built into our container runtime. If we don't use this with Docker's `buildx`, the image is used as a cache for new builds only by default. To avoid any cached layer within this build process, we used `--no-cache`, and this ensures the complete execution of each step defined in the Dockerfile.

This lab showed you that you can prepare your images for any available architecture by using a unified Dockerfile and executing the build process with the `--platform` argument.

Summary

In this chapter, we learned how to create container images for applications. We started with an overview of CoW filesystems, which are the base for creating container images using layers. We looked at different methods to build images, along with their pros, cons, and examples. Using Dockerfiles is the best method because it provides a reproducible way of creating images by using different steps, written in order in these files. We provided a quick overview of the most important directives we can use in Dockerfiles and the command line and the arguments for using them. As the container-image-building process can be tricky, we presented some advanced features and practices we can use to improve our workflow in terms of speed, reusability, and quality.

In the next chapter, we will provide a quick overview of image registries, learn how to store and tag our images in them, and learn how to improve integrity and security by signing and scanning container images.

3

Sharing Docker Images

Sharing container images is key to being able to run your applications anywhere. You've built your application components on your workstation or laptop, and now you are ready to move them to different platform stages. This chapter will cover how images will be stored and shared with other users or orchestration platforms. We will also review various methods for signing container images to improve security in our development workflow. We will also learn how to use content scanners to find any possible security issues in our container images. By the end of this chapter, you will be ready to deliver secure and trusted images to production.

In this chapter, we will cover the following main topics:

- Container image registries and repositories
- Improving security by signing container images
- Analyzing container image content by using image scanners

This chapter will teach you about the different tools and techniques used to deliver secure images, which will really improve the use of software containers in your projects.

Technical requirements

You can find the labs for this chapter at `https://github.com/PacktPublishing/Containers-for-Developers-Handbook/tree/main/Chapter3`, where you will find some extended explanations, omitted in the chapter's content to make it easier to follow. The *Code In Action* video for this chapter can be found at `https://packt.link/JdOIY`.

Container image registries and repositories

In *Chapter 1, Modern Infrastructure and Applications with Docker*, we discussed why software containers have become so popular. In *Chapter 2, Building Docker Images*, we learned how we can create containers by using them as templates and building container images. Before deep-diving into container execution, we will learn, in this chapter, how to store and manage container images by using registries.

What is a registry?

A **registry** is a service where container images can be stored. This storage can be delivered as a service by cloud providers or on-premises by deploying your own registry. Cloud registries require zero maintenance from you; you just need to manage the usual housekeeping of unused images. Docker Hub (`docker.io`) is probably the most common service of this kind, but we also have Google's Container Registry (`gcr.io`) and similar services from Red Hat, GitHub, and GitLab, among others. Container images have become the new artifacts to run applications, and nowadays, many cloud code repositories include their own image registry services.

Nowadays, most registries can ingest and manage Docker's Image Manifest V2 and Open Container Initiative specifications, and therefore, we can use any of these formats for our images. These image schemas specify how image layers and metadata will be associated for use in our container runtime.

At the end of this chapter, we will run our own local registry and verify how images are stored.

Container runtimes completely manage all the actions we do with software containers or their images; hence, your client will just tell your runtime to download images before executing a container or upload them once built.

Understanding how image registries work

Before continuing to understand how image registries work, we should introduce here the complete container image naming convention. Images will always be referenced by using the following `name:tag` format:

```
full_FQDN_registry_name[:registry_port]/[username or team]/image_
repository:image_tag
```

This means that all image references must contain the registry where they are stored. You can omit this while your container runtime works with local images, but any time you use any remote image, you should always use the registry's **Fully Qualified Domain Name** (**FQDN**). Usually, registries are published on port 443, using **TLS/SSL**. This port will be used by default by container runtimes to connect to registries, but we can use any other port by specifying it when we refer to its images. **Image repository** is the term we commonly use to refer to the software repository where images are stored, but it is important to understand that images follow the same nomenclature used for code repositories, and hence, different tags are used to identify images within repositories.

Container images may be associated with the users or teams that own the images in the registries. This is common for these kinds of services, as the user is the one who pushed the image, which may be public or private. If you are part of a group, your image may be accessible to the members of the group. This role-based access may differ from one registry to another; you should ask your administrators or read the appropriate documentation for your registry. Docker images published in Docker Hub are shown as `docker.io/library`, and all their public repositories are published under this root group, similar to the schema presented at the beginning of this subsection.

Conversely, although tags help identify images, we learned in *Chapter 2, Building Docker Images*, that one image can have multiple tags; hence, a tag does not uniquely identify a specific image. **Image digests**, however, uniquely identify a set of layers for each image, and these really differ between images. To ensure that we use the correct image, we should use its **digest**.

Tagging images is not always easy; you must ensure, as a developer, that anyone can follow your work by using appropriate tags that reference the release on which users are running. Remember that you can add as many labels as you need to include additional information that is relevant for users. I always try to follow the code release schema, using X.Y.Z to represent major versions, minor versions, and fixes. It is a good idea to include a label regarding the commit in your code that generated the image artifact. This will help you track problems and also improve the application's life cycle. Automation will really help you to implement a workflow and follow your own tagging and labeling logic schema.

Whenever we pull an image from a registry, we get its digest within its metadata, as we can see in the following example, by executing docker image pull alpine:3.17.2:

```
Windows PowerShell                                         —    □    ×
PS C:\> docker image pull alpine:3.17.2
3.17.2: Pulling from library/alpine
63b65145d645: Already exists
Digest: sha256:69665d02cb32192e52e07644d76bc6f25abeb5410edc1c7a81a10ba3f0efb90a
Status: Downloaded newer image for alpine:3.17.2
docker.io/library/alpine:3.17.2
PS C:\> docker image ls --digests
REPOSITORY    TAG        DIGEST
                   IMAGE ID        CREATED        SIZE
alpine        3.17.2     sha256:69665d02cb32192e52e07644d76bc6f25abeb5410edc1c7a81a
10ba3f0efb90a    b2aa39c304c2    3 weeks ago    7.05MB
PS C:\> docker image pull docker.io/alpine:3.17.2
3.17.2: Pulling from library/alpine
Digest: sha256:69665d02cb32192e52e07644d76bc6f25abeb5410edc1c7a81a10ba3f0efb90a
Status: Image is up to date for alpine:3.17.2
docker.io/library/alpine:3.17.2
PS C:\>
```

Figure 3.1 – An image digest can be easily recovered after pulling an image from remote registries

Depending on which container runtime you use, a default container registry can be set. The Docker container runtime will use docker.io by default, which is why we can execute docker image pull alpine:3.17.2 to download an alpine:3.17.2 image, as shown in *Figure 3.1*. The digest in both cases is the same; hence, they are the same image. However, we can create a new image with this same name in our system. If we first delete the previously downloaded image (names are unique within a container runtime) and create a new one, it will have a completely different digest. This will happen even if we used the same Dockerfile because digests will also integrate the dates for each execution done during the build process.

> **Important note**
>
> We can download all images from a repository by using the `--all-tags` argument, as in the following example: `docker image pull --all-tags alpine`. This will download all images inside the `alpine` repository.

We can review which registries are included by default in Docker Desktop by using `docker info`. The following screenshot shows an example output using **Windows PowerShell**:

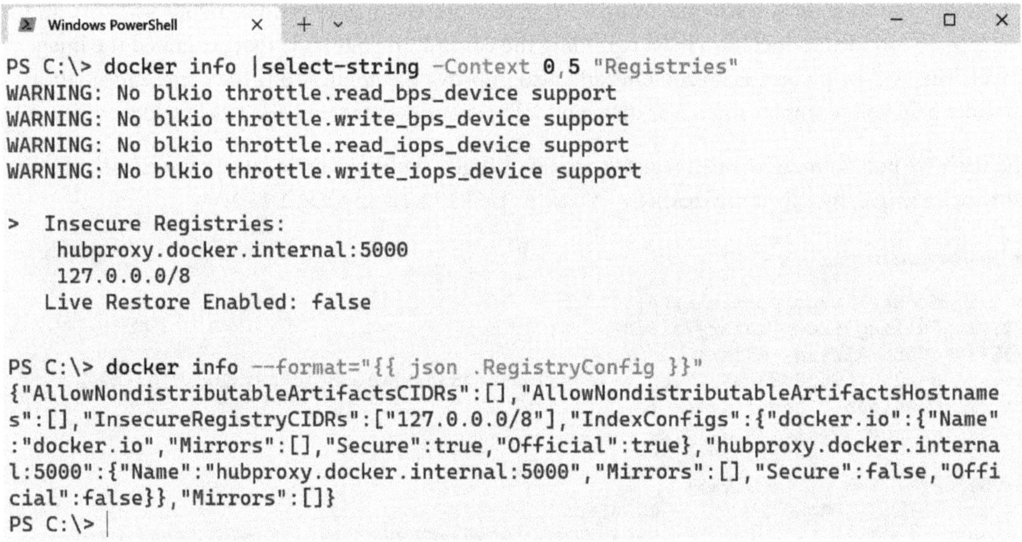

```
Windows PowerShell            ×   + ∨                              –   □   ×

PS C:\> docker info |select-string -Context 0,5 "Registries"
WARNING: No blkio throttle.read_bps_device support
WARNING: No blkio throttle.write_bps_device support
WARNING: No blkio throttle.read_iops_device support
WARNING: No blkio throttle.write_iops_device support

>   Insecure Registries:
    hubproxy.docker.internal:5000
    127.0.0.0/8
    Live Restore Enabled: false

PS C:\> docker info --format="{{ json .RegistryConfig }}"
{"AllowNondistributableArtifactsCIDRs":[],"AllowNondistributableArtifactsHostname
s":[],"InsecureRegistryCIDRs":["127.0.0.0/8"],"IndexConfigs":{"docker.io":{"Name"
:"docker.io","Mirrors":[],"Secure":true,"Official":true},"hubproxy.docker.interna
l:5000":{"Name":"hubproxy.docker.internal:5000","Mirrors":[],"Secure":false,"Offi
cial":false}},"Mirrors":[]}
PS C:\>
```

Figure 3.2 – Local container image registries used by our container runtime

In this example, the `localhost` registry and `hubproxy.docker.internal:5000` are shown. The local registry is used to store images locally. Desktop clients such as **Docker Desktop** and **Rancher Desktop** will show local images graphically:

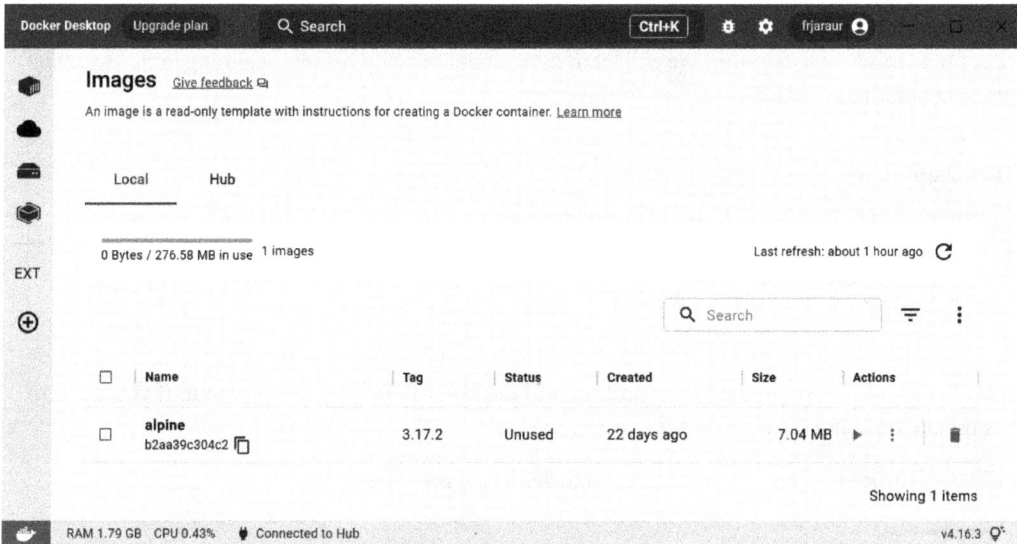

Figure 3.3 – An overview of Docker Desktop local images

In fact, Docker offers integration with your Docker Hub account. You can also view your remote images:

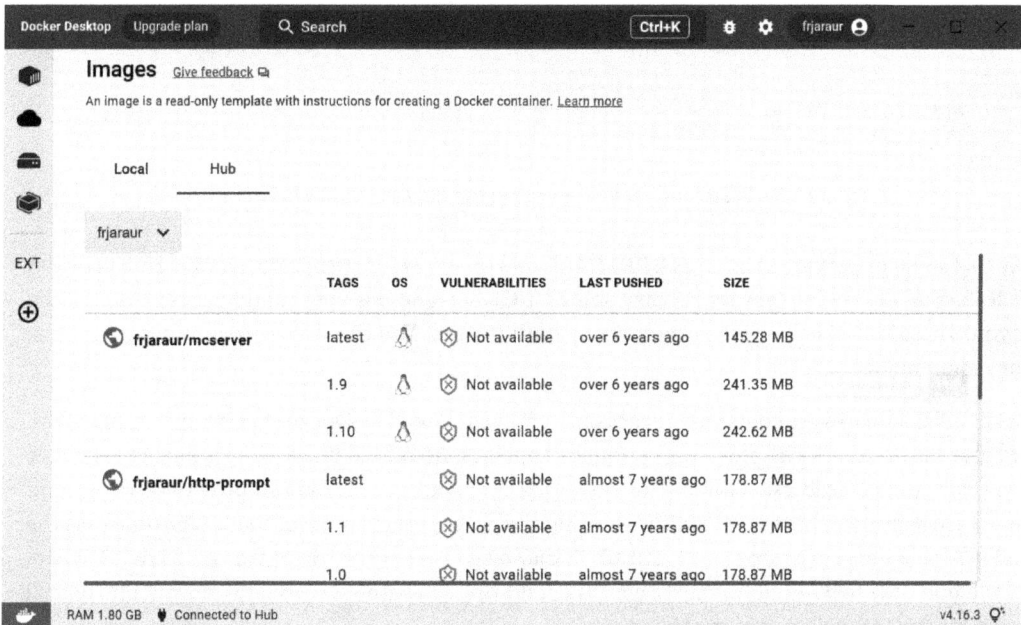

Figure 3.4 – An overview of Docker Desktop Docker Hub remote images

This interface also allows you to download and review the vulnerabilities found in the image's content if you pay for a Docker subscription. We will learn more about security content scanning in the *Scanning image content for vulnerabilities* section in this chapter.

> **Important note**
>
> Your container runtime may require some specific configurations to allow new container registries. All container image registries are expected to use **HTTPS**. Insecure registries are those that don't use secure protocols or use certificates that aren't trusted. Please add any registry with these characteristics to the `insecure-registries` list.

Registries will usually require a login, and we will use their FQDN to access them. If you need to access private repositories, an account will be required.

We will use `docker search` to find repositories based on a string:

```
PS C:\> docker search docker.io/nginx --limit 5
NAME                              DESCRIPTION                                STARS   OFFICIAL   AUTOMATED
nginx                             Official build of Nginx.                   18172   [OK]
bitnami/nginx                     Bitnami nginx Docker Image                 151                [OK]
ubuntu/nginx                      Nginx, a high-performance reverse proxy & we…  79
bitnami/nginx-ingress-controller  Bitnami Docker Image for NGINX Ingress Contr…  23                [OK]
kasmweb/nginx                     An Nginx image based off nginx:alpine and in…  4
PS C:\> docker search docker.io/nginx --limit 5
NAME                              DESCRIPTION                                STARS   OFFICIAL   AUTOMATED
nginx                             Official build of Nginx.                   18172   [OK]
bitnami/nginx                     Bitnami nginx Docker Image                 151                [OK]
ubuntu/nginx                      Nginx, a high-performance reverse proxy & we…  79
bitnami/nginx-ingress-controller  Bitnami Docker Image for NGINX Ingress Contr…  23                [OK]
kasmweb/nginx                     An Nginx image based off nginx:alpine and in…  4
PS C:\>
```

Figure 3.5 – Searching for images using the docker search command

Note that some images are marked as `OFFICIAL` and others as `AUTOMATED`. Docker Hub provides this feature as, depending on your registry, some CI/CD integrations may build images automatically for you. These values help mark automatically built images. You can use their automation if it fits your requirements.

Docker official images are container images built and maintained by Docker following best practices. They also ensure security by providing continuous updates. All official image code is publicly available on GitHub, and you can use it directly or customize it to your needs. You can provide feedback and contact Docker if you have any issues with using them. A great example of these repositories is the one for Alpine images (`https://github.com/alpinelinux/docker-alpine/`), where you will find code to build all Alpine images. The following screenshot shows how can you find official images for any technology you may need:

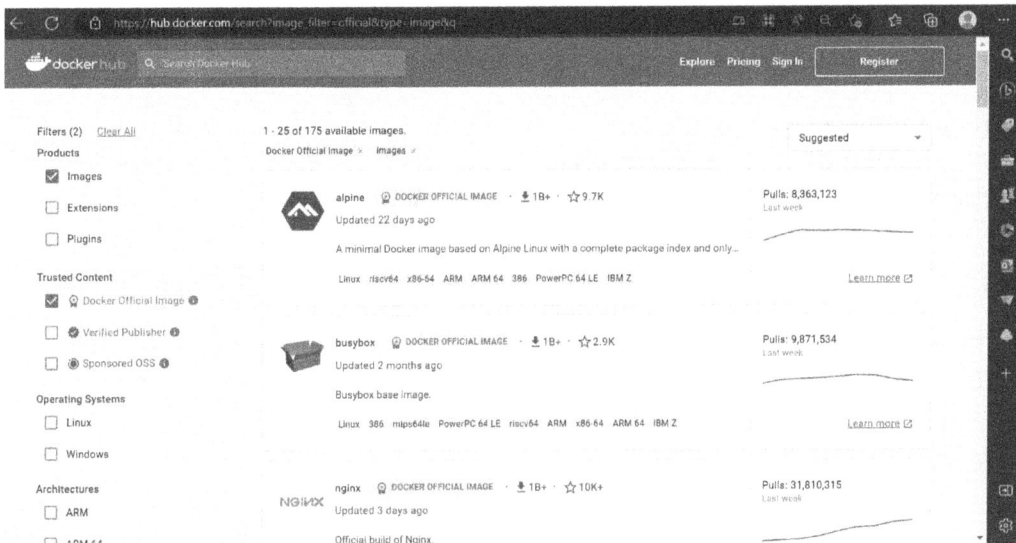

Figure 3.6 – An overview of official images available in Docker Hub

Software vendors and open source community projects also offer container images, prepared and maintained by them. The best method to integrate third-party projects or components in your applications will always be using images already prepared for you by Docker, a verified publisher, or sponsored open source software providers. These images are quite well documented, and you will be able to customize container behavior by using arguments and environment variables. You will avoid a lot of problems by using these images, instead of having to create your own from scratch. In the *Labs* section, we will review a very commonly used PostgreSQL database example to understand this better.

Searching in repositories

The Docker client provides certain repository search features, but they are not sufficient when you try to find specific images. You will probably use the Docker Hub web interface to perform fine-grained searches. Image registries publish their API, and we can use `curl` or any other web interface with arguments to find these images. Alternatively, we can use `skopeo` instead. This tool allows us to filter specific tags when we search inside a repository.

As mentioned previously in this section, registries provide an HTTP API that we can use to query specific repositories, tags, and so on, but it is not easy to use. We can use, for example, `curl https://myregistry.com:5000/v2/_catalog` to list all the images in the catalog. It is recommended to use `skopeo` instead because it will give you a clear and easy command line, especially when working with authenticated registries (certificates and login are required). It is available for different Linux flavors in their package repositories. You can follow the instructions for your distribution, or WSL, here: `https://github.com/containers/skopeo/blob/main/install.md`.

Unfortunately, `skopeo` isn't available in the Ubuntu 20.04 LTS package repository. We can install a newer WSL distribution by executing `wsl --install -d Ubuntu-22.04` in a Windows PowerShell terminal. Once the new Ubuntu 22.04 WSL distribution is ready, we can install the package by using `sudo apt-get update -qq && sudo apt-get install skopeo -qq`. Then, we can integrate the container runtime from either Docker Desktop or Rancher Desktop by navigating to **Settings | Resources | WSL Integration** or **File | Preferences | WSL**. The following screenshot shows the interface in Docker Desktop:

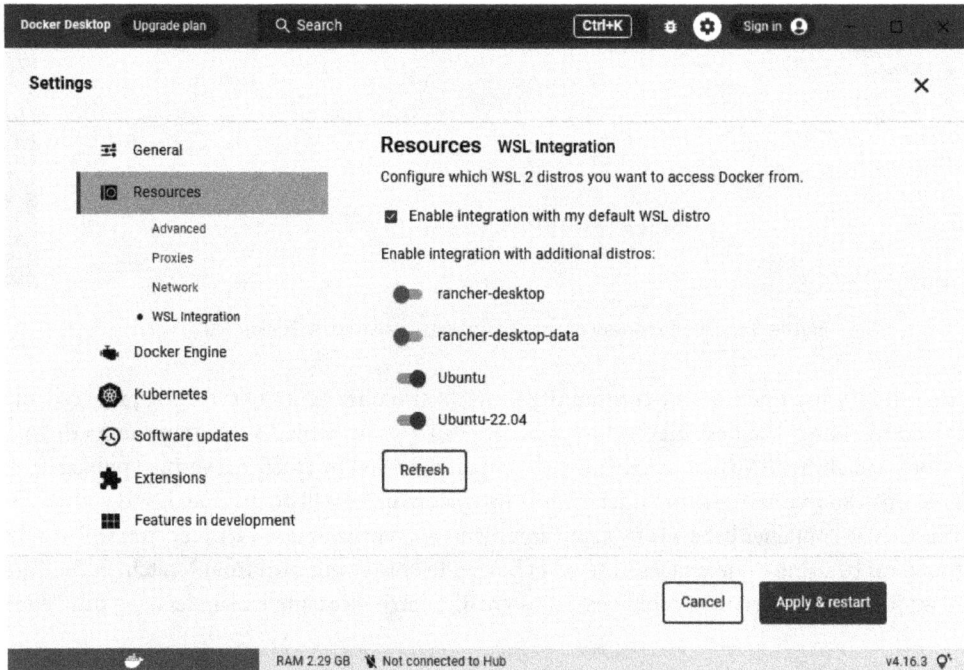

Figure 3.7 – Enabling Docker Desktop integration in Ubuntu 22.04 LTS WSL

With `skopeo`, we can easily list all tags included in a repository, as shown in the following example:

```
$ skopeo  list-tags docker://docker.io/frjaraur/colors
{
    "Repository": "docker.io/frjaraur/colors",
    "Tags": [
        "1.0",
        "1.1",
        "1.2",
        "1.5",
        "latest"
    ]
}
```

We can even inspect remote image information by using `skopeo inspect`:

```
$ skopeo inspect docker://docker.io/frjaraur/colors:1.0 \
--format="{{ .Digest }}"
sha256:cb7c1e49bcac66663aafea571ce5a6e6626e387c43b4836cc4d9e4c0e5d-
9faff
```

Instead of installing `skopeo` locally, we can use Red Hat's official image by using `docker container run`:

```
$ docker container run --rm quay.io/skopeo/stable \
inspect docker://docker.io/frjaraur/colors:1.0 \
--format="{{ .Digest }}"
Unable to find image 'quay.io/skopeo/stable:latest' locally
latest: Pulling from skopeo/stable
1a72627e77ed: Already exists
...
Digest: sha256:23f4b378c4aff49621e90289b33daf133462824b5e-
ba603b0834e25cb83a97ca
Status: Downloaded newer image for quay.io/skopeo/stable:latest
sha256:cb7c1e49bcac66663aafea571ce5a6e6626e387c43b4836cc4d9e4c0e5d-
9faff
```

Hopefully, this has given you a new perspective on using software containers. We can package tools in container images and use them instead of having to install software on our computers.

Now that we know how to store and reference images from registries, it is time to learn how we can improve the ownership and security of our images by signing them.

Improving security by signing container images

As we mentioned in the previous section, a digest is the only way we can validate which image we are really using. In this section, we will review how we can improve this by signing images. This will really ensure we use the right images, as we can check the signature and verify the ownership of every image.

We are going to analyze and learn about the Docker methodology used to sign images, but there are other methods available. We will use **Cosign** in the *Labs* section, which seems easier and integrates very well with the Kubernetes container orchestrator.

Docker created **Docker Content Trust** some years ago to integrate digital signatures in the container images management workflow and associate signatures with image tags. We will be able to have repositories with signed and unsigned images, for example, for local tests before moving on to a new stage. You, as a developer, create your images and decide which ones should be signed.

The signing process is based on a list of different keys that will be used to include meta-information in your registered image. Some of these keys are interactively managed, while others are calculated during the execution. These include the following:

- A `root` key is always used to start the signing process. You will be asked for this key, but you can also include it in your environment as a variable. It is very important to mention that if you lose this key, you will need to re-sign all your images, as there isn't a known process to recover them. Always have a backup of this key; losing it can be a real problem in a production environment because your old signatures will not be valid.

- Each repository will also have its own key. This key will be used to sign images in a specific repository, so losing it will only affect the images there. Nonetheless, you should keep it safe and have a backup.

- The `timestamp` key is added automatically to the final signature. This ensures security because each signature will always be completely different.

You will be asked to create a passphrase for the root key, and each repository key will be created randomly. Docker provides its own password manager under your user's `~/.docker` directory.

> **Important note**
>
> To back up your keys, prepare a `tar.gz` file with the contents of `~/.docker/trust/private`. We can execute the following command line:
>
> ```
> $ umask 077; tar -zcvf private_keys_backup.tar.gz ~/.docker/
> trust/private; umask 022
> ```

We can enable Docker Content Trust in our client by setting the `DOCKER_CONTENT_TRUST=1` environment variable. This will enable Content Trust for any new command executed, which means that only signed images will be enabled in your environment. If we just need to enable Content Trust for a specific command, we can use the `--disable-content-trust=false` argument.

Let's try enabling Content Trust for an image by setting the `DOCKER_CONTENT_TRUST` variable. In this example, we perform the following steps:

1. We first pull the `docker.io/busybox:latest` image:

   ```
   $ docker image pull busybox
   Using default tag: latest
   latest: Pulling from library/busybox
   Digest: sha256:7b3ccabffc97de872a30dfd234fd972a66d247c8cf-
   c69b0550f276481852627c
   Status: Image is up to date for busybox:latest
   docker.io/library/busybox:latest
   ```

2. Now, we enable DOCKER_CONTENT_TRUST=1 and download the same image again:

```
$ export DOCKER_CONTENT_TRUST=1
$ docker image pull busybox
Using default tag: latest
Pull (1 of 1): busybox:latest@sha256:7b3ccabffc97de872a30dfd-
234fd972a66d247c8cfc69b0550f276481852627c
docker.io/library/busybox@sha256:7b3ccabffc97de872a30dfd234f-
d972a66d247c8cfc69b0550f276481852627c: Pulling from library/
busybox
Digest: sha256:7b3ccabffc97de872a30dfd234fd972a66d247c8cf-
c69b0550f276481852627c
Status: Image is up to date for busybox@sha256:7b3ccabff-
c97de872a30dfd234fd972a66d247c8cfc69b0550f276481852627c
Tagging busybox@sha256:7b3ccabffc97de872a30dfd234fd972a66d247c-
8cfc69b0550f276481852627c as busybox:latest
docker.io/library/busybox:latest
```

3. I have prepared my own busybox image, docker.io/frjaraur/busybox-un-trusted:0.1. Let's see what happens if we try to pull an untrusted image when Docker Content Trust is enabled:

```
$ docker pull docker.io/frjaraur/busybox-untrusted:0.1
Error: remote trust data does not exist for docker.io/frjaraur/
busybox-untrusted: notary.docker.io does not have trust data for
docker.io/frjaraur/busybox-untrusted
```

This means that no valid signature is found, but if we disable Content Trust, we can pull the image without a problem:

```
$ export DOCKER_CONTENT_TRUST=0
$ docker pull docker.io/frjaraur/busybox-untrusted:0.1
0.1: Pulling from frjaraur/busybox-untrusted
Digest: sha256:907ca53d7e2947e849b839b1cd258c98fd3916c-
60f2e6e70c30edbf741ab6754
Status: Downloaded newer image for frjaraur/busybox-un-
trusted:0.1
docker.io/frjaraur/busybox-untrusted:0.1
```

In fact, we can't run any container using this untrusted image, which we can easily verify:

```
$ docker run -ti --disable-content-trust=false\
docker.io/frjaraur/busybox-untrusted:0.1
docker: Error: remote trust data does not exist for docker.io/
frjaraur/busybox-untrusted: notary.docker.io does not have trust
data for docker.io/frjaraur/busybox-untrusted.
See 'docker run --help'.
```

Conversely, if we run a trusted image, everything works as expected:

```
$ docker run -ti --disable-content-trust=false \
docker.io/busybox:latest ls -ld /tmp
drwxrwxrwt    2 root      root          4096 Jan  3 22:44 /tmp
```

4. We will now review the signatures in these images by executing the `docker trust` command line:

```
$ docker trust inspect \
--pretty docker.io/busybox:latest
Signatures for docker.io/busybox:latest


SIGNED TAG    DIGEST
SIGNERS
latest        7b3ccabffc97de872a30dfd234fd972a66d247c8cf-
c69b0550f276481852627c    (Repo Admin)

Administrative keys for docker.io/busybox:latest

  Repository Key:       02d15c99120886f6e02b4b0186522bc-
72d21a339ec35fad8af0a1b4a47c871d2
  Root Key:       074cad59e43e13b440b11d1b5521e20aa8633f-
c8f3928720590268895711d0c6
$ docker trust inspect --pretty docker.io/frjaraur/busybox-un-
trusted:0.1
No signatures or cannot access docker.io/frjaraur/busybox-un-
trusted:0.1
```

No signatures are included in `frjaraur/busybox-untrusted:0.1`; hence, this image cannot be used in a container runtime in which Docker Content Trust is enabled.

5. Let's start the signing process by creating a key for our user by using `docker trust key generate`:

```
$ docker trust key generate frjaraur
Generating key for frjaraur...
Enter passphrase for new frjaraur key with ID ceb39cd:
Repeat passphrase for new frjaraur key with ID ceb39cd:
Successfully generated and loaded private key. Corresponding
public key available: /home/frjaraur/frjaraur.pub
```

Our key is stored under `~/.docker/trust/private/`:

```
$ ls -lart ~/.docker/trust/private/
total 12
drwx------ 4 frjaraur frjaraur 4096 Mar  5 17:43 ..
```

```
-rw------- 1 frjaraur frjaraur  420 Mar  5 17:50 ceb39cd48cf78d-
478ffef211cc9da3e97ff9912ae60585254d6dc661076d0d2a.key
drwx------ 2 frjaraur frjaraur 4096 Mar  5 17:50 .
```

6. Now, we can sign our image. We will just retag the original busybox image with docker. io/frjaraur/busybox-trusted:0.1; hence, both images are the same, but the signature is related to the tag:

```
$ docker image tag busybox docker.io/frjaraur/busy-
box-trusted:0.1
$ docker image push docker.io/frjaraur/busybox-trusted:0.1
The push refers to repository [docker.io/frjaraur/busy-
box-trusted]
b64792c17e4a: Mounted from frjaraur/busybox
0.1: digest: sha256:907ca53d7e2947e849b839b1cd258c98fd3916c-
60f2e6e70c30edbf741ab6754 size: 528
Signing and pushing trust metadata
You are about to create a new root signing key passphrase. This
passphrase
will be used to protect the most sensitive key in your signing
system. Please
choose a long, complex passphrase and be careful to keep the
password and the
key file itself secure and backed up. It is highly recommended
that you use a
password manager to generate the passphrase and keep it safe.
There will be no
way to recover this key. You can find the key in your config
directory.
Enter passphrase for new root key with ID dfbeee2:
Repeat passphrase for new root key with ID dfbeee2:
Enter passphrase for new repository key with ID 9cfa33d:
Repeat passphrase for new repository key with ID 9cfa33d:
Finished initializing "docker.io/frjaraur/busybox-trusted"
Successfully signed docker.io/frjaraur/busybox-trusted:0.1
```

During the signing process, the root key and repository passphrases are requested. Remember that if you lose your root key, you will need to re-sign all your images, and users will need to pull new images because your old signature will be invalid.

We have signed and pushed the image to the Docker Hub registry, but this process can be executed using two steps. We can sign the container image by using docker trust sign docker.io/frjaraur/busybox-trusted:0.1 and then push it with docker image push docker.io/frjaraur/busybox-trusted:0.1.

7. Now, we can easily review our image signature:

```
$ docker trust inspect \
--pretty docker.io/frjaraur/busybox-trusted:0.1
Signatures for docker.io/frjaraur/busybox-trusted:0.1
SIGNED TAG    DIGES
T                                                        S
IGNERS
0.1           907ca53d7e2947e849b839b1cd258c98fd3916c-
60f2e6e70c30edbf741ab6754    (Repo Admin)
Administrative keys for docker.io/frjaraur/busybox-trusted:0.1
  Repository Key:        9cfa33df6e6b93596416b06bb82198a46be-
fb94479bbf5b0d92e73a213a30126
  Root Key:        f802546452481df2edc8b9670d30638e079164e7dc7187b-
698cd275d894531f4
```

It is important to say here that you will need a registry server with **Notary**. This project is responsible for managing your signatures. It is a client-server application that runs alongside your registry to integrate the signing part. Docker Hub already integrates a Notary server with their registry, which is why we are able to integrate our signatures in our image metadata.

If you are planning to use a local registry with Content Trust, you will need to also run and integrate a Notary server. You can learn more about Notary at `https://github.com/notaryproject/notary`.

We can revoke our signatures for a specific container image tag by using `docker trust revoke`. Signatures can be delegated to other users by sharing our public key. The integration of image signatures in your platform really depends on the registries you use. We have learned about the process for Docker Hub, but other solutions may implement different commands and options. We will show you how to use Cosign, which manages a different type of signature, but it also works to implement a good, secure supply chain, ensuring an image's provenance and ownership.

Now that we know how to ensure the uniqueness of container images by using their digests and signatures, we will continue to secure our images by implementing image content vulnerability scanning.

Scanning image content for vulnerabilities

Container images can be securely stored in registries, and we can track their provenance and ownership by reviewing their digests and signatures. It would be great if we could trust all the files included in the image layers. There are many solutions that can check whether any of the files included inside the image layers are somehow vulnerable to any reported issues or exploits that can affect your application's integrity. However, this requires new tools and effort.

Image scanning can be implemented in either your local development environment or the remote registries where images are eventually stored and shared. Most content scanners use well-known public and community-supported databases of known vulnerabilities and exploits. These will give us a list of **Common Vulnerabilities and Exposures** (**CVEs**) to compare against our content. Each binary or library in this list is identified by its digest, and it is easy to find out whether a file in our image layers is considered vulnerable in this list.

The Docker container runtime scanning facility is integrated into the client command line and can be executed with either images or Dockerfiles, which is very interesting because we can get an overview of possible problems in our images even before they are built.

All of Docker Hub's official images provide a pretty descriptive report of the security scanner results. This is also available for **verified publisher** (software vendors) and **sponsored OSS** (open source community-supported projects) images. We can access these reports from the **Tags** section on any of these images. The following screenshot shows the summary report for each tag in Postgres:

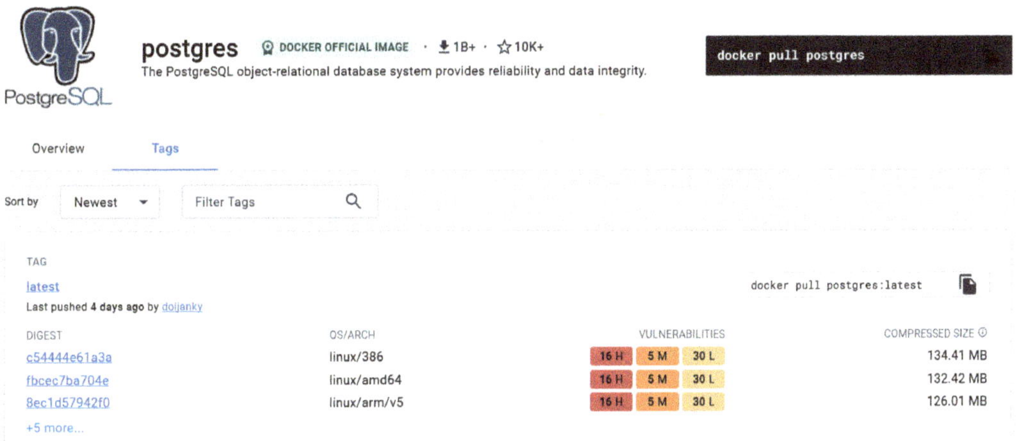

Figure 3.8 – A review of the vulnerabilities for the official Postgres image in Docker Hub

We can deep-dive by clicking on any tag. This will show us the CVE number associated with the vulnerability detected. The following screenshot shows some CVEs found for **stdlib 1.18.2**, included in the **postgres:latest** image layers:

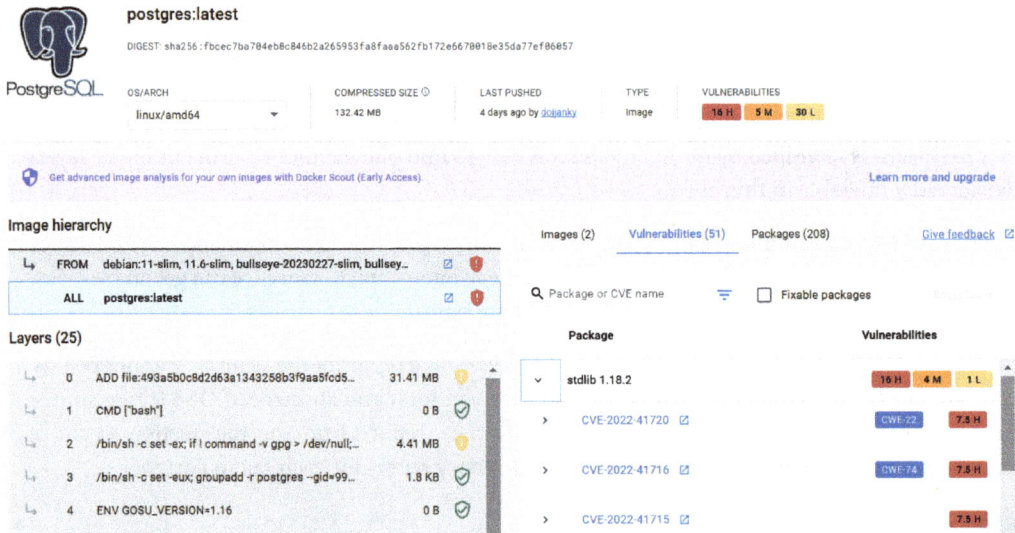

Figure 3.9 – Deep-diving inside the current vulnerabilities in the postgres:latest image

This report gives us a good idea of the health of Docker Hub's images. It is recommended to always review the scanner report of the official images we will use in our developments.

The Docker command-line scanning implementation uses the **Snyk engine**. It is accessible by executing `docker scan`. Here is an example of the execution against **postgres:latest**:

```
frjaraur@antares:~$ docker scan --accept-license docker.io/postgres:latest --severity high

Testing docker.io/postgres:latest...

Package manager:    deb
Project name:       docker-image|docker.io/postgres
Docker image:       docker.io/postgres:latest
Platform:           linux/amd64
Base image:         postgres:15.2-bullseye

✗ Tested 147 dependencies for known vulnerabilities, no vulnerable paths found.

According to our scan, you are currently using the most secure version of the selected base image

For more free scans that keep your images secure, sign up to Snyk at https://dockr.ly/3ePqVcp

-------------------------------------------------------

Testing docker.io/postgres:latest...

Package manager:    gomodules
Target file:        /usr/local/bin/gosu
Project name:       github.com/tianon/gosu
Docker image:       docker.io/postgres:latest

✗ Tested 3 dependencies for known vulnerabilities, no vulnerable paths found.

For more free scans that keep your images secure, sign up to Snyk at https://dockr.ly/3ePqVcp

Tested 2 projects, no vulnerable paths were found.
```

Figure 3.10 – Using the docker scan facility in our local environment

In the preceding screenshot, we used the `--accept-license` argument to use a non-interactive execution; otherwise, we would be asked to accept the agreement with the Snyk service. We also included `--severity` for filtering and only showed critical vulnerabilities. In this example, Snyk does not show any critical vulnerabilities, which differs from Docker Hub's website. This may depend on base image scanning. The Snyk scanner only provides 10 scans a month, which probably won't be enough. You can sign up to increase this limit or even subscribe to their service to increase the features available, avoiding any limitations. In this chapter's *Labs* section, we will learn how to use another scanning tool, **Trivy**, which will be executed as a container to avoid any installation in our working environment.

We have learned how to work with image registries and ensure security by scanning and signing images. In the next section, we will review some easy labs to put into practice the concepts we've discussed.

Labs

The following labs will provide examples to put into practice the concepts and procedures learned in this chapter. We will use Docker Desktop as the container runtime and WSL 2 (or your Linux/macOS Terminal) to execute the commands described.

The first step for all labs is always to download the most updated version of this book's GitHub repository at `https://github.com/PacktPublishing/Containers-for-Developers-Handbook. git`. To do this, simply execute `git clone https://github.com/PacktPublishing/ Containers-for-Developers-Handbook.git` to download all its content. If you have already downloaded it before, ensure you have the newest version by executing `git pull` inside the directory.

All commands presented in these labs will be executed within the `Containers-for-Developers- Handbook/Chapter3` directory.

Deploying and using your own local registry

In this first lab, we will deploy a simple unauthenticated and untrusted (HTTP, not HTTPS) registry. We will use the currently available Docker official registry image, which is `registry:2.8.1` at the time of writing. We can review its vulnerabilities by navigating to `https://hub.docker.com/ layers/library/registry/2.8.1/images/sha256-a001a2f72038b13c1cbee7 cdd2033ac565636b325dfee98d8b9cc4ba749ef337?context=explore`:

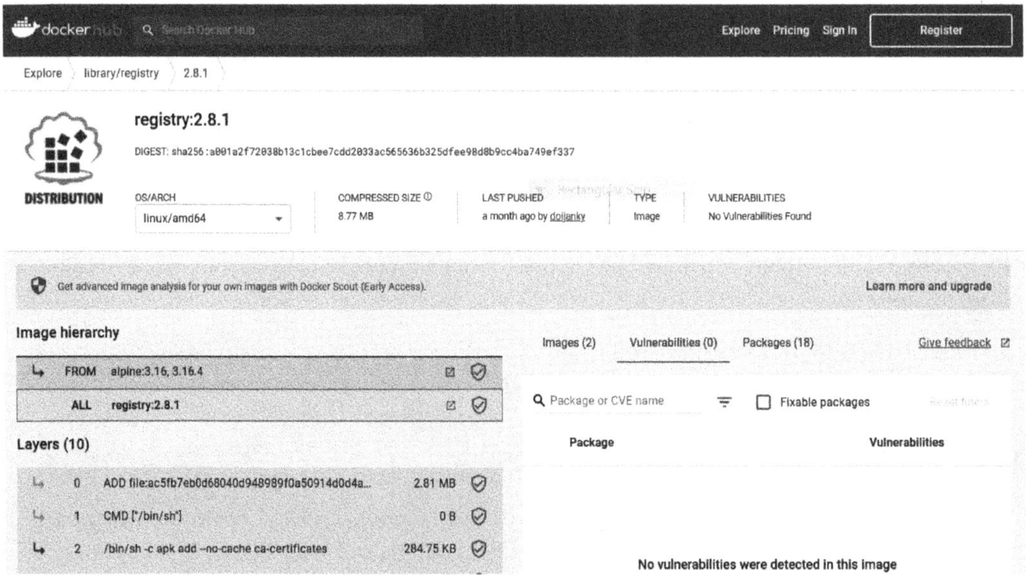

Figure 3.11 – An overview of the official Docker registry:2.8.1 vulnerabilities

We then perform the following steps:

1. Pull the `docker.io/registry:2.8.1` image:

    ```
    $ docker image pull docker.io/registry:2.8.1
    ...
    Digest: sha256:3f71055ad7c41728e381190fee5c4cf9b8f7725839dcf-
    5c0fe3e5e20dc5db1faStatus: Downloaded newer image for regis-
    try:2.8.1
    docker.io/library/registry:2.8.1
    ```

2. Now, review its **CMD, ENTRYPOINT, VOLUME**, and **EXPOSE** keys. These will show us which command will be executed, the port that will be used, and which directory will be used for persistent data:

    ```
    $ docker image inspect docker.io/registry:2.8.1 \
    --format="{{ .Config.Cmd }} {{.Config.Entrypoint }} {{.Config.
    Volumes }} {{.Config.ExposedPorts }}"
    [/etc/docker/registry/config.yml] [/entrypoint.sh] map[/var/lib/
    registry:{}] map[5000/tcp:{}]
    ```

    Port 5000 will be published, and a custom script will be launched with a configuration file as the argument. The /var/lib/registry directory will be used for our images; hence, we will map it to a local folder in this lab.

If you've already downloaded this book's GitHub repository, change to the `Chapter3` folder and follow the next steps from there. If you haven't, please download the repository to your computer by executing `git clone https://github.com/PacktPublishing/Containers-for-Developers-Handbook.git`. We will remove the long path in the following prompts.

3. Create a directory for registry data, and execute a container using the registry image pulled previously:

```
Chapter3$ mkdir registry-data
Chapter3$ docker container run -P -d \
--name myregstry \
-v $(pwd)/registry-data:/var/lib/registry \
registry:2.8.1
```

This command executed a container in the background, publishing the image's defined port, `5000`. It also used the directory we created to store all our images, by using `$(pwd)` to get the current directory. Adding volumes to a container requires the use of the directory's full path.

As we identified our new container as `myregistry`, we can easily review its status:

```
$ docker container ls
CONTAINER ID    IMAGE             COMMAND                   CREATE
D            STATUS         PORTS                  NAMES
1c7b40ed71d0    registry:2.8.1    "/entrypoint.sh /etc…"    7
minutes ago    Up 7 minutes   0.0.0.0:32768->5000/tcp    myregis-
try
```

From this output, we get our host's port mapped to the container's port, `5000`. We used `-P` to allow the container runtime to choose any port available to publish the application's port; therefore, this port may be different in your environment:

```
$ curl -I 0.0.0.0:32768
HTTP/1.1 200 OK
Cache-Control: no-cache
Date: Sat, 11 Mar 2023 09:18:53 GMT
```

We are now ready to start using this local registry, published on port 32768 (in my example environment).

4. Let's download an `alpine` container image and upload it to our registry. First, we need to pull this image:

```
Chapter3$ docker pull alpine
Using default tag: latest
...
Status: Downloaded newer image for alpine:latest
docker.io/library/alpine:latest
```

5. Now, we retag the image to our repository, published and accessible locally as `localhost:32768`:

    ```
    Chapter3$ docker image tag alpine localhost:32768/alpine:0.1
    ```

6. We can list the local images before pushing them to the local registry:

    ```
    Chapter3$ docker image ls |grep "alpine"
    alpine                         latest      b2aa39c304c2   3 weeks
    ago     7.05MB
    localhost:32768/alpine         0.1         b2aa39c304c2   3 weeks
    ago     7.05MB
    ```

    Both images are identical; we use a second tag to name the same image.

7. Now, we push it to our `localhost:32768` registry:

    ```
    Chapter3$ docker image push localhost:32768/alpine:0.1
    The push refers to repository [localhost:32768/alpine]
    7cd52847ad77: Pushed
    0.1: digest: sha256:e2e16842c9b54d985bf1ef9242a313f36b856181f-
    188de21313820e177002501 size: 528
    ```

    As you can see, everything works as if we were pushing to the Docker Hub registry. The only difference here is that we didn't have to log in and our registry uses HTTP. We can manage this by adding an NGINX web server as a frontend, behind the registry server.

8. Let's now review how images are distributed in our filesystem:

    ```
    Chapter3$ ls -lart registry-data/docker/registry/v2/
    total 16
    drwxr-xr-x 3 root root 4096 Mar  6 19:55 repositories
    drwxr-xr-x 3 root root 4096 Mar  6 19:55 ..
    drwxr-xr-x 3 root root 4096 Mar  6 19:55 blobs
    drwxr-xr-x 4 root root 4096 Mar  6 19:55 .
    ```

    There are two different directories. The `repositories` directory manages the metadata for each image repository, while the `blobs` directory stores all the layers from all container images.

    The `blobs` directory is distributed in many other directories to be able to manage an enormous number of layers:

    ```
    Chapter3$ ls -lart registry-data/docker/registry/v2/blobs/
    sha256/
    total 20
    drwxr-xr-x 3 root root 4096 Mar  6 19:55 63
    drwxr-xr-x 3 root root 4096 Mar  6 19:55 ..
    drwxr-xr-x 3 root root 4096 Mar  6 19:55 e2
    drwxr-xr-x 3 root root 4096 Mar  6 19:55 b2
    drwxr-xr-x 5 root root 4096 Mar  6 19:55 .
    ```

9. Now, we will push a new image into our registry. We again retag the original `alpine:latest` image as `localhost:32768/alpine:0.2`:

    ```
    Chapter3$ docker image tag alpine localhost:32768/alpine:0.2
    ```

    This means that we have a new tag for the original Alpine image; hence, we expect that only metadata should be modified.

10. Let's push the image and review the filesystem changes:

    ```
    $ docker image push localhost:32768/alpine:0.2
    The push refers to repository [localhost:32768/alpine]
    7cd52847ad77: Layer already exists
    0.2: digest: sha256:e2e16842c9b54d985bf1ef9242a313f36b856181f-
    188de21313820e177002501 size: 528
    ```

    Note that our `localhost:32768` registry says that the image layers already exist.

11. We can list the content of the registry again:

    ```
    Chapter3$ ls -lart registry-data/docker/registry/v2/blobs/
    sha256/
    total 20
    drwxr-xr-x 3 root root 4096 Mar  6 19:55 63
    drwxr-xr-x 3 root root 4096 Mar  6 19:55 ..
    drwxr-xr-x 3 root root 4096 Mar  6 19:55 e2
    drwxr-xr-x 3 root root 4096 Mar  6 19:55 b2
    drwxr-xr-x 5 root root 4096 Mar  6 19:55 .
    ```

12. The `blobs` directory wasn't changed, but let's review the `repositories` directory, where the image metadata is managed:

    ```
    Chapter3$ ls -lart registry-data/docker/registry/v2/
    repositories/alpine/_manifests/tags/
    total 16
    drwxr-xr-x 4 root root 4096 Mar  6 19:55 0.1
    drwxr-xr-x 4 root root 4096 Mar  6 19:55 ..
    drwxr-xr-x 4 root root 4096 Mar  6 19:59 0.2
    drwxr-xr-x 4 root root 4096 Mar  6 19:59 .
    ```

    A new folder was created to reference the layers already included inside the `blobs` directory for both tags `0.1` and `0.2`. Let's now push a new image with some changes.

13. We will now create a modified version of the original `alpine.latest` image by using it as the base image in a new build process. We will use an on-the-fly build by piping a Dockerfile:

    ```
    Chapter3$ cat <<EOF | docker build -t \
    localhost:32768/alpine:0.3 -
    FROM docker.io/alpine:latest
    ```

```
RUN apk add --update nginx
EXPOSE 80
CMD ["whatever command"]
EOF
```

This is a different way of building images using a Dockerfile. In this case, we can't use the image content, and therefore, copying files wouldn't work, but it is fine for this example. We create a new image using Unix pipes to avoid the creation of a file. This way, we create an image on the fly:

```
Chapter3$ cat <<EOF | docker build -t \
localhost:32768/alpine:0.3 -
FROM> FROM docker.io/alpine:latest
> RUN apk add --update nginx
> EXPOSE 80
> CMD ["whatever command"]
> EOF
[+] Building 1.3s (6/6) FINISHED
...
=> [1/2] FROM docker.io/library/alpine:latest
0.0s
=> [2/2] RUN apk add --update nginx
1.2s
...
=> => writing image sha256:e900ec26c76b9d779b-
c3d6a7f828403db07daea66c85b5271ccd94e12b460
ccd                                    0.0s
=> => naming to localhost:32768/alpine:0.3
```

14. We now push this new image and review the directories:

```
Chapter3$ docker push localhost:32768/alpine:0.3
The push refers to repository [localhost:32768/alpine]
33593eed7b41: Pushed
7cd52847ad77: Layer already exists
0.3: digest: sha256:1bf4c7082773b616fd2247ef9758dfec9e3084ff-
0d23845452a1384a6e715c40 size: 739
```

As you can see, one new layer is pushed.

15. Now, we'll review the local folders, where the image registry stores data in our host:

```
Chapter3$ ls -lart registry-data/docker/\
registry/v2/repositories/alpine/_manifests/tags/
total 20
drwxr-xr-x 4 root root 4096 Mar  6 19:55 0.1
drwxr-xr-x 4 root root 4096 Mar  6 19:55 ..
drwxr-xr-x 4 root root 4096 Mar  6 19:59 0.2
drwxr-xr-x 4 root root 4096 Mar  6 20:08 0.3
```

```
drwxr-xr-x 5 root root 4096 Mar  6 20:08 .
 Chapter3$ ls -lart registry-data/docker/registry/v2/blobs/
sha256/
total 32
drwxr-xr-x 3 root root 4096 Mar  6 19:55 63
drwxr-xr-x 3 root root 4096 Mar  6 19:55 ..
drwxr-xr-x 3 root root 4096 Mar  6 19:55 e2
drwxr-xr-x 3 root root 4096 Mar  6 19:55 b2
drwxr-xr-x 3 root root 4096 Mar  6 20:08 c1
drwxr-xr-x 3 root root 4096 Mar  6 20:08 e9
drwxr-xr-x 3 root root 4096 Mar  6 20:08 1b
drwxr-xr-x 8 root root 4096 Mar  6 20:08 .
```

As a result of this new push, new folders are created under both the `repositories` and `blobs` locations.

We have seen how images are stored and managed inside our registry. Let's now move on to a new lab, in which we will review how to sign images with a different tool, **Cosign**.

Signing images with Cosign

For this new lab, we will use a new tool, Cosign, which can be easily downloaded in different formats:

1. We will install Cosign by downloading its binary:

    ```
    Chapter3$ mkdir bin
    Chapter3$ export PATH=$PATH:$(pwd)/bin
    Chapter3$ curl -sL -o bin/cosign https://github.com/sigstore/
    cosign/releases/download/v2.0.0/cosign-linux-amd64
    Chapter3$ chmod 755 bin/*
    Chapter3$ cosign --help
    A tool for Container Signing, Verification and Storage in an OCI
    registry.
    Usage:
      cosign [command]
    ...
    ```

2. Once installed, we will use Cosign to create our key pair to sign images. We will use `--output-key-prefix` to ensure our keys have an appropriate name:

    ```
    Chapter3$ cosign generate-key-pair \
    --output-key-prefix frjaraur
    Enter password for private key:
    Enter password for private key again:
    Private key written to frjaraur.key
    Public key written to frjaraur.pub
    ```

Use your own name for your key. You will be asked for a password. Use your own, and remember that this will be required to sign any image. This will create your public and private keys:

```
Chapter3$ ls -l
total 12
-rw------- 1 frjaraur frjaraur  649 Mar  7 19:51 frjaraur.key
-rw-r--r-- 1 frjaraur frjaraur  178 Mar  7 19:51 frjaraur.pub
```

3. We will add a new name and tag to the image, and after that, we will push it:

```
Chapter3$ docker tag localhost:32768/alpine:0.3 \
localhost:32768/alpine:0.4-signed
Chapter3$ docker push localhost:32768/alpine:0.4-signed
The push refers to repository [localhost:32768/alpine]
dfdda8f0d335: Pushed
7cd52847ad77: Layer already exists
0.4-signed: digest: sha256:f7ffc0ab458dfa9e-
474f656afebb4289953bd1196022911f0b4c739705e49956 size: 740
```

4. Now, we can proceed to sign the image:

```
Chapter3$ cosign sign --key frjaraur.key \
localhost:32768/alpine:0.4-signed
Enter password for private key:
WARNING: Image reference localhost:32768/alpine:0.4-signed uses
a tag, not a digest, to identify the image to sign.
    This can lead you to sign a different image than the
intended one. Please use a
    digest (example.com/ubuntu@sha256:abc123...) rather than tag
    (example.com/ubuntu:latest) for the input to cosign. The
ability to refer to
    images by tag will be removed in a future release.
        Note that there may be personally identifiable informa-
tion associated with this signed artifact.
        This may include the email address associated with the
account with which you authenticate.
        This information will be used for signing this artifact
and will be stored in public transparency logs and cannot be
removed later.
By typing 'y', you attest that you grant (or have permission to
grant) and agree to have this information stored permanently in
transparency logs.
Are you sure you would like to continue? [y/N] y
tlog entry created with index: 14885625
Pushing signature to: localhost:32768/alpine
```

Note the warning message. As we learned in *Chapter 2, Building Docker Images*, only the image digest really ensures image uniqueness, and in this example, we use the tags to reference the image we are signing. We should use the digest to improve the signing process and ensure that we sign the right image for production, but for this example, we can continue.

5. We can now verify the signature associated with the image:

```
Chapter3$ cosign verify --key frjaraur.pub \
localhost:32768/alpine:0.4-signed
Verification for localhost:32768/alpine:0.4-signed --
The following checks were performed on each of these signatures:
  - The cosign claims were validated
  - Existence of the claims in the transparency log was verified
offline
  - The signatures were verified against the specified public
key
[{"critical":{"identity":{"docker-reference":"lo-
calhost:32768/alpine"},"image":{"docker-manifest-di-
gest":"sha256:f7ffc0ab458dfa9e474f656afebb4289953bd-
1196022911f0b4c739705e49956"},"type":"cosign container image
signature"},"optional":{"Bundle":{"SignedEntryTimestamp":"MEU-
CIQCFALeoiF8cs6zZjRCFRy//ZFujGalzoVg1ktPYFIhVqAIgI94xz+dCIVI-
jyAww1SUcDG22X4tjNGfbh4O4d+iSwsA=","Payload":{"body":"eyJhcGl-
WZXJzaW9uIjoiMC4wLjEiLCJraW5kIjoiaGFzaGVkcmVrb3JkIiwic3BlYyI....
RDZz09In19fX0=","integratedTime":1678215719,"logIndex-
":14885625,"logID":"c0d23d6ad406973f9559f3ba2d1ca01f84147d8ff-
c5b8445c224f98b9591801d"}}}}]
```

6. We can use `cosign triangulate` to verify whether an image is signed:

```
Chapter3$ cosign triangulate localhost:32768/\
alpine:0.4-signed
localhost:32768/alpine:sha256-f7ffc0ab458dfa9e-
474f656afebb4289953bd1196022911f0b4c739705e49956.sig
```

This hash is the digest referenced:

```
Chapter3$ docker image ls --digests |grep "0.4-signed"
localhost:32768/alpine        0.4-signed    sha256:f7ff-
c0ab458dfa9e474f656afebb4289953bd1196022911f-
0b4c739705e49956    c76f61b74ae4    24 hours ago    164MB
```

7. Let's review what happens if we now remove the signature by renaming (change name and/or tag) the older original `localhost:32768/alpine:0.3` image:

```
Chapter3$ docker tag localhost:32768/alpine:0.3 \
localhost:32768/alpine:0.4-signed
```

8. Now, we push it again:

```
Chapter3$ docker push localhost:32768/\
alpine:0.4-signed
The push refers to repository [localhost:32768/alpine]
33593eed7b41: Layer already exists
7cd52847ad77: Layer already exists
0.4-signed: digest: sha256:1bf4c7082773b616fd2247ef9758d-
fec9e3084ff0d23845452a1384a6e715c40 size: 739
```

9. We can now verify the newly pushed image again:

```
Chapter3$ cosign verify --key frjaraur.pub \
localhost:32768/alpine:0.4-signed
Error: no matching signatures:
 main.go:69: error during command execution: no matching signa-
tures:
```

This means that the new image doesn't have a signature. We maintained the image repository and tag, but no signature is attached. We changed the image, and this also changed the signature. Orchestrators such as Kubernetes can improve application security by validating the image signatures, by executing ValidatingWebHook. This will ensure that only signed images (we can also include specific signatures) will be available to create containers.

Now that we know how to improve security by signing and verifying their signatures, we can go a step further by using security scanners to review any possible vulnerability in their content.

Improving security by using image content vulnerability scanners

For this lab, we will use **Trivy**, from Aquasec. It is a very powerful security scanner for file content, misconfigurations, and even Kubernetes resources. It will really help you in your daily DevOps tasks as well as as a developer. We will create a custom Trivy image for offline usage by including the online database. With this example, we will also learn how to manage cached content inside our images.

Inside the Chapter3 folder, you will find the trivy directory, with a Dockerfile ready for you to build the aforementioned custom image for this lab:

1. First, we will verify the digest for the latest stable version of the docker.io/trivy image by using skopeo:

```
Chapter3$ skopeo inspect \
docker://aquasec/trivy:0.38.2-amd64|grep -i digest
    "Digest": "sha256:8038205ca56f2d88b93d-
804d0407831056ee0e40616cb0b8d74b0770c93aaa9f",
```

2. We will use this digest to ensure we have the right base image for our custom `trivy` image. We will move inside the `trivy` folder to build our new image. Review the Dockerfile's content, and write down the appropriate hash for your base image:

```
FROM aquasec/trivy:0.38.2-amd64@sha256:8038205ca56f2d88b93d-
804d0407831056ee0e40616cb0b8d74b0770c93aaa9f
LABEL MAINTAINER "frjaraur at github.com"
LABEL TRIVY "0.38.2-amd64"
ENV TRIVY_CACHE_DIR="/cache"   \
    TRIVY_NO_PROGRESS=true
RUN TRIVY_TEMP_DIR=$(mktemp -d) \
   && trivy --cache-dir $TRIVY_CACHE_DIR image --download-db-only
\
   && tar -cf ./db.tar.gz -C $TRIVY_CACHE_DIR/db metadata.json
trivy.db
ENV TRIVY_SKIP_DB_UPDATE=true
RUN chmod 777 -R /cache
USER nobody
```

3. We will now build our image:

```
Chapter3/trivy$ docker build -t \
localhost:32768/trivy:custom-0.38.2 . --no-cache
[+] Building 23.5s (7/7) FINISHED
=> [internal] load build definition from Dockerfile
0.1s
...
=> => writing image sha256:de8c7b30b715d05ab3167f6c8d66ef-
47f25603d05b8392ab614e8bb8eb70d4b3            0.1s
=> => naming to localhost:32768/trivy:custom-0.38.2
0.0s
```

4. Now, we are ready to run a scan of any image available remotely. We will test our scanner with the latest `python:alpine` image available in Docker Hub. We will only scan for content vulnerability:

```
Chapter3/trivy$ docker run -ti \
localhost:32768/trivy:custom-0.38.2 \
image python:alpine --scanners vuln \
--severity HIGH,CRITICAL
2023-03-08T20:49:21.927Z        INFO    Vulnerability scanning
is enabled
2023-03-08T20:49:26.865Z        INFO    Detected OS: alpine
2023-03-08T20:49:26.865Z        INFO    Detecting Alpine vulner-
abilities...
2023-03-08T20:49:26.869Z        INFO    Number of language-spe-
cific files: 1
```

```
2023-03-08T20:49:26.869Z          INFO     Detecting python-pkg
vulnerabilities...
python:alpine (alpine 3.17.2)
Total: 1 (HIGH: 1, CRITICAL: 0)
| Library    | Vulnerability | Severity | Installed
Version | Fixed Version |                                 Titl
e                                 |
| libcom_err | CVE-2022-1304 | HIGH     | 1.46.5-r4          |
1.46.6-r0    | e2fsprogs: out-of-bounds read/write via crafted
filesystem |
  https://avd.aquasec.com/nvd/cve-2022-1304                  |
```

We filtered on HIGH and CRITICAL severities only to avoid any non-critical output. We used the default table format for the output, but it is possible to use the JSON format, for example, to include the vulnerability scanner in automated tasks.

Image scanning will really help us decide which releases to use, or even fix issues in our images by understanding the vulnerabilities included in our base images. Scanning processes will usually be included in your building pipelines to ensure that your workflow does not produce images with vulnerabilities that can be easily managed.

Summary

In this chapter, we learned how to store container images in registries, using appropriate repositories and tags for our application components. You, as a developer, must provide the logic names, tags, and required information to your images to allow users to run your applications correctly. Labels will also allow you to include any relevant information that can help you track code changes and how they apply to your application's processes.

In addition, it is critical to ensure a secure supply chain for our image artifacts. We learned that digests provide uniqueness, but that is not enough. We can include signatures to inform users about the provenance and ownership of the images we create, but signing does not guarantee the health of the files included inside our image layers. We will include content vulnerability scanning in our build process. This will allow us to review and verify whether the images we use to create our projects contain any security problems. Knowing the vulnerabilities in our images will help us to improve our application life cycle. Although fixing all vulnerabilities can be hard or even impossible, it is key to understand the possible issues we have to manage in our projects.

Now that we have a good base to create and share our application images, using the best techniques and improved security, it's time to move on to the next chapter, in which we will learn how to run software containers and the command line for different features and container runtimes.

4

Running Docker Containers

Software containers are the new standard application artifacts for modern platforms. In the previous chapters, we learned to create software container images and share them with other developers or services. In this chapter, we will learn how to effectively work with containers. We will understand the main Docker containers' objects and how to manage them using the appropriate command-line actions and options. Understanding the container network model and how to manage persistent data is key for working with containers. We will also cover the concepts for managing both. At the end of this chapter, we will review some very important maintenance tasks you should know about so that you can manage your environment.

In this chapter, we will cover the following topics:

- Understanding Docker software container objects
- Learning about using the command line to work with containers
- Limiting container access to host resources
- Managing container behavior
- Container runtime maintenance tasks

Technical requirements

This book teaches you how to use software containers to improve your application's development. The labs for this chapter can be found at `https://github.com/PacktPublishing/Docker-for-Developers-Handbook/tree/main/Chapter4`. Here, you will find some extended explanations that have been omitted in this chapter's content to make it easier to follow. The *Code In Action* video for this chapter can be found at `https://packt.link/JdOIY`.

Let's begin this chapter by introducing the most important Docker container objects.

Understanding Docker software container objects

Container runtimes usually work by following a client-server model. We interact with the runtime by using a client command line such as `docker`, `nerdctl`, or `crictl`, depending on the backend. The runtime itself is responsible for managing different objects or resources, which can easily be manipulated by interacting with it. Before we learn how to interact with and manage software containers, we will learn about the different objects that are managed by the container runtime in this section. All commands or actions will be related to them, to either create, remove, or modify their properties. We learned about container images in *Chapter 1*, *Modern Infrastructure and Applications with Docker*, and *Chapter 2*, *Building Docker Images*, where we also learned how to build them. Let's start by reviewing these well-known objects, which are common within all container runtimes:

- **Container images**: These objects are also referred to as **container artifacts**. They are the base for creating a container because they contain all the files, integrated into different layers, that will be included inside a container's filesystem. The images also contain the meta-information required for running a container, such as the processes that will run internally, the ports that will be exposed externally, the volumes that will be used to override the container's filesystem, and so on. You, as a developer, will create and use a lot of images for your applications. Please take the best practices for security that were reviewed in *Chapter 2*, *Building Docker Images*, into account.

- **Containers**: When we use a container image and run a container using it, we are telling the container runtime to execute the processes defined in the image's meta-information and use the image layers to provide a base filesystem for these processes. Kernel features such as cgroups and namespaces are also provided to isolate containers. This makes it possible to run different containers in the same hosts in a secure way. None of them will see each other unless specifically declared. The container's filesystem layer will be added on top of the image layers in read and write mode. All layers below the container layer will be used in read-only mode, and the files that have been modified or created will be managed using the features of CoW filesystems.

- **Networks**: Containers always run in isolation within kernel namespaces. We can share some of the underlying operating system's namespaces, such as for networks, processes, IPCs, and so on, but this feature should only be used in very special use cases (for example, for monitoring host resources). By default, each container will run with its own virtual interface and IP address, and this interface will be linked to a specially created `docker0` bridge interface at the host level, although other network options and drivers can be used. The container runtime manages IP addresses with an internal IPAM, and a NAT is used to allow container access to the real host's attached network. Network objects allow us to create different bridge interfaces and attach containers to them, isolating containers running within different networks.

- **Volumes**: CoW filesystems may impact the application's behavior. If your processes change a lot of files or any file must persist throughout a container's life cycle, a volume must be used to override CoW filesystem management. We use volumes to store files outside of the container's layers. These can be folders in our host system, remote filesystems, or even external block devices. Different drivers can be used and, by default, volumes will be locally available in each underlying host as folders.

All these objects will be identified by unique IDs and we will use either their names or IDs to refer to them. Common actions for creating and removing them will be available in our client command line. We can also list them and inspect their properties, and we can use Go template formatting, as we learned in *Chapter 2*, *Building Docker Images*.

Container orchestrators will have their own set of objects or resources (in Kubernetes). We will review them in *Chapter 7*, *Orchestrating with Swarm*, and *Chapter 8*, *Deploying Applications with the Kubernetes Orchestrator*, respectively.

Before we review these new objects, let's remember that containers are processes that run on top of hosts thanks to container runtimes. These processes run isolated from each other by using the special kernel features of the hosts. Container images will provide the base filesystems for these processes and we will use a client command line to interact with them.

Containers are considered stateless and ephemeral, although they exist on the underlying host. Applications running within containers should be prepared to run anywhere, and their state and data should be managed out of the container's life cycle. What if we need to store an application's data or its status? We can use the volume objects to persist data and processes' states when containers are removed or a new container is created using the same data. If we are working in a distributed or orchestrated environment, sharing these volumes is critical, and we need to use external volumes to attach the data to the containers wherever it's needed.

Depending on the container runtime we use, the location of the files related to containers may change, but in the case of a Docker container runtime, we expect to have all images and container layers and their files under `/var/lib/docker` or `c:\ProgramData\docker`. This may seem completely different on new desktop environments, such as Docker Desktop and Rancher Desktop. In these environments, we will use WSL or the Windows command line to execute the client and interact with a container runtime. The runtime runs in a different WSL environment; hence, you will not be able to reach its data path. To review the current data path from your client, you can use `docker info` if you are using Docker as the container runtime:

```
$ docker info --format="{{ .DockerRootDir }}"
/var/lib/docker
```

If you use `containerd` directly, the data root path will be located under the `/var/lib/containerd` directory, but in both cases, you will not be able to access these folders in desktop environments because client access uses a pipe connection to access the container runtime remotely. All the objects' meta-information will be stored under this `DockerRootDir` path and we will be able to retrieve the objects' properties by using the container runtime client with appropriate commands.

If you are using WSL2 with Docker Desktop, two WSL instances will have been created: `docker-desktop` and `docker-desktop-data`. The second one is used for mounting all data inside your own WSL instance (`ubuntu-22.04` in my case, but it may differ for you). This is possible thanks to the integration with Docker Desktop. We can find all the Docker container content inside the `\\wsl.localhost\docker-desktop-data\data\docker` directory. The following PowerShell screenshot shows my environment's data:

```
Windows PowerShell
PS C:\Users\frjaraur> wsl --list --running
Windows Subsystem for Linux Distributions:
docker-desktop (Default)
Ubuntu-22.04
docker-desktop-data
PS C:\Users\frjaraur> dir \\wsl.localhost\docker-desktop-data\data\docker


    Directory: \\wsl.localhost\docker-desktop-data\data\docker


Mode                 LastWriteTime         Length Name
----                 -------------         ------ ----
d-----         3/12/2023     4:39 PM                trust
d-----         3/12/2023     4:39 PM                image
d-----         3/12/2023     4:39 PM                plugins
d-----         3/12/2023     4:39 PM                swarm
d-----         3/18/2023     7:55 PM                volumes
d-----         3/18/2023     7:55 PM                overlay2
d-----         3/18/2023     7:55 PM                runtimes
d-----         3/12/2023     5:07 PM                buildkit
d-----         3/12/2023     6:37 PM                containers
d-----         3/18/2023     7:59 PM                tmp
d-----         3/12/2023     4:39 PM                network


PS C:\Users\frjaraur>
```

Figure 4.1 – Docker container runtime objects data inside the docker-desktop-data WSL instance

Now that we know about the main objects that are managed by the container runtime, we can review the command-line options we have for managing and interacting with them.

Learning about using the command line to work with containers

In this section, we will learn how to manage containers. We will use the Docker command line, provided by the Docker client (the `docker-client` package or WSL integrated into Docker Desktop environments). All the command-line actions we are going to discuss in this chapter will be similar for other clients, such as `nerdctl` or `podman`, although in the latter case, it does not use a `containerd` daemon.

The Docker client sends actions to the Docker daemon via an API every time we retrieve information about Docker objects. Clients can use **SSH**, **HTTP/HTTPS**, or direct **sockets** (or **pipes** in Microsoft operating systems).

First, we will start with the actions that are common and available to all container runtime objects:

- `create`: All the container runtime objects can be created and destroyed. This doesn't apply to container images because we will use `docker image build` to start a building process to create them. All objects will automatically receive an ID, and in some cases, such as with containers, an automated random name will also be added. If we want to assign a defined name, we can use the `--name` argument.

- `list`: This action will show all the objects in the defined category; for example, `docker image list` will retrieve all the images available locally in our container runtime. As we learned in *Chapter 2, Building Docker Images*, we have options for filtering and formatting the output using **Go templates**. When we list the available containers, only those currently running will be listed by default. To also include already stopped containers, we should use the `--all` argument. If we only need the object identifiers, we can use the `--quiet` option. This can be very useful for piping the output to another command.

> **Important note**
>
> You may notice that we can also use `docker container ps` or the shorter version, `docker ps`, to list containers. Containers are processes that run in our host with kernel features providing isolation; hence, it seems appropriate to use this argument as if we were listing processes. By default, only running processes (or containers) will be listed, and we will have to use the `--all` argument to show stopped ones as well.

- `inspect`: Inspecting objects will allow us to retrieve all the information related to the defined object. By default, all objects' data will be presented in JSON format, but we can also use the `--format` argument to format the output.

- `remove`: All objects can also be removed. We will use their IDs or names to delete them. In some cases, internal dependencies may appear. For example, we can't remove a container image if any existing container is using it. To avoid these dependencies, we can use the `--force` argument.

These actions are common to all container runtime-managed objects, but when we deep dive into containers, more are available. Next, we will review the current actions for containers, but first, we have to understand that when we create an object, we prepare all the required configurations. This means that creating a container will prepare the container to run, but the container will be stopped. Let's see this feature in action with a quick example:

```
$ docker container create --name test alpine
f7536c408182698af04f53f032ea693f1623985ae12ab0525f7fb4119c8850d9
$ docker container inspect test --format="{{ .Config.Cmd }}"
[/bin/sh]
$ docker container ls
CONTAINER ID    IMAGE      COMMAND     CREATED     STATUS     PORTS      NAMES
```

Our Docker container runtime has just created a container, but it is not running, although it is still in our host system.

We can remove this container and check it again. We will not be able to retrieve any value from this object because it will now not exist:

```
$ docker container rm test
test
$ docker container inspect test --format="{{ .Config.Cmd }}"
Error: No such container: test
```

Now, we can continue reviewing the actions for containers, starting with the action for actually starting a container after its creation:

- `start`: This action requires a previously created container object to exist. The container runtime will execute the defined container object's processes with its host's isolation and defined attached resources.

- `run`: This action will **create** and **start** a container in one go. The container runtime will start a container and it will attach our terminal to the container's main process. This process runs in the **foreground** unless we use the `--detach` argument; in this case, the container will run in the background and our terminal will be detached from the container. We can use `--interactive` and `--tty` to execute the current container in interactive mode and use a pseudo-terminal; this way, we can actively interact with the container's main process. Containers will run using all the parameters defined in their configuration, such as usernames, defined kernel namespaces, volumes, networks, and so on. Some of these definitions may be modified by adding different arguments to our command line.

- `stop`: Containers can be stopped. This action will ask the container runtime to send a stop signal (`SIGTERM`) to the container's main process and it will wait 10 seconds (by default) before sending a kill signal (`SIGKILL`) if the process is still alive.

- `kill`: Killing a container will directly ask the container runtime to send a `SIGKILL` signal. You, as a developer, should prepare your applications to die correctly in case `SIGTERM` or `SIGKILL` is received. Ensure your files are closed correctly and no unmanaged process continues running after the main process has died.

- `restart`: This action will be used to stop and start the container. We can ask the container runtime to always restart our container whenever our main process dies.

- `pause/unpause`: Containers can be paused. This will make the container runtime inform the kernel to remove any CPU time from the container's processes. This is important because paused containers can be used to share container resources, such as volumes and namespaces.

> **Important note**
> You may get stuck and be unable to exit the container's main process standard and error output if you run certain processes in the foreground without the `--interactive` argument. To avoid this situation, you can use the *CTRL + P + Q* keyboard sequence.

Now that have learned about the most important actions for managing containers, let's review some of the arguments we can use to modify the container's behavior:

- `--name`: Each container will be identified by a unique ID, but a name will always be assigned. This name will be random and composed of two strings. An internal database will be used to generate them and the final concatenated string will be unique. We can avoid this behavior by using `--name` and passing a string of our choice, but remember, container names must be unique and you will not be able to reuse this name.

- `--restart`: As mentioned before, we can ask the container runtime to manage the container's life cycle for us. By default, containers will not restart if the main process dies, but we can use strings such as `on-failure`, `always`, or `unless-stopped` to define whether the container should start in case of failure (any exit code other than 0), always, or just in case we didn't effectively stop the container, respectively. We can also ensure that the Docker runtime does not manage the life cycle of the container by not using any specific string or command.

- `--entrypoint`: This option will allow us to override the container image's defined entry point (main process). It is very important to understand that anyone can change your image's entry point, executing whatever binary or script is available in your image's layers; therefore, it is critical to strictly include the files required by your application.

- `--env`: We can add new environment variables by using this argument or `--env-file`. In this case, a file with a key-value format will be included to add a set of variables.

- `--expose`: By default, only the ports defined in the container's image will be exposed, but it is possible to add new ones if some modifications of the container's behavior require new ports or protocols.

- `--user`: This argument allows us to modify the user who effectively executes the container's main process. You must ensure that your application can run if you change the container's user; this may be critical if you run your applications inside Kubernetes. In *Chapter 8, Deploying Applications with the Kubernetes Orchestrator*, we will learn whether we can improve our application's security by using **security contexts**.

- `--publish` and `--publish-all`: These arguments allow us to publish one port (we can use the arguments multiple times to add multiple ports) or all the image's defined exposed ports. This will make the application accessible from outside of the container's network using NAT. Random host ports will be used to publish your applications unless you define specific ports during container execution.

- `--memory` and `--cpus`: These options, among others, will allow us to manage the amount of memory and CPU resources that will be attached to the container. We can also include the host's GPUs by using `--gpus`.

Now that we have had an overview of the most important arguments that are used with containers, we will take a look at some critical options for securing our workloads:

- `--cap-add`: This option allows us to specifically add some kernel capabilities to our processes inside the container's execution. Kernel capabilities are a set of privileges associated with a superuser, which the system provides in a fine-grained way. By default, a container runtime does not allow *privileged* containers to run with all available capabilities. Container runtimes allow only a subset of all available capabilities by default (the currently available capabilities can be reviewed at `https://man7.org/linux/man-pages/man7/capabilities.7.html`). The Docker runtime, for example, allows 14 capabilities (`https://docs.docker.com/engine/reference/run/#runtime-privilege-and-linux-capabilities`), which will probably be enough for your applications to run, but if your applications need some specific capability, such as the permission to manage network interfaces using `NET_ADMIN`, you should add that using the `--cap-add NET_ADMIN` argument. Adding capabilities may be useful for modifying your current kernel behavior if your application needs some special features. You, as a developer, should inform the relevant parties about the special privileges needed by your applications because capabilities may be dropped in secure container orchestrator environments. Inform your DevOps or cluster administrator teams about your special requirements.

- `--cap-drop`: This option, in contrast, is used to remove certain capabilities. This may be very useful, for example, if we need to remove the possibility of changing file ownership inside a container's life cycle, which we can do using `--cap-drop CHWON`, or remove the ability to send raw network packets, for example, ICMP, which is done using `--cap-drop NET_RAW`. You may find secure environments where all capabilities are dropped.

> **Important note**
>
> Both `--cap-drop` and `--cap-add` can be used with the `ALL` argument, which means that all capabilities will be dropped or added, respectively. It is very important for you, as a developer, to test the possible issues that may appear if you drop all available capabilities. This will help you prepare your applications for secure environments.

- `--privileged`: This argument will provide all capabilities and avoid any resource limitations. You should avoid using this option for your application containers. Take time to review which capabilities and resources are required for your application and apply them. Overriding all the process limits in production is a bad idea and should be applied only to specific application containers, for example, to monitor your infrastructure. In these specific cases, you may need extra resources or be able to access all the host's capabilities, processes, and so on to manage applications from the containers themselves.

- `--disable-content-trust`: This option will disable any Docker Content Trust verification; hence, any signature or image source check will be omitted.

- --read-only: Executing containers in **read-only** mode will ask the container runtime to present the root filesystem inside the container in read-only mode. This feature improves the security of your applications. All changes to files must be managed outside of the container's life cycle using volumes. It is important to test your containers with this feature because it is quite probable that you may need to set this option for production environments. This applies to the entire root filesystem; hence, if your application needs to write, for example, in the /tmp directory, you need to set up a volume attached to this path to allow this interaction.

- --security-opt: Some extended security measures may need extra options, for example, for setting up a different seccomp profile or specifying SELinux options. In general, this option allows us to modify Linux security modules' behavior.

Now that we know how to run containers and the most important options, let's review how to limit and include the underlying host's resources.

Limiting container access to host resources

In this section, we will learn how to limit hosts' resources inside containers, but first, we will take a look at the container network model and how to use volumes to override container storage.

Network isolation

Network isolation is provided by assigning a network namespace to each container; hence, virtualized IP addresses are added to containers. These assignments are provided from a pool of defined IP addresses managed by the container runtime's internal IPAM. This is the way that internal IP addresses are assigned, but all virtual interfaces are associated by default with a bridged host interface. Each container runtime will create and manage its own bridge interface. Docker will use docker0 by default. This interface is created during Docker daemon installation and all IP containers' interfaces will be associated with docker0. Different drivers can be used to extend this default behavior, for example, to attach network VLANs to containers directly.

> **Important note**
> By default, the following network plugins are available: bridge, host, ipvlan, macvlan, null, and overlay.

By default, a fresh Docker container runtime installation creates three different interfaces. We can review them by listing the default network objects after installation:

```
$ docker network list
NETWORK ID      NAME      DRIVER    SCOPE
490f99141fa4    bridge    bridge    local
b984f74311fa    host      host      local
25c30b67b7cd    none      null      local
```

All containers will run using the `bridge` interface by default. Every time we create a container, a virtual interface is created in the host, attached to `docker0`. All egress and ingress traffic will go through this interface. In this scenario, it is very important to understand that all containers attached to this common bridge will see each other. Let's see how this happens in this example:

```
$  docker container create --name one alpine sleep INF
116220a54ee1da127a4b2b56974884b349de573a4ed27e2647b1e780543374f9
$ docker container inspect one --format='{{ .NetworkSettings.IPAddress }}'
```

This container does not receive an IP address until it runs. We can now execute it and review the IP address again:

```
$ docker container start one
one
$ docker container inspect one --format='{{ .NetworkSettings.IPAddress }}'
172.17.0.2
```

The container runtime manages the IP assignment and we can verify the network segment that was used:

```
$ docker network inspect bridge --format='{{ .IPAM }}'
{default map[] [{172.17.0.0/16  172.17.0.1 map[]}]}
```

Let's verify what happens when another container runs attached to the network bridge interface:

```
$ docker container ls
CONTAINER ID    IMAGE       COMMAND         CREATED         STATU
S          PORTS       NAMES
116220a54ee1    alpine      "sleep INF"     12 minutes ago  Up 8
minutes                 one
$ docker container run -ti alpine ping -c 3 172.17.0.2
PING 172.17.0.2 (172.17.0.2): 56 data bytes
64 bytes from 172.17.0.2: seq=0 ttl=64 time=1.148 ms
64 bytes from 172.17.0.2: seq=1 ttl=64 time=0.163 ms
64 bytes from 172.17.0.2: seq=2 ttl=64 time=0.165 ms
--- 172.17.0.2 ping statistics ---
3 packets transmitted, 3 packets received, 0% packet loss
round-trip min/avg/max = 0.163/0.492/1.148 ms
```

The second container executed three pings to the first one's IP address and it was reachable. Both containers run in the same network segment, associated with the same bridge interface. This default behavior can be managed by setting certain network object's keys during their creation, such as `com.docker.network.bridge.enable_icc`, which manages the isolation between containers in the same network (additional information can be found at `https://docs.docker.com/engine/reference/commandline/network_create`).

We will use --network to define the networks to which containers should attach.

The none network can be used to initialize and run containers without any networking capabilities. This can be interesting when we run certain tasks that don't require any network traffic – for example, managing data stored in a volume.

We can share the host's network namespace by using the host network. When we attach a container to this network, it will use the host's IP interfaces. We can verify this behavior by executing docker container run --rm --network=host alpine ip address show. The following screenshot shows the output of this command, showing the interfaces inside a container using the host network:

```
Ubuntu 22.04.2 LTS
frjaraur@sirius:~$ docker container run --rm --network=host alpine ip address show
1: lo: <LOOPBACK,UP,LOWER_UP> mtu 65536 qdisc noqueue state UNKNOWN qlen 1000
    link/loopback 00:00:00:00:00:00 brd 00:00:00:00:00:00
    inet 127.0.0.1/8 scope host lo
       valid_lft forever preferred_lft forever
    inet6 ::1/128 scope host
       valid_lft forever preferred_lft forever
2: tunl0@NONE: <NOARP> mtu 1480 qdisc noop state DOWN qlen 1000
    link/ipip 0.0.0.0 brd 0.0.0.0
3: sit0@NONE: <NOARP> mtu 1480 qdisc noop state DOWN qlen 1000
    link/sit 0.0.0.0 brd 0.0.0.0
4: docker0: <BROADCAST,MULTICAST,UP,LOWER_UP> mtu 1500 qdisc noqueue state UP
    link/ether 02:42:7a:37:1f:5e brd ff:ff:ff:ff:ff:ff
    inet 172.17.0.1/16 brd 172.17.255.255 scope global docker0
       valid_lft forever preferred_lft forever
    inet6 fe80::42:7aff:fe37:1f5e/64 scope link
       valid_lft forever preferred_lft forever
7: eth0@if6: <BROADCAST,MULTICAST,UP,LOWER_UP,M-DOWN> mtu 1500 qdisc noqueue state UP
    link/ether 1a:bd:cb:58:97:37 brd ff:ff:ff:ff:ff:ff
    inet 192.168.65.4 peer 192.168.65.5/32 scope global eth0
       valid_lft forever preferred_lft forever
    inet6 fe80::18bd:cbff:fe58:9737/64 scope link
       valid_lft forever preferred_lft forever
9: vethfd492ad@if8: <BROADCAST,MULTICAST,UP,LOWER_UP,M-DOWN> mtu 1500 qdisc noqueue master docker0 state UP
    link/ether f6:84:28:4d:98:13 brd ff:ff:ff:ff:ff:ff
    inet6 fe80::f484:28ff:fe4d:9813/64 scope link
       valid_lft forever preferred_lft forever
frjaraur@sirius:~$
```

Figure 4.2 – The network interfaces of our host, included inside a running container

Here, we can see that docker0 and the previous container's interface are inside the new container. The host network is used for monitoring and security applications – when we need access to all the host interfaces to manage or retrieve their traffic statistics, for example.

> **Important note**
> We used the --rm argument to remove the container right after its execution. This option is very useful for testing and executing quick commands inside containers.

Understanding custom networks

We can also create custom networks, as with any other container runtime objects. We can use `docker network create <NETWORK_NAME>` for such tasks; a new bridge network interface will be created by default. These new network interfaces will be similar to the `docker0` interface, but some important features are added:

- Each custom network is isolated from the others, using a completely different network segment. A new bridge interface will be created for each custom network and the network the segment will be associated with. All containers running attached to this network will see each other, but they won't reach the ones attached to any other network, including the default bridge. This also works the opposite way; hence, containers attached to a network will only see those working on the same network.

- Custom networks can be dynamically attached. This means that containers can be attached and detached by using `docker network connect <CONTAINER>` and `docker network disconnect <CONTAINER>`. This behavior can't be reproduced in the default bridge network.

- An internal DNS is provided for each custom network. This means that all containers that are attached can be accessed by using their names. Hence, network discovery is provided, and each time a new container runs attached to this network, a new entry is added to the internal DNS. However, remember that DNS names will only be accessible internally in the defined network. Default bridge networks can also access containers via their names if we use the `--link` argument. This way, we can link containers together to make them work as if they were using a DNS, but this will only work for the containers included in this argument; no other containers will be seen by their names.

Let's see a quick example by creating a new network using `docker network create`. We will also define a name, its scope, and the associated subnet:

```
$ docker network create --subnet 192.168.30.0/24 mynetwork
43ee9a8bde09de1882c91638ae7605e67bab0857c0b1ee9fe785c2d5e5c9c3a7
$ docker network inspect mynetwork --format='{{ .IPAM }}'
{default map[] [{192.168.30.0/24   map[]}] }
$ docker run --detach  --name forty \
--network=mynetwork alpine sleep INF
3aac157b4fd859605ef22641ea5cc7e8b37f2216f0075d92a36fc7f62056e2da
$ docker container ls
CONTAINER ID    IMAGE       COMMAND         CREATED         STATU
S           PORTS       NAMES
3aac157b4fd8    alpine      "sleep INF"     10 seconds ago  Up 8
seconds                 forty
116220a54ee1    alpine      "sleep INF"     2 hours ago     Up 2
hours                   one
```

Now, let's try to access the container attached to the created custom network:

```
$ docker run   --rm --network=mynetwork alpine ping \
-c 1 forty
PING forty (192.168.30.2): 56 data bytes
64 bytes from 192.168.30.2: seq=0 ttl=64 time=0.222 ms
  --- forty ping statistics ---
1 packets transmitted, 1 packets received, 0% packet loss
round-trip min/avg/max = 0.222/0.222/0.222 ms
```

It is accessible by its name, but let's try this again with the container attached to the default network:

```
$ docker run   --rm --network=mynetwork alpine ping \
-c 1 one
ping: bad address 'one'
```

It is not accessible by the DNS name. Let's verify whether the network is available:

```
$ docker container inspect one \
--format='{{ .NetworkSettings.IPAddress }}'
172.17.0.2
$ docker run   --rm --network=mynetwork alpine ping -c 1 172.17.0.2
PING 172.17.0.2 (172.17.0.2): 56 data bytes
--- 172.17.0.2 ping statistics ---
1 packets transmitted, 0 packets received, 100% packet loss
```

All packets are lost. The bridge network is not accessible from the custom one we created, although both use the host's interfaces. Each network is attached to its own bridge network, but we can attach a container to both networks using docker connect <NETWORK> <CONTAINER>:

```
$ docker network connect mynetwork one
```

Now, we have a container attached to the custom and default bridge networks:

```
$ docker exec -ti one ip address show|grep inet
    inet 127.0.0.1/8 scope host lo
    inet 172.17.0.2/16 brd 172.17.255.255 scope global eth0
    inet 192.168.30.3/24 brd 192.168.30.255 scope global eth1
```

Therefore, we can now reach the containers in the custom network:

```
$ docker exec -ti one ping -c 1 forty
PING forty (192.168.30.2): 56 data bytes
64 bytes from 192.168.30.2: seq=0 ttl=64 time=0.199 ms
  --- forty ping statistics ---
1 packets transmitted, 1 packets received, 0% packet loss
round-trip min/avg/max = 0.199/0.199/0.199 ms
```

Some options can modify the default networking behavior inside containers. Let's see some of them for running containers or creating networks:

- `--add-host`: This option allows us to include some external hosts in `host:ip` format to make them available as if they were included in the DNS.

- `--dns`, `--dns-search`, and `--dns-option`: These options allow us to modify the DNS resolution for the container. By default, the container runtime will include its current DNS, but we can change this behavior.

- `--domainname`: We can set the container's domain name to something other than the default one.

- `--ip`: Although it is quite important to use default dynamic IP address mappings, we may prefer to assign a specific IP address to the container. Use this option with care as you can't reuse IP addresses.

- `--hostname`: By default, each container will use the container's ID as its name, but we can change this behavior by using this option.

- `--link`: This option will allow us to attach two or more containers by using this option multiple times. It is quite similar to the `--add-host` option, but in this case, we will use it to attach a container to a DNS name in `CONTAINER_NAME:DNS_ALIAS` format to make it accessible via its DNS name.

- `--network-alias`: Sometimes, we need a container to be known in the network with multiple names. This option allows us to add a DNS alias to the container.

- `--subnet` and `--ip-range`: This option is available for networks and allows us to modify the internal IP's assignation. We can also modify the default gateway for each network by using the `--gateway` argument (by default, the lowest IP address will be used).

In the next section, we'll learn how volumes work.

Managing persistent data with containers

Applications running in containers must be prepared to run in any host. We can even go further and say that we should be able to run them in the cloud or on-premises environments. However, containers' life cycles should not include process states and data. **Volumes** will help us manage data outside of containers' life cycles and hosts (if we're using remote storage solutions such as NAS). In this section, we will learn how container runtimes manage local volumes. Volumes will allow containers to access hosts' filesystems or remote filesystems.

Local volumes will be expected to be located by default under the `DockerRootDir` path in the container runtime's host. This will work for **unnamed volumes**, which are created automatically by a container runtime whenever a `VOLUME` key is declared in any image's meta-information, and **named volumes**, which are created or declared by users during a container's execution.

We can also use any local filesystem (**bind mount**) and host's memory by declaring a tmpfs volume (this is very interesting when we need the fastest possible storage backend). In the case of bind mounts, any directory or file from the host's filesystem can be included inside a container. But a problem arises here: whenever we move an application to another host, any expected location related to the host may be different. To avoid such a situation, it is recommended to use external storage, presented on other hosts at the same time or whenever a container needs to run. This is especially relevant for clusters, where a pool of nodes can run your containers. At this point, we will just discuss having data outside of a container's life cycle; in *Chapter 10*, *Leveraging Application Data Management in Kubernetes*, we will discuss how to manage data in these more complicated scenarios.

Let's deep dive a bit and describe the local volume types:

- **Unnamed volumes**: These are the volumes that are created automatically for you whenever you run a container from an image with some VOLUME definition. These volumes are used to override the container's filesystem, but we, as users, are not responsible for its content. In other terms, whenever a new container runs, a new volume will be created dynamically and new data will be used. If we need data to persist between executions, we have to define a volume by ourselves. The container runtime manages the unnamed volumes completely. We can remove them using the docker volume rm command but we will need to stop and remove the associated containers first. Let's run a quick example with the postgres:alpine image, which uses a dynamic unnamed volume to override the container's filesystem for the /var/lib/postgresql/data directory:

```
$ docker pull postgres:alpine -q
docker.io/library/postgres:alpine
$ docker image inspect postgres:alpine \
--format="{{ .Config.Volumes }}"
map[/var/lib/postgresql/data:{}]
$ docker run -d -P postgres:alpine
27f008dea3f834f85c8b8674e8e30d4b4fc6c643df5080c62a14b63b5651401f
$ docker container inspect 27f008dea3 \
--format="{{ .Mounts }}"
[{volume 343e58f19c66d664e92a512ca2e8bb201d8787bc62bb9835d5b2d-
5ba46584fe2 /var/lib/docker/volumes/343e58f19c66d664e92a-
512ca2e8bb201d8787bc62bb9835d5b2d5ba46584fe2/_data /var/lib/
postgresql/data local  true }]
frjaraur@sirius:~$ docker volume ls
DRIVER     VOLUME NAME
local      343e58f19c66d664e92a512ca2e8bb201d8787bc62bb9835d5b2d-
5ba46584fe2
```

In this example, a local volume is created without a name, and it is used by a container running a PostgreSQL database. Any data created in the database will be stored inside the newly created volume.

- **Named volumes:** These volumes have a definition and we can use `docker volume create` to create them or just include a volume name when we run a container. If this volume already exists, the container runtime will attach it to the container, and if isn't already present, it will be created. Container runtimes allow us to extend their volumes' functionality by using different plugins; this will allow us to use NFS, for example. Let's run the previous example using a defined volume; we will use `DATA` as the name for this new volume:

```
$ docker run -d -P \
-v DATA:/var/lib/postgresql/data postgres:alpine
ad7dde43bfa926fb7afaa2525c7b54a089875332baced7f86cd3709f04629709
$ docker container inspect ad7dde43bf \
--format="{{ .Mounts }}"
[{volume DATA /var/lib/docker/volumes/DATA/_data /var/lib/post-
gresql/data local z true }]
$ docker volume ls
DRIVER     VOLUME NAME
local      343e58f19c66d664e92a512ca2e8bb201d8787bc62bb9835d5b2d-
5ba46584fe2
local      DATA
```

In this example, `DATA` is the name of the volume, and we will be able to reuse this volume whenever we remove the `postgresql` container and create a new one.

Data will be persisted in the volumes in both examples but the named volume will allow us to manage the data most conveniently.

> **Important note**
>
> It is very important to understand that named and unnamed (dynamic) volumes use our host's storage. You must take care of volumes forgotten in your filesystem; we will review some techniques for this in the *Container runtime maintenance tasks* section later in this chapter.

- **Bind mounts:** In this case, we will include a host's directory or file inside our container. We will practice using this type of volume in the *Labs* section.

- **In-memory or tmpfs volumes:** These volumes can be used to override the container's storage, providing fast storage, such as the host's memory. This can be very useful for storing a small amount of data that changes quite often, such as statistics. It also can be very dangerous if you don't limit the amount of memory for use.

> **Important note**
>
> Volumes can be used in read-only mode to preserve any existing data. This is very useful for presenting data from the host that you want to ensure remains unchanged, such as operating system files. We can also manipulate the ownership and permissions seen inside the container's filesystem for any mounted volume. It is also common to use `--volumes-from` to share volumes between containers.

As you have probably noticed, we used the `-v` or `--volumes` argument to add volumes to our container at runtime. We can use this argument multiple times and use the `--volume SOURCE_VOLUME:FULL_DESTINE_PATH[:ro][:Z]` format, where `SOURCE_VOLUME` can be any of the previously described types (you have to use the full path for the shared directory when using bind mounts). A volume can be mounted in read-only mode, which is very interesting when you provide configurations to your container, and we can force SELinux usage if needed with the Z option. However, we can also use volumes with the `--mount` argument, which provides an extended version of the volume-mounting options in key-value format:

- `type`: We specify the type of the mount (`bind`, `volume`, or `tmpfs`).
- `source` (or `src`): This key is used to define the source of the mount. The value can either be a name (for named volumes), a full path (for bind mounts), or empty (for unnamed volumes).
- `target` (or `dst`): This key defines the destination path where the volume will be presented.

We can also include the **read-only** value or even extend the volume mount behavior by using specific options and adding `volume-opt`.

Now that we have learned how to include the host's network and filesystems inside containers, let's continue by accessing the CPU and memory, among other resources.

Limiting access to host hardware resources

Sharing host resources within containers is the key to the container model, but this requires being able to limit how they access these resources. In *Chapter 1*, *Modern Infrastructure and Applications with Docker*, we learned that resource isolation is provided by cgroups.

If our host runs out of memory or CPU, all containers running on top will be affected. That's why is so important to limit access to the host's resources.

By default, containers run without any limits; hence, they can consume all the host's resources. You, as a developer, should know the resources that are required by all your application components and limit the resources provided to them. We will now review the arguments we can pass to the container runtime to effectively limit the container's access to resources:

- `--cpus`: This argument allows us to define the number of CPUs provided to the container's main process. This value depends on the number of CPUs available to the host. We can use decimals to indicate a subset of the total number of CPUs. This value guarantees the number of CPUs that can be used to run the container's process.
- `--memory`: We can set the maximum memory available to the container's processes. When this limit is reached, the host's kernel will kill the container's main process by executing a runtime called **Out-Of-Memory-Killer (OOM-Killer)**. We can avoid this task by using `--oom-kill-disable`; however, it is not recommended as you may leave your host without any protection if too much memory is consumed.

> **Important note**
>
> Container runtimes provide more options for managing CPU and memory resources available for any container. It is possible to even limit the access to block devices' **input/output (I/O)** operations or modify the default kernel's scheduling behavior. We have just reviewed the most important ones to help you understand how we can limit access to the host's resources. You can review the full options at `https://docs.docker.com/config/containers/resource_constraints/`.

Extending access to host resources

Containers run on top of our hosts thanks to container runtimes. By default, the host's CPU and memory are provided thanks to cgroups. Volumes are provided inside containers using different types (dynamic unnamed volumes, named volumes, host binds, or tmpfs volumes) and the network is provided using kernel namespaces. We can use other Docker client arguments to integrate or modify kernel namespaces:

- `--ipc`: This argument allows us to modify the IPC behavior (shared memory segments, semaphores, and message queues) inside containers. It is quite common to include the host's IPC by using `--ipc host` for monitoring purposes.

- `--pid`: This option is intended to set the PID kernel namespace. By default, containers run with their own process trees, but we can include other containers' PIDs by using `--pid container:CONTAINER_NAME`. We can also include the underlying host's PIDs tree by using `--pid host`. This can be very interesting if your application needs to monitor the host's processes.

- `--userns`: We can create a different kernel user's namespace and include it inside containers. This allows us to map different user IDs to the processes running inside the containers.

Other interesting options allow us to include different host devices:

- `--device`: This option allows us to include the host's devices inside containers. Processes running inside the containers will see these devices as if they were directly connected to the container. We can use this option to mount block devices (`--device=/dev/sda:/dev/xvdc`), sound devices (`--device=/dev/snd:/dev/snd`), and so on.

- `--gpus`: We can include **graphic processing units (GPUs)** inside containers. This is only possible if your application is prepared for working with modern GPUs and your host provides some of these devices. We can include a defined number of GPUs or all of them at once by using `--gpus all`.

In this section, we learned how to limit access to different hosts' resources. In the next section, we will learn how to manage containers running in our host and their behavior.

Managing container behavior

The container runtime will help us understand containers' behavior by providing a set of options for reviewing process logs, copying files to and from containers, executing processes inside them, and so on.

The following actions allow us to interact with container processes and filesystems:

- exec: We can attach new processes to the containers' namespaces by using docker container exec. This option will allow us to run any script or binary included in the container's filesystems or mounted columns.

- attach: When a container is running in the background, detached from the container runtime client's command line, we can attach its output by using this action. We will attach the Docker client to the container's main process; hence, all the output and errors will be shown in our terminal. Take care with this option because you should detach from the container's main process output to free your terminal. Do not use the *Ctrl + C* keyboard combination because this will send a SIGNINT signal to the container's main process and it will probably be stopped. You can detach from the container's process by using *Ctrl + P + Q*.

- cp: Sometimes, we need to retrieve some files from the container for debugging, for example, certain errors. We can use the cp action to copy files to/from a container. Remember that you can use volumes to provide files or directories to containers; the cp action should be used with small files because it uses the container runtime and your client to retrieve the files from the containers or send them from the local client.

- logs: Retrieving the logs from containers is key to understanding how your applications work. A container has a main process and this process's STDOUT and STDERR streams are the output we receive by using the logs action. We can use --follow to attach to the container's process output continuously and --tail to retrieve only a set of lines. It is possible to filter logs by dates by using --since and --until.

> **Important note**
>
> If you need to execute an interactive session within a container, it is important to include --interactive and --tty. These options ask the container runtime to prepare a pseudo-terminal and interactive session attached to the binary defined in the exec action; for example, we will use docker container exec --ti CONTAINER /bin/bash to execute a *bash* shell inside a defined container.

We can also use a running container to create a container image. We learned about this feature in *Chapter 2, Building Docker Images*. Creating images from containers does not provide a reproducible recipe and it is not a good method for creating images, but in certain situations, you may need the container's content to review the files included.

We can use docker container commit to create an image from the container's layer and export all the layers by using docker container export. This action will only store the files included in the container. It does not include any meta-information because we are only working with the content.

Another action that's quite interesting for quickly debugging the file changes made by the container's processes is diff. This action allows us to retrieve the changes that are created in the container's layer by comparing all its files with the image's layers. Let's review this action with a quick example:

```
$ docker container run --name test alpine touch /tmp/TESTFILE
$ docker container diff test
C /tmp
A /tmp/TESTFILE
```

As we can see from the command's output, the /tmp directory was changed (indicated by C) and a file was added, /tmp/TESTFILE (indicated by A).

Now that we have had a good overview of how to interact with containers and obtain information for debugging our applications, let's learn some housekeeping tasks that will help us maintain our container's environment healthily.

Container runtime maintenance tasks

In this section, we are going to take a quick look at some housekeeping actions that are available for maintaining our container runtime.

Maintaining the right amount of storage available in your host is a very important task. Depending on your container runtime, you may have to prune certain objects instead of using a general tool. This happens, for example, with the containerd client called nerdctl. If you are using Rancher Desktop, you will need to specifically remove unnecessary objects per category. Let's review how this can be done with the Docker client using docker system prune. But before you prune your system and clean old objects, you should first understand where the disk space has been used.

To review the actual amount of disk that's been allocated to different objects, we can use the docker system df command:

```
$ docker system df
TYPE            TOTAL    ACTIVE    SIZE      RECLAIMABLE
Images          5        4         1.564GB   11.1MB (0%)
Containers      17       0         729.3MB   729.3MB (100%)
Local Volumes   2        2         0B        0B
Build Cache     13       0         1.094GB   1.094GB
```

This command shows the space used by images, containers, and local volumes on your system. We can obtain quite descriptive information by adding the --verbose argument, which will show us exactly the amount of space that's used by every object in our host. Specific sections for each object category will show the space we will free up after removing those objects (only object headers and one object line are shown as an example; you can access the full output at `https://github.com/PacktPublishing/Docker-for-Developers-Handbook/blob/main/Chapter4/Readme.md`):

```
$ docker system df --verbose
Images space usage:
 REPOSITORY               TAG              IMAGE ID         CREATE
D        SIZE        SHARED SIZE   UNIQUE SIZE    CONTAINERS
localhost:32768/trivy    custom-0.38.2    bdde1846d546    2 weeks ago
1.3GB       7.05MB        1.293GB           4
Containers space usage:
CONTAINER ID   IMAGE                                     COMMAN
D                LOCAL VOLUMES    SIZE       CREATED        STATU
S                NAMES
df967027f21a   alpine                                   "touch /tmp/TEST-
FILE"    0                0B       47 hours ago   Exited (0) 47 hours
ago      test
Local Volumes space usage:
VOLUME NAME                                                      LI
NKS     SIZE
DATA
1          0B
343e58f19c66d664e92a512ca2e8bb201d8787bc62bb9835d5b2d-
5ba46584fe2    1        0B
 Build cache usage: 1.094GB
 CACHE ID       CACHE TYPE      SIZE       CREATED        LAST
USED       USAGE       SHARED
1ipx4a3h7x8j   regular         4.05MB     2 weeks ago    2 weeks
ago      2           true
```

This output gives us a good idea of how are we running our environment. You, as a developer, will probably have a lot of cached layers if you are building your images locally. These layers will help speed up your build processes but all the layers that have not been used for a long time can be removed.

Let's take a look at how images are distributed in our example:

```
REPOSITORY               TAG              IMAGE ID         CREATED
SIZE        SHARED SIZE   UNIQUE SIZE    CONTAINERS
localhost:32768/trivy    custom-0.38.2    bdde1846d546    2 weeks ago
1.3GB       7.05MB        1.293GB           4
localhost:32768/alpine   0.3              a043ba94e082    2 weeks ago
11.1MB      7.05MB        4.049MB           0
registry                 2.8.1            0d153fadf70b    6 weeks ago
24.15MB     0B            24.15MB           1
postgres                 alpine           6a35e2c987a6    6 weeks ago
243.1MB     7.05MB        236MB             2
alpine                   latest           b2aa39c304c2    6 weeks ago
7.05MB      7.05MB        0B                10
```

All the images based on `alpine` share `7.05MB`; hence, using common base images will help you save a lot of storage and is a good practice.

The `CONTAINERS` section will help us find possible problems because we don't expect to have much space in containers. Remember that containers are intended to be ephemeral, and persistent data should be maintained outside of their storage. Application logs should be redirected to either volumes or `STDOUT/STDERR` (this is the recommended option). Therefore, the space used by containers should be minimal, only consisting of runtime modifications that shouldn't persist. In our example, we can see a couple of containers with several megabytes of usage:

```
737aa47334e2    localhost:32768/trivy:custom-0.38.2    "trivy image pyth
on:…"    0                365MB    2 weeks ago    Exited (0) 2 weeks
ago        infallible_mirzakhani
f077c99cb082    localhost:32768/trivy:custom-0.38.2    "trivy image pyth
on:…"    0                365MB    2 weeks ago    Exited (0) 2 weeks
ago        sharp_kowalevski
```

In both cases, the `trivy` database is probably included in the container's layer (we used `trivy` and updated its database during the build process for these images).

We also have a few volumes (dynamic and named) present, but no data was stored because we didn't add any data to the database example.

And finally, we can see the cache section in the output of the `docker system df -verbose` command, where we will find the shared layers used in the `buildx` processes.

The objects' disk usage shown by `docker system df` is a representation of the physical space distributed in `/var/lib/docker` (the default being `rootDir`).

Pruning container objects

Once we know how our host's storage is distributed, we can proceed with cleaning unused objects. We will use `docker system prune` to clean all unused objects in one go. It will try to free disk space by removing objects from different categories. We can include the volumes by using the `--volumes` argument. The `system prune` command will remove the following:

- **All dangling images (by default)**: Image layers not referenced by any container image.
- **All unused images (using the --all argument)**: Images not referenced by any container (running or stopped in our system).
- **All stopped containers (by default)**: By default, all stopped containers will be removed (those with an *exited* status). This will remove the containers' layers.
- **All unused volumes (using --volumes)**: Volumes are not used by any container.
- **All unused networks (by default)**: Networks with no containers attached.
- **All dangling cache layers (by default)**: All layers that are not referenced in any build process.

You will always be asked to confirm this action as it can't be undone:

```
$ docker system prune
WARNING! This will remove:
  - all stopped containers
  - all networks not used by at least one container
  - all dangling images
  - all dangling build cache
 Are you sure you want to continue? [y/N] y
Deleted Containers:
df967027f21a15e473d236a9c30fa95d5104a8a180a91c3ca9e0e117bdeb6400
...
Deleted Networks:
test1
...
Deleted build cache objects:
1cmmyj0xgul6e37qdrwjijrhf
...
 Total reclaimed space: 1.823GB
```

A summary of the reclaimed space is shown after the cleaning process.

For each category of objects, we can execute these pruning processes:

- docker container prune

- docker image prune

- docker buildx prune

- docker network prune

These actions will only clean specific objects, which may be very useful if you don't want to change other objects.

All pruning options can be filtered using the appropriate --filter argument. These are some of the most common filters:

- until: We use a timestamp argument to only remove containers that were created before a date.

- label: This will help us filter which objects from a category will only be removed. Multiple labels can be used, separated by commas. Labels can be filtered by their existence or absence and we can use keys and values for fine-grained selections.

> **Important note**
>
> If you are planning to schedule prune processes, you may need to use `--force` to execute them in a non-interactive way.

Before you move on to the next section, it is important to know that your containers' logs will also be present in your host system.

Configuring container runtime logging

Container logging options for your system depend on your container runtime. In this section, we will quickly review some of the options available for a Docker container runtime as this is the one that has the most advanced options. You will probably find options for logging using JSON format, but Docker also provides other logging drivers.

By default, the Docker daemon will use the `json-file` logging driver, but we can change this behavior in Docker's `daemon.json` file. This driver uses more disk space than others and that's why it is recommended to use the local logging driver for local development. We can use our host's system logs in Linux environments by configuring `syslog` or `journald` drivers, but if we need to send our containers logs to an external application, we will probably use `gelf` (a commonly used standard) or `splunk` drivers, although there are also some drivers specific for cloud environments.

We can configure some housekeeping options by adding specific keys to the `daemon.json` file. Here is an example that will keep the logs' size under 20 MB:

```
{
    "log-driver": "json-file",
    "log-opts": {
          "max-size": "20m",
          "max-file": "10",
    }
}
```

We will apply this configuration and restart our Docker container runtime. We can make these changes in Docker Desktop:

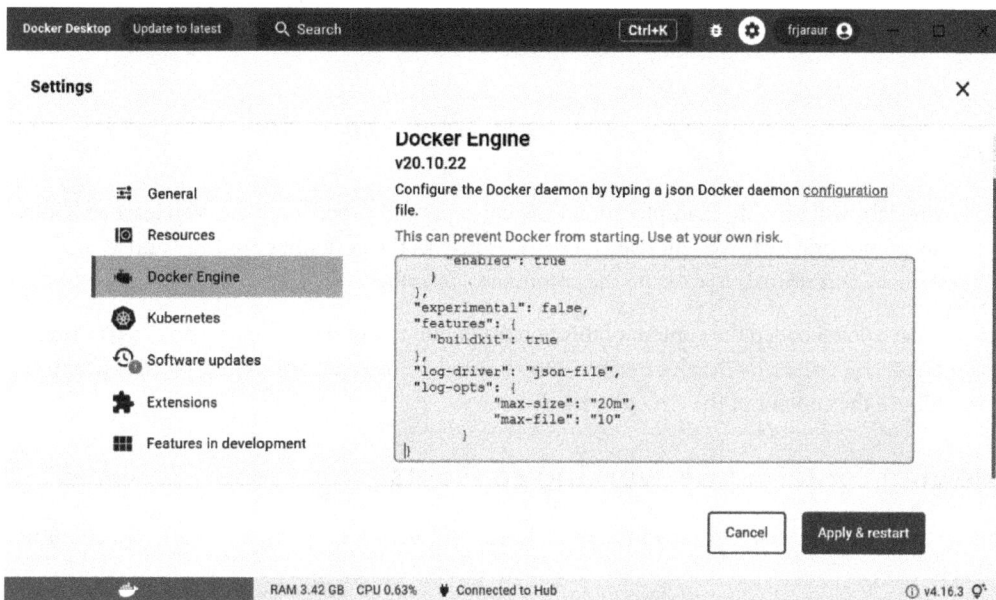

Figure 4.3 – The available Docker daemon settings in Docker Desktop (the embedded daemon.json file configured in our environment)

It is possible to define a specific logging driver for each container, although it is preferred to define a common one for your entire environment:

```
$ docker run \
        --log-driver local --log-opt max-size=10m \
        alpine echo hello world
```

Before we complete this section, we should talk about the different logging strategies we can use in our environments:

- **Local logging**: You will probably use local logging when developing your applications. These logs will be removed whenever you remove a container and will always be managed by the container runtime. These are only present locally on your computer desktop, laptop, or server.

- **Volumes**: Using a container's external storage will allow us to ensure that logs persist between executions; although these logs may be attached to the host's storage, which will keep them locally only. If you want to keep these logs available in other servers just in case you move your containers (or execute new ones with the same attached volume), you will need to use external storage solutions such as NAS or SAN for your volumes.

- **External logging ingestion**: This should be your choice for production. Your application may send your logs directly from your code to an external logs ingestion solution or you may configure your container runtimes to send them directly for you. This will help you keep a homogeneous environment if your applications run in containers.

In the next section, we will review some of the content we learned about in this chapter by executing some labs.

Labs

The following labs will provide examples to put the concepts and procedures that you learned about in this chapter into practice. We will use Docker Desktop as the container runtime and WSL2 (or your Linux/macOS Terminal) to execute the commands described.

Ensure you have downloaded the content of this book's GitHub repository from `https://github.com/PacktPublishing/Docker-for-Developers-Handbook.git`. For this chapter's labs, we will use the content of the `Chapter4` directory.

Reviewing container networking concepts

In this section, we will review some of the most important networking topics we learned about in this chapter:

1. First, we will run a container in the background, which will be used as a reference in other steps. We will run a simple `sleep` command:

    ```
    $ docker container run -d --name one alpine sleep INF
    025e24a95b6939e025afda09bb9d646651025dfecc30357732e629aced18e66b
    ```

2. Now that container `one` is running, we will run a second one directly with the `ping` command. We will use `one` as the name to test the default bridge network DNS's existence:

    ```
    $ docker container run -ti --rm \
    --name two alpine ping -c1 one
    ping: bad address 'one'
    ```

    We verified that the default bridge network doesn't include a DNS because `one` can't be resolved, but let's verify whether communications exist. We used the `--rm` argument to delete the container right after its execution.

3. Let's verify the container's IP address by using the `inspect` action:

    ```
    $ docker container inspect one \
    --format="{{ .NetworkSettings.IPAddress }}"
    172.17.0.2
    ```

    Let's test whether container `two` can reach container `one`:

    ```
    $ docker container run -ti --rm --name two \
    --add-host one:172.17.0.2 alpine ping -c1 one
    PING one (172.17.0.2): 56 data bytes
    64 bytes from 172.17.0.2: seq=0 ttl=64 time=0.116 ms
    ```

```
   --- one ping statistics ---
  1 packets transmitted, 1 packets received, 0% packet loss
  round-trip min/avg/max = 0.116/0.116/0.116 ms
```

As expected, both containers see each other because they are running in the default bridge network. Let's remove the reference container so that we can test this again using a custom network:

```
$ docker container rm --force one
one
```

4. We can repeat the same steps using a custom network, but first, we will create the new testnet network and review its IPAM configuration:

```
$ docker network create testnet
582fe354cf843270a84f8d034ca9e152ac4bffe47949ce5399820e81fb0ba555
$ docker network inspect testnet --format="{{ .IPAM.Config }}"
[{172.18.0.0/16  172.18.0.1 map[]}]
```

And now we start our reference container attached to this network:

```
$ docker container run -d --net testnet --name one alpine sleep
INF
027469ad503329300c5df6019cfe72982af1203e0ccf7174fc7d0e242b7999aa
```

> **Important note**
>
> This can also be done by using docker network connect NETWORK CONTAINER if the container is already running (for example, if we reused the container from previous steps and attached it to the bridge network, we would have been able to also connect the new custom network).

Now, let's review the IP addresses that were assigned to the containers in this custom network:

```
$ docker network inspect testnet \
--format="{{ .Containers }}"
map[027469ad503329300c5df6019cfe72982af1203e0ccf7174fc7d0e-
242b7999aa:{one cc99284ffccb5705605075412b0a058bc58ec2ff5738efb-
d8d249a45bc5d65df 02:42:ac:12:00:02 172.18.0.2/16 }]
```

Now, let's verify the DNS resolution inside this custom network by executing a new container attached with a ping command with the first container's name as the target:

```
$ docker container run -ti --rm --name two \
--net testnet alpine ping -c1 one
PING one (172.18.0.2): 56 data bytes
64 bytes from 172.18.0.2: seq=0 ttl=64 time=0.117 ms
--- one ping statistics ---
1 packets transmitted, 1 packets received, 0% packet loss
round-trip min/avg/max = 0.117/0.117/0.117 ms
```

As we expected, DNS resolution and communications work when we use a custom network (which is also attached to the docker0 bridge interface by default).

Access to container services

In this lab, we will use the created custom network and run a simple NGINX web server:

1. Let's run a new container using the nginx:alpine image, attached to the custom network. Notice that we didn't use --it (interactive and pseudo-terminal attached) arguments because we will not interact with the NGINX process:

```
$ docker container run -d --net testnet \
--name webserver nginx:alpine
1eb773889e80f06ec1e2567461abf1244fe292a53779039a7731bd85a0f500b8
```

We can verify the running containers by using docker container ls or docker ps:

```
$ docker ps
CONTAINER ID    IMAGE           COMMAND              CREATE
D               STATUS          PORTS       NAMES
1eb773889e80    nginx:alpine    "/docker-entrypoint...."   4 minutes
ago    Up 4 minutes    80/tcp      webserver
027469ad5033    alpine          "sleep INF"          23
minutes ago    Up 23 minutes            one
```

2. We are now in our reference container, where we can install the curl package and test the connection to the web server running in the custom network:

```
$ docker container exec -ti one /bin/sh
/ # ps -ef
PID    USER     TIME   COMMAND
    1 root      0:00 sleep INF
    7 root      0:00 /bin/sh
   26 root      0:00 ps -ef
/ # apk add --update --no-cache curl
fetch https://dl-cdn.alpinelinux.org/alpine/v3.17/main/x86_64/
APKINDEX.tar.gz
...
OK: 9 MiB in 20 packages
/ # curl webserver -I
HTTP/1.1 200 OK
...
```

Now that we are executing a shell within the reference container, we can verify that the reference container's hostname is the container's ID by default before exiting:

```
/ # hostname
027469ad5033
/ # exit
```

3. We're removing the `webserver` container because we are going to modify its main page by using a **bind mount** volume:

    ```
    $ docker container rm -fv webserver
    ```

    We will now create the `data` directory in the current path, and we will just create the `index.html` file by using a simple `echo` command:

    ```
    $ mkdir $(pwd)/data
    $ echo "My webserver" >data/index.html
    ```

    Now, we can execute the `webserver` container again, but this time, we will add the created directory as a volume so that we can include our `index.html` file:

    ```
    $ docker container run -d --net testnet -v $(pwd)/data:/usr/
    share/nginx/html \
    --name webserver nginx:alpine
    b94e7a931d2fbe65fab58848f38a771f7f66ac8306abce04a3ac0ec7e0c5e750
    ```

4. Now, let's test the `webserver` service again:

    ```
    $ docker container exec -ti one curl webserver
    My webserver
    ```

    If we run a new `webserver` container using the same volume, we will obtain the same result because this directory provides persistency for static content:

    ```
    $ docker container run -d --net testnet -v $(pwd)/data:/usr/
    share/nginx/html \
    --name webserver2 nginx:alpine
    $ docker container exec -ti one curl webserver2
    My webserver
    ```

    We can now change the content of the `index.html` file and verify the result:

    ```
    $ echo "My webserver 2" >data/index.html
    $ docker container exec -ti one curl webserver2
    My webserver 2
    ```

Notice that we can change the static content with the container in a running state. If your application manages static content, you will be able to verify the changes online while developing, but this may not work for your application if your processes read the information while they start. In these cases, you will need to restart/recreate your containers.

5. Finally, let's remove the second web server:

```
$ docker container rm -fv webserver2
webserver2
```

Notice that we used the -fv argument to force-remove the container (stop it if it was running) and the associated volumes (in this case, we used a bind mount, which will never be removed by the container runtime, so don't worry about this type of mount). Let's also launch our web server by using the extended mount definition just to understand its usage:

```
$ docker container run -d --net testnet \
-name webserver \
--mount type=bind,source=$(pwd)/data,target=/usr/share/nginx/
html \
nginx:alpine
b2446c4e77be587f911d141238a5a4a8c1c518b6aa2a0418e574e89dc135d23b
$ docker container exec -ti one curl webserver
My webserver 2
```

6. Now, let's test the behavior of a named volume:

```
$ docker container run -d --net testnet -v WWWROOT:/usr/share/
nginx/html --name webserver nginx:alpine
fb59d6cf6e81dfd43b063204f5fd4cdbbbc6661cd4166bcbcc58c633fee26e86
$ docker container exec -ti one curl webserver
<!DOCTYPE html>
<html>
<head>
<title>Welcome to nginx!</title>
...
<h1>Welcome to nginx!</h1>
```

As you can see, NGINX's default page is shown, but we can copy our index.html page to the webserver's WWWROOT named volume by using the cp action:

```
$ docker cp data/index.html webserver:/usr/share/nginx/html
```

Let's test this once more to verify the changes:

```
$ docker container exec -ti one curl webserver
My webserver 2
```

As we have seen, we can manage persistent data inside containers using volumes and we can copy some content inside them using docker cp (you can use the same command to retrieve the container's content). We also tested all the internal communications; we didn't expose any service outside of the container runtime environment. Let's remove both webserver containers if they still exist:

```
$ docker rm -f webserver webserver2
```

Now, let's move on to the next lab.

Exposing applications

In this lab, we will expose the application's containers outside of the container runtime's internal networks:

1. We will use the `--publish-all` or `-P` argument to publish all the image's defined exposed ports:

    ```
    $ docker container run -d \
    --net testnet -P -v WWWROOT:/usr/share/nginx/html \
    --name webserver nginx:alpine
    dc658849d9c34ec05394a3d1f41377334261283092400e0a0de4ae98582238a7
    $ docker ps
    CONTAINER ID    IMAGE           COMMAND              CREATE
    D               STATUS          PORTS                NAMES
    dc658849d9c3    nginx:alpine    "/docker-entrypoint...."   14
    seconds ago     Up 12 seconds   0.0.0.0:32768->80/tcp    webserver
    027469ad5033    alpine          "sleep INF"          23 hours
    ago      Up 23 hours                          one
    ```

    You may have noticed that a NAT port is created. Consecutive host ports will be used within the port range of `32768` and `61000`.

2. Now, let's check the service from our host. We can use `localhost`, `127.0.0.1`, or `0.0.0.0` as the IP address because we didn't specify any of the host's IP addresses:

    ```
    $ curl 127.0.0.1:32768
    My webserver 2
    ```

3. Now, let's use the `host` network.

    As we already have `curl` installed on one container, we can use `commit` for these changes to prepare a new image for running new containers with `curl`:

    ```
    $ docker container commit one myalpine
    sha256:6732b418977ae171a31a86460315a83d13961387d-
    aacf5393e965921499b446e
    ```

    > **Important note**
    >
    > If you are using the `host` network in Linux directly, you will be able to connect directly to your container's ports, even if they aren't exposed. This doesn't work in WSL environments directly, but you can use this behavior in cluster environments.

4. Now, we can use this new container by connecting it to the host's network to verify how the network changed:

    ```
    $ docker container run -d --net host \
    --name two myalpine sleep INF
    885bcf52115653b05645ee10cb2862bab7eee0199c0c1b99e367d8329a8cc601
    ```

5. If we use `inspect`, we will notice that no IP addresses will be associated with the container via the container runtime. Instead, all the host's network interfaces will be attached to the containers:

```
$ docker container exec \
-ti two ip add show|grep "inet "
    inet 127.0.0.1/8 scope host lo
    inet 172.17.0.1/16 brd 172.17.255.255 scope global docker0
    inet 192.168.65.4 peer 192.168.65.5/32 scope global eth0
    inet 172.18.0.1/16 brd 172.18.255.255 scope global
br-582fe354cf84
```

6. However, you should notice that DNS container resolution doesn't work in the host network:

```
$ docker container exec -ti two curl webserver -I
curl: (6) Could not resolve host: webserver
```

7. Let's retrieve the web server's IP address to access it via the host network container:

```
$ docker container inspect webserver --format="{{ .NetworkSet-
tings.Networks.testnet.IPAddress }}"
172.18.0.3
$ docker container exec -ti two curl 172.18.0.3
My webserver 2
```

Next, we will review how to limit access to the host's hardware resources and how exceeding the memory limit will trigger the execution of the OOM-Killer kernel process.

Limiting containers' resource usage

In this lab, we will review how to limit the host's memory inside a container:

1. First, we will create a custom image, including the `stress-ng` application:

```
$ cat <<EOF|docker build -q -t stress -
FROM alpine:latest
RUN apk add --update --no-cache stress-ng
EOF
sha256:4158ba4e466c974dee2a13ebc5a32462b38f687b28004a2dd79ca-
f97ae764a08
```

2. Now, we can test how `stress-ng` works by using just one worker process and a maximum memory capacity of 1,024 MB:

```
$ docker run -d --name stress stress stress-ng \
--vm-bytes 1024M --fork 1 -m 1
2bea5bcba3a9609e0f47b7c27b24fde9767a75764b4a9bb628ba696f569da001
```

3. Let's use `docker stats` to retrieve the current container's resource usage:

```
$ docker stats --no-stream
CONTAINER ID   NAME       CPU %      MEM USAGE / LIMIT      MEM
%      NET I/O      BLOCK I/O    PIDS
2bea5bcba3a9   stress     219.77%    1.017GiB /
9.236GiB   11.01%    836B / 0B   0B / 0B     5
```

We can wait a few seconds and run this command again or simply execute `docker stats` to retrieve the statistics continuously:

```
$ docker stats --no-stream
CONTAINER ID   NAME       CPU %      MEM USAGE / LIMIT      MEM
%      NET I/O      BLOCK I/O    PIDS
2bea5bcba3a9   stress     217.13%    1.015GiB /
9.236GiB   10.99%    906B / 0B   0B / 0B     5
```

4. Now, let's kill the current `stress` container and run it again, limiting its access to the host's memory:

```
$ docker container kill stress
stress
$ docker run -d --name stress-limited \
--memory 128M stress stress-ng --vm-bytes 1024M \
--fork 1 -m 1
238e34215885f5fc20b0ff157f17b18e6559720c7453064a1c7aedb9cb635284
```

5. Now, we will execute the `stats` action again continuously (to show the output in this book, we executed it using `--no-stream` a few times) and we can verify that although `stress-ng` runs a process with 1,024 MB, the container never uses that amount of memory:

```
$ docker stats --no-stream
CONTAINER ID   NAME              CPU %      MEM USAGE /
LIMIT    MEM %      NET I/O      BLOCK I/O    PIDS
ff3f4797af43   stress-limited    166.65%    125.1MiB /
128MiB   97.74%    1.12kB / 0B   0B / 0B     4
```

Wait a few seconds and execute it again:

```
$ docker stats --no-stream
CONTAINER ID   NAME              CPU %      MEM USAGE /
LIMIT    MEM %      NET I/O      BLOCK I/O    PIDS
ff3f4797af43   stress-limited    142.81%    127MiB /
128MiB   99.19%    1.12kB / 0B   0B / 0B     5
```

As we expected, the memory usage is limited. You can verify what happened by reviewing the current host's system log. The container runtime uses cgroups to limit the container's use of resources and the kernel launched the OOM-Killer feature to kill the processes that were consuming more memory than expected:

```
$ dmesg|grep -i oom
[22893.337110] oom_reaper: reaped process 19232 (stress-ng), now
anon-rss:0kB, file-rss:0kB, shmem-rss:32kB
[22893.915193] stress-ng invoked oom-killer: gfp_mask=0xcc0(GFP_
KERNEL), order=0, oom_score_adj=1000
[22893.915221]  oom_kill_process.cold+0xb/0x10
[22893.915307] [  pid  ]   uid  tgid total_vm      rss pgtables_
bytes swapents oom_score_adj name
[22893.915316] oom-kill:constraint=CONSTRAINT_MEMCG,no-
demask=(null),cpuset=ff3f4797af43980e7ea223c-
bee27b39921dbaa84b61f22d8dcfb409347ba4a5a,mems_allowed=0,oom_
memcg=/docker/ff3f4797af43980e7ea223cbee27b39921db
```

This kernel feature is killing the `stress-ng` worker processes, but it launches more (this is the normal `stress-ng` behavior, but your applications may die if OOM-Killer is asked to destroy your processes).

6.   We will finish this lab by simply removing the used containers:

```
$ docker container rm --force stress-limited
stress-limited
```

We can now move on to the next lab, in which we will learn how to limit the use of privileged users in our processes if they are not needed.

Avoiding the use of root users inside containers

This quick lab will show you how to run an NGINX web server without `root`. But first, we will review what happens when you change the default NGINX environment:

1.   First, let's review the user that a default `nginx:alpine` image will use by simply executing a new web server:

```
$ docker container run -d --publish 8080:80 \
--name webserver nginx:alpine
cbcd52a7ca480606c081edc63a59df5b6a237bb2891a4f4bb2ae68f9882fd0b3
$ docker container ls
CONTAINER ID    IMAGE           COMMAND                     CREATE
D          STATUS          PORTS                   NAMES
cbcd52a7ca48    nginx:alpine    "/docker-entrypoint.…."    7 seconds
ago    Up 6 seconds    0.0.0.0:8080->80/tcp    webserver
```

As expected, it is running and served in the host's port 8080:

```
$ curl 0.0.0.0:8080 -I
HTTP/1.1 200 OK
...
```

2. Now, let's retrieve its logs:

```
$ docker logs webserver
/docker-entrypoint.sh: /docker-entrypoint.d/ is not empty, will
attempt to perform configuration
...
2023/03/31 19:26:57 [notice] 1#1: start worker process 33
172.17.0.1 - - [31/Mar/2023:19:28:18 +0000] "GET / HTTP/1.1" 200
615 "-" "curl/7.81.0" "-"
```

The log shows the request we make to the published service and the output of NGINX's main process (standard output and error). We can limit the number of lines shown by using `--tail 2` (this will show only the last two lines of the container's logs):

```
$ docker logs webserver --details \
--timestamps --tail 2
2023-03-31T19:29:35.362006700Z  172.17.0.1 - - [31/
Mar/2023:19:29:35 +0000] "GET / HTTP/1.1" 200 615 "-"
"curl/7.81.0" "-"
2023-03-31T19:29:39.427574300Z  172.17.0.1 - - [31/
Mar/2023:19:29:39 +0000] "HEAD / HTTP/1.1" 200 0 "-"
"curl/7.81.0" "-"
```

Notice that we also used `--timestamp` to show the container runtime's included timestamp. This can be very useful when the running application does not provide any timestamp.

By default, NGINX writes to `/var/log/nginx/access.log` and `/var/log/nginx/error.log`. It is very interesting to learn how this container's image developers set the processes up to write to `/dev/stdout` and `/dev/stderr`. You can learn more at `github.com/nginxinc/docker-nginx/blob/73a5acae6945b75b433caf-d0c9318e4378e72cbb/mainline/alpine-slim/Dockerfile`. An extract of the currently important lines is shown here:

```
# forward request and error logs to docker log collector
    && ln -sf /dev/stdout /var/log/nginx/access.log \
    && ln -sf /dev/stderr /var/log/nginx/error.log \
```

3. Now, let's check the user running this instance:

```
$ docker exec -ti webserver id
uid=0(root) gid=0(root) groups=0(root),1(bin),2(dae-
mon),3(sys),4(adm),6(disk),10(wheel),11(floppy),20(di-
alout),26(tape),27(video)
```

4. As we already discussed in this chapter, running containers as non-`root` should always be preferred, so let's remove this container and create a new safer one (without `root`):

```
$ docker container rm webserver --force -v
webserver
```

5. We will try changing the current 0 user (`root`) to a common `1000` ID:

```
$ docker container run -d --publish 8080:80 \
--name webserver  --user 1000 nginx:alpine
6fce3675a104ca658454d33bfa5f38fb48a0c7f71defd56caf70886c94c82e89
```

As we expected, issues appear because this image hasn't been set up to be run by a non-`root` user:

```
$ docker logs webserver
...
nginx: [warn] the "user" directive makes sense only if the
master process runs with super-user privileges, ignored in /etc/
nginx/nginx.conf:2
2023/04/01 11:36:03 [emerg] 1#1: mkdir() "/var/cache/nginx/
client_temp" failed (13: Permission denied)
nginx: [emerg] mkdir() "/var/cache/nginx/client_temp" failed
(13: Permission denied)
```

6. We can try to modify this image behavior by adding a volume for the problematic path, but even with this change, it will not work. NGINX should avoid using port 80 because it is system-restricted; if this port must be used in your environment, special capabilities such as NET_BIND_SERVICE should be added. Instead of changing the current image behavior, we will use a new image from NGINX, Inc.:

```
$ docker search nginxinc
NAME                                            DESCRIP-
TION                                  STARS OFFICIAL    AUTO-
MATED
nginxinc/nginx-unprivileged                     Unprivileged NGINX
Dockerfiles                        90
...
```

You can find this image and its information at https://hub.docker.com/r/nginx-inc/nginx-unprivileged#!.

7. Let's pull the image from Docker Hub and review the ports and the user used:

```
$ docker image pull nginxinc/nginx-unprivileged:alpine-slim -q
docker.io/nginxinc/nginx-unprivileged:alpine-slim
```

Run `docker inspect` to do so:

```
$ docker image inspect \
nginxinc/nginx-unprivileged:alpine-slim \
--format="{{ .Config.ExposedPorts }} {{ .Config.User }}"
map[8080/tcp:{}] 101
```

Now, let's run a container by publishing port 8080 on our host's port 8080. Notice that we used the `--publish` option, which allows us to even use a specific IP address from our host in `IP:host_port:container_port` format:

```
$ docker container run -d --publish 8080:8080 --name webserver
nginxinc/nginx-unprivileged:alpine-slim
369307cbd5e8b74330b220947ec41d4f263ebfe7727efddae3efbcc3a1610e5e
$ docker container ps
CONTAINER ID    IMAGE                                        COM
MAND                    CREATED           STATUS        PORT
S                       NAMES
369307cbd5e8    nginxinc/nginx-unprivileged:al-
pine-slim    "/docker-entrypoint.…"    15 seconds ago    Up 13
seconds    0.0.0.0:8080->8080/tcp    webserver
```

8. Let's test our web server again and review the logs:

```
$ curl 0.0.0.0:8080  -I
HTTP/1.1 200 OK
...
 $ docker logs --tail 2 webserver
2023/04/01 11:40:29 [notice] 1#1: start worker process 32
172.17.0.1 - - [01/Apr/2023:11:41:36 +0000] "HEAD / HTTP/1.1"
200 0 "-" "curl/7.81.0" "-"
```

9. Now, let's review the web server's user:

```
$ docker exec webserver id
uid=101(nginx) gid=101(nginx) groups=101(nginx)
```

As we expected, this `webserver` application runs using a non-privileged user and it's safer than the one running as `root`. You, as the developer, must prioritize the usage of non-privileged users in your applications to improve the components' security.

Cleaning the container runtime

To finish this chapter's labs, we will quickly clean up all the objects that were created during the labs by using a combination of commands:

1. Kill all the running containers (we can also remove them using a single line, but we will kill them before using the `prune` action):

```
$ docker ps -q|xargs docker kill
6f883a19a8f1
3e37afe57357
369307cbd5e8
```

Here, we piped two commands. The first command retrieves the list of all the running containers, but the `-q` argument is used to only show the containers' IDs. Then, we piped the result using the `xargs` command to `docker kill`. This combination kills all the running containers.

2. Now, we can use `docker system prune` to remove all the objects that were created. We will use `--all` to remove all the unused images and the volumes by adding `--volumes` (you will be asked for confirmation):

```
$ docker system prune --all --volumes
WARNING! This will remove:
  - all stopped containers
  - all networks not used by at least one container
  - all volumes not used by at least one container
  - all images without at least one container associated to them
  - all build cache
 Are you sure you want to continue? [y/N] y
...
Total reclaimed space: 49.95MB
```

After a few seconds, your system will be clean, and you can start the next chapter's labs without old objects. We can verify this by executing `docker system df`:

```
$ docker system df
TYPE            TOTAL       ACTIVE      SIZE        RECLAIMABLE
Images          0           0           0B          0B
Containers      0           0           0B          0B
Local Volumes   0           0           0B          0B
Build Cache     0           0           0B          0B
```

In these labs, we covered almost all the content that we reviewed in this chapter. You may find additional information in the GitHub repository that's been prepared for this chapter: `https://github.com/PacktPublishing/Docker-for-Developers-Handbook/tree/main/Chapter4`.

Summary

In this chapter, we learned how to run containers and manage their behavior. We also reviewed how we can limit access to the host's resources by applying different kernel features. Different techniques allow us to interact with containers while they are running, and we can use them to retrieve important information about the applications running inside. By the end of this chapter, we also learned about some simple commands that will help us keep our environments free of old unused containers' objects.

Now that we know how to create container images, store them, and run containers using them, we can move on to the next chapter, where we will learn how to run applications by using multiple containers that interact with each other.

5

Creating Multi-Container Applications

This book guides you step by step along the path of developing your applications using containers. In previous chapters, we learned how to create container images, how to share them, and, finally, how to run application processes within containers. In this chapter, we will go a step further by running applications using multiple containers. This is the method you would probably use for developing your applications, running different interconnected components, sharing information, and publishing only the frontend processes to users. By the end of this chapter, you will be able to build, deliver, and run applications by using a composition of multiple containers managed all at once with a newly learned command line.

This chapter will cover the following topics:

- Installing and using Docker Compose
- Introducing the Docker Compose file syntax
- Building and sharing multi-container applications
- Running and debugging multi-container applications
- Managing multiple environments with Docker Compose

Technical requirements

We will use open source tools for building, sharing, and running an application composed of multiple containers. The labs for this chapter will help you understand the content presented, and they are published at `https://github.com/PacktPublishing/Containers-for-Developers-Handbook/tree/main/Chapter5`. The *Code In Action* video for this chapter can be found at `https://packt.link/JdOIY`.

Installing and using Docker Compose

Docker Compose is a tool developed by Docker Inc. to help developers create, deliver, and run applications with multiple components running within containers. This tool may come with your Docker container runtime distribution or have to be installed separately. If you are using tools such as **Podman**, you will also have available equivalent command-line tools.

Docker Compose, developed in 2014 as an open source project, aims to manage multiple containers based on YAML definitions. This command line will talk directly with the Docker container runtime API. This means that all containers managed by a `docker-compose` file will run together on top of the same container runtime, hence on the same host. Understanding this is key because you will need third-party tools and configurations if you need to provide high availability for your applications. We can think of Docker Compose as a single-node container orchestrator.

If you are using Docker Desktop, you will notice that Docker Compose is already available for you. It should be integrated into your WSL environment if you checked the **Enable integration** option in your Docker Desktop (this option should be already checked for all previous chapters' labs and examples in your environment). You can verify this by quickly accessing your Docker Desktop's settings by navigating to **Settings** | **Resources** | **WSL Integration**:

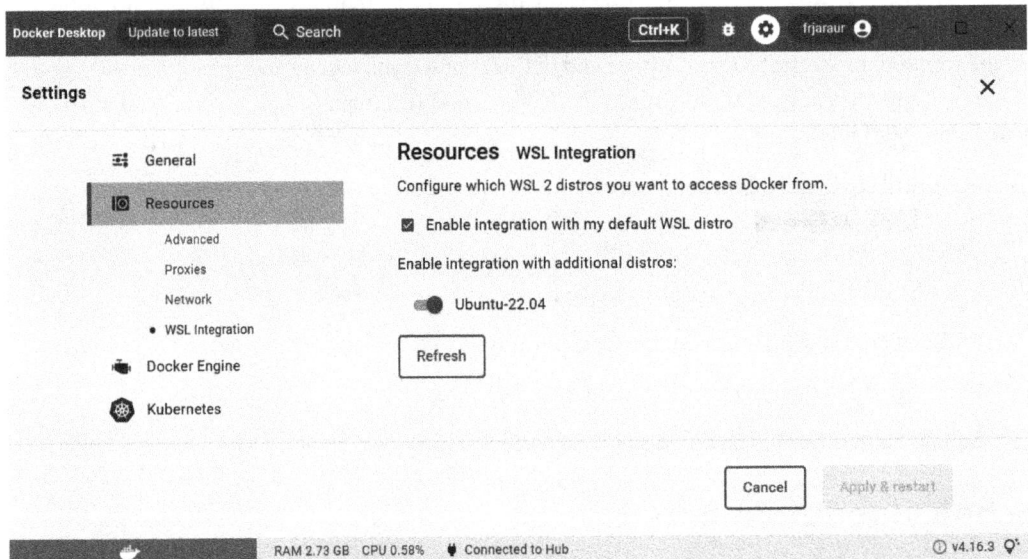

Figure 5.1 – Docker Desktop WSL Integration settings

Then, you can open a terminal in your WSL environment and simply execute `which docker-compose`:

```
$ which docker-compose
/usr/bin/docker-compose
```

Docker Desktop installs a modern Docker CLI environment, and this includes a docker-compose built-in link. You can verify this by simply retrieving the related information as follows:

```
$ docker compose --help
Usage:  docker compose [OPTIONS] COMMAND
Docker Compose
Options:
...
Commands:
  build        Build or rebuild services
...
    version    Show the Docker Compose version information
Run 'docker compose COMMAND --help' for more information on a command.
```

Therefore, we can use either docker-compose or docker compose for running compose commands.

If you are using the Docker container runtime directly on your computer and your client environment does not include this built-in link, you will need to properly install docker-compose binaries. You can use any of the following procedures:

- **As a Python module**: You can install the docker-compose module by using the Python package installer (pip install docker-compose). This will install the latest Python-based docker-compose release (1.29.2):

  ```
  $ pip install docker-compose
  Defaulting to user installation because normal site-packages is
  not writeable
  Collecting docker-compose
    Downloading docker_compose-1.29.2-py2.py3-none-any.whl (114
  kB) —— 114.8/114.8 KB 3.9 MB/s eta 0:00:00
  ...
  Successfully installed attrs-22.2.0 bcrypt-4.0.1 certifi-
  2022.12.7 cffi-1.15.1 charset-normalizer-3.1.0 docker-6.0.1
  docker-compose-1.29.2 dockerpty-0.4.1 docopt-0.6.2 idna-3.4
  jsonschema-3.2.0 packaging-23.0 paramiko-3.1.0 pycparser-2.21
  pynacl-1.5.0 pyrsistent-0.19.3 python-dotenv-0.21.1
  requests-2.28.2 texttable-1.6.7 urllib3-1.26.15 websocket-
  client-0.59.0
  ```

  However, this method will be deprecated as newer docker-compose binaries are built using the Go language. We can check the version currently installed by using the –version argument:

  ```
  $ docker-compose --version
  docker-compose version 1.29.2, build unknown
  ```

- **As a system package**: Depending on your client's operating system, you will find different options for installing the docker-compose package. We will show you the steps for Ubuntu 22.04, which provides the required package and its dependencies:

```
$ sudo apt-get install -qq docker-compose
Preconfiguring packages ...
Selecting previously unselected package pigz.
...
Setting up docker-compose (1.29.2-1) ...
Processing triggers for dbus (1.12.20-2ubuntu4.1) ...
Processing triggers for man-db (2.10.2-1) .
```

As you can see, this method also installs the latest Python-based version of docker-compose:

```
$ file /usr/bin/docker-compose
/usr/bin/docker-compose: Python script, ASCII text executable
```

Docker Compose v1 will be deprecated in June, 2023. We should use at least Docker Compose v2 and an appropriate command-line release based on Go. This release may be installed automatically with Docker Desktop, as mentioned at the beginning of this section, or by using Docker Compose as a Docker client plugin.

- **As a plugin**: This method is preferred if you are not using Docker Desktop. You will need to configure the package report for your operating system (this may be already configured in your environment if you installed the Docker container runtime and its client). You can follow the specific instructions for your Linux distribution published at https://docs.docker.com/engine/install. Having configured the Docker Inc. repository, we can use the distribution-specific package manager to install docker-compose-plugin. We will show you the following Ubuntu process as an example:

```
$ sudo apt-get install docker-compose-plugin -qq
Selecting previously unselected package docker-compose-plugin
...
Unpacking docker-compose-plugin (2.17.2-1~ubuntu.22.04~jammy)
...
Setting up docker-compose-plugin (2.17.2-1~ubuntu.22.04~jammy)
...
$ docker compose version
Docker Compose version v2.17.2
```

As you may have noticed, this method installs a modern docker-compose version compatible with both Docker Compose v2 and v3.

Important note

It is possible to install `docker-compose` directly by downloading its binary from the project's GitHub repository. You can use the following link for further instructions: `https://docs.docker.com/compose/install/linux/#install-the-plugin-manually`.

Once we install `docker-compose` by following any of the methods described, we are ready to quickly review the main features available:

- We can build multiple images, code blocks, and Dockerfiles, which may be separated into different folders. This is very useful for automating the construction of all application components at once.

- Sharing the applications' container image components is easier with `docker-compose`, as all images will be pushed at once.

- We can start and stop applications based on multiple containers with `docker-compose`. All components will run at the same time by default, although we can define dependencies between components.

- All applications' standard errors and output will be available from a single command, which means that we can access all application logs at once. This will be very useful when debugging interactions between multiple components at the same time.

- Provisioning and decommissioning environments are very easy using `docker-compose`, as all required application components will be created and removed by using simple actions such as `docker compose create` and `docker compose rm`.

- **Volumes** and **networks** will be managed for all application containers. This makes the use of `docker-compose` perfect for easily sharing data and isolating inter-process communication. We will publish only specific application processes, while others will be kept internal, hidden from the users.

- We will use **projects** to isolate one application from another. Therefore, projects will allow us to run the same application multiple times (with its own set of container objects, such as processes, volumes, networks, and configurations). By default, the name of the current directory will be used as the project name if none is specified. The project name will be used as a prefix for all the objects created. Therefore, container names created for a project, for example, will follow the `<PROJECT>-<SERVICE_NAME>` syntax. We can retrieve the list of running projects by using `docker-compose ls`. This command will show you all running `docker-compose` projects with their Compose YAML file definitions.

- By using **profiles**, we will be able to define which objects should be created and managed. This can be very useful under certain circumstances – for example, we can define one profile for production and another for debugging. In this situation, we will execute `docker-compose --profile prod up --detach` to launch our application in production, while using `--profile debug` will run some additional components/services for debugging. We will use the `profile` key in our Compose YAML file to group services, which can be added to multiple profiles. We will use a string to define these profiles and we will use it later in the `docker-compose` command line. If no profile is specified, `docker-compose` will execute the actions without using any profile (objects with no profile will be used).

The following list shows the main actions available for `docker-compose`:

- `config`: This action will check and show a review of the Compose YAML file. It can be used in combination with `--services` or `--volumes` arguments to retrieve only these objects. As mentioned before, `--profile` can be used to specifically retrieve information about a certain set or group of objects.

- `images`: This shows the images defined in our Compose YAML file. This will be useful if you are wondering whether images will need to be built or may already be present in your environment.

- `build`: This action makes `docker-compose` a great tool, even if you are planning to deploy your applications on a container orchestration cluster such as Kubernetes, as we are able to build all our application components' container images with just one command. Images created using `docker-compose` will include the project's name in its name; hence, they will be identified as `<PROJECT_NAME>-<SERVICE_NAME>`. A Dockerfile should be included in all component directories, although we can override the building of certain images by specifying an image repository directly. Remember all the content we learned about tagging images in *Chapter 3, Shipping Docker Images*. We can modify the build context and the Dockerfile filename using the `context` and `dockerfile` keys, respectively. If our Dockerfile contains various targets, we can define which one will be used for building the service's image by using the `target` key. Arguments can also be passed to the build process to modify the environment by using the `args` key with a list of the key-value pairs that should be included.

- `pull/push`: The images defined can be downloaded all at once and the build definitions can also be pushed to remote registries once your images are created.

- `up`: This action is equivalent to executing `docker run` for each component/service defined in our Compose YAML file. By default, `docker compose up` will start all the containers at once and our terminal will attach to all containers' outputs, which may be interesting for testing but not for production (our terminal will be stuck attached to the processes, and we must use *Ctrl + P + Q* to detach from them). To avoid this situation, we should use the `-d` or `--detach` argument to launch our containers in the background. `docker-compose` also supports the `run` action, but this is generally used for running specific services one at a time.

- down: This action, as expected, does the opposite of up; it will stop and remove all the containers that are running. It is important to understand that new containers will be created if they were previously removed by using this action. Any persistent data must be stored outside of the container's life cycle. To completely remove your application, remember to always remove the associated volumes. We can add the --volumes argument to force the removal of any associated volume.

- create/run/start/stop/rm: All these actions are equivalent to the ones we learned about in *Chapter 4*, *Running Docker Containers*, but in this case, they will apply to multiple containers at once.

- ps: As we are running multiple containers for a project, this action will list all associated containers. Containers' performances can be reviewed by using docker-compose top, which is an extension of the docker stats command we learned in *Chapter 4*, *Running Docker Containers*.

- exec: This option allows us to execute a command attached to one of the containers (in this case, a project's service).

- logs: We can use docker-compose logs to retrieve all the project's container logs. This is very useful for retrieving all application logs by using a single point of view and just one command. The container output will be separated by colors and all the filter options learned about in *Chapter 4*, *Running Docker Containers*, as well as by --follow, which continuously follows all of them. We can retrieve just one service log by adding the service name as an argument.

> **Important note**
>
> Although you will usually execute docker-compose actions against all containers, it is possible to specify one service at a time by adding the specific service name, docker-compose <ACTION> <SERVICE>. This option is extensible to almost all commands and very useful for debugging purposes when things go wrong with some containers.

Now that we know how to install docker-compose and what features we may expect, we can learn how to create applications using it.

Introducing the Docker Compose file syntax

We will use docker-compose with a YAML file, in which we will define all the services, volumes, and networks that will run together and be managed as components of an application. The YAML file used should follow the **Compose application model** (more information is available at https://github.com/compose-spec/compose-spec/blob/master/spec.md). This model will distribute application components in **services** and their intercommunication with each other using **networks**. These networks provide the isolation and abstraction layers for our application containers. Services will store and share their data by using **volumes**.

Services may need additional configurations, and we will use **config** and **secret** resources to add specific information to manage the application's behavior. These objects will be mounted inside our containers, and processes running inside them will use the provided configurations. Secrets will be used to inject sensitive data and the container runtime will treat them differently.

As discussed earlier in this chapter, Compose v1 will be deprecated soon, and you should migrate to at least Compose v2. Your files may need some changes. You can verify this by reviewing the documentation at `https://docs.docker.com/compose/compose-file/compose-versioning`. The Compose application model specification merges the object definitions from v2 and v3.

Now, let's deep dive into the Docker Compose YAML file definition keys.

YAML file definition keys

By default, the `docker-compose` command will look for `docker-compose.yaml` or `compose.yaml` files in the current directory (you can use either `.yaml` or `.yml` extensions). Multiple Compose files can be used at the same time, and the order in which they appear will define the final file specification to use. Values will be overridden by the latest ordered file. We can also use variables that can be expanded in runtime by setting our environment variables. This will help us use a general file with variables for multiple environments.

The basic schema of a Compose YAML file will be presented as follows:

```
services:
     service_name1:
           <SERVICE_SPECS>
  ...
     service_nameN:
           <SERVICE_SPECS>
volumes:
     volume_name1:
           <VOLUME_SPECS>
  ...
     volume_nameN:
           <VOLUME_SPECS>
networks:
     network_name1:
           <NETWORK_SPECS>
  ...
     network_nameN:
           <NETWORK_SPECS>
```

Each service will need at least a container image definition or a directory where its Dockerfile is located.

Let's review the Compose syntax with an example file:

```yaml
version: "3.7"

services:
  # load balancer
  lb:
    build: simplestlb
    image: myregistry/simplest-lab:simplestlb
    environment:
      - APPLICATION_ALIAS=simplestapp
      - APPLICATION_PORT=3000
    networks:
      simplestlab:
          aliases:
          - simplestlb
    ports:
      - "8080:80"
  db:
    build: simplestdb
    image: myregistry/simplest-lab:simplestdb
    environment:
        - "POSTGRES_PASSWORD=changeme"
    networks:
        simplestlab:
        aliases:
          - simplestdb
    volumes:
      - pgdata:/var/lib/postgresql/data
  app:
    build: simplestapp
    image: myregistry/simplest-lab:simplestapp
    environment:
      - dbhost=simplestdb
      - dbname=demo
      - dbuser=demo
      - dbpasswd=d3m0
    networks:
        simplestlab:
        aliases:
            - simplestapp
    depends_on:
      - lb
      - db
```

```
volumes:
  pgdata:

networks:
  simplestlab:
    ipam:
      driver: default
      config:
        - subnet: 172.16.0.0/16
```

The first line is used to identify the Compose syntax version used. Currently, the `version` key is only informative, added for backward compatibility. If some keys are not allowed in the current Compose release, we will be warned, and those keys will be ignored. At the time of writing this book, Compose YAML files do not require this `version` key.

This Compose YAML file contains three service definitions: `lb`, `db`, and `app`. All of them have an `image` key, which defines the image repository that will be used for creating each service. We also have a `build` key, which defines the directory that will be used for building the defined image. Having both keys will allow us to create the required image with the defined name before executing the service. As you may have noticed, we have defined dependencies for the `app` service. This service depends on the `lb` and `db` services; hence, their containers must be running and healthy before any `app` container starts. Health checks defined in each container image will be used to verify the healthiness of the container's processes. That's why you, as a developer, should define the appropriate health checks for your application's components.

> **Important note**
>
> Although, in this example, we used the `depends_on` key, it is very important to include the management of different component dependencies in our application's code. This is important because the `depends_on` key is only available in Compose YAML files. When you deploy your applications in Docker Swarm or Kubernetes, the dependencies can't be managed in the same way. Compose manages dependencies for you, but this feature does not exist in orchestrated environments and your applications should be prepared for that. You may, for example, verify the connectivity with your database component before executing certain tasks, or you might manage the exceptions in your code related to the loss of this connection. Your application component may need several components, and you should decide what your application has to do if one of them is down. Key application components should stop your code in case full application functionality breaks.

In this example, we also defined one volume, `pgdata`, and a network, `simplestlab`. The `volumes` and `networks` sections allow us to define objects to be used by the containers. Each defined service should include the volumes and networks that should be attached to the service's containers. Containers associated with a service will be named after the service name, including the project as a prefix. Each container is considered an instance for the service and will be numbered; hence, the final container will be named `<PROJECT_NAME>-<SERVICE_NAME>-<INSTANCE_NUMBER>`.

We can have more than one instance per service. This means that multiple containers may run for a defined service. We will use `--scale SERVICE_NAME=<NUMBER_OF_REPLICAS>` to define the number of replicas that should be running for a specific service.

> **Important note**
>
> As mentioned before, dynamic names will be used for the service containers, but we can use the `container_name` key to define a specific name. This may be interesting for accessing a container name from other containers, but this service wouldn't be able to scale because, as you already know, container names are unique for each container runtime; thus, we cannot manage replicas in this situation.

Compose YAML files allow us to overwrite all the keys defined in the container images. We will include them inside each `services` definition block. In the presented example, we have included some environment variables for all services:

```
...
services:
  lb:
    environment:
      - APPLICATION_ALIAS=simplestapp
      - APPLICATION_PORT=3000
...
  db:
    environment:
      - "POSTGRES_PASSWORD=changeme"
...
  app:
    environment:
      - dbhost=simplestdb
      - dbname=demo
      - dbuser=demo
      - dbpasswd=d3m0
...
```

As you may notice, these environment variables define some configurations that will change the application components' behavior. Some of these configurations contain sensitive data, and we can use additional Compose objects such as `secrets`. Non-sensitive data can be written using `config` objects.

For these objects, an additional key will be used at the root level:

```
...services:
  app:
...
    configs:
      - source: appconfig
        target: /app/config
        uid: '103'
        gid: '103'
        mode: 0440
volumes:
...
networks:
...
configs:
  appconfig:
    file: ./appconfig.txt
```

In this example, we changed all the app component environment variables for a config object, which will be mounted inside the container.

> **Important note**
>
> By default, config object files will be mounted in /<source> if no target key is used. Although there is a short version for mounting config object files inside service containers, it is recommended to use the presented long format, as it allows us to specify the complete paths for both source and target, as well as the file's permissions and ownership.

Secret objects are only available in swarm mode. This means that even if you are just using a single node, you must execute docker swarm init to initialize a single-node Swarm cluster. This will allow us to create secrets, which are stored as cluster objects by the Docker container engine. Compose can manage these objects and present them in our service's containers. By default, secrets will be mounted inside containers in the /run/secrets/<SECRET_NAME> path, but this can be changed, as we will see in the following example.

First, we create a secret with the database password, used in the db service, by using docker secret create:

```
$ printf "mysecretdbpassword" | docker secret create postgres_pass -
dzr8bbh5jqgwhfidpnrq7m5qs
```

Then, we can change our Compose YAML file to include this new `secret`:

```
...
  db:
    build: simplestdb
    image: myregistry/simplest-lab:simplestdb
    environment:
        - POSTGRES_PASSWORD_FILE: /run/secrets/postgres_pass
    secrets:
    - postgres_pass
...
secrets:
  postgres_pass:
    external: true
```

In this example, we created the secret using the standard output and we used `external: true` to declare that the secret is already set and the container runtime must use its key store to find it. We could have used a file instead as the source. It is also common to integrate some files as `secrets` inside containers by adding them in the following format:

```
secrets:
  my_secret_name:
    file: <FULL_PATH_TO_SECRET_FILE>
```

The main difference here is that you may be using a plain text file as a secret that will be encrypted by the Docker container runtime and mounted inside your containers. Anyone with access to this plain text file will read your secrets. Using the standard output increases security because only the container runtime will have access to the `secret` object. In fact, the Docker Swarm store can also be encrypted, adding a new layer of security.

Now that we understand the basic Compose YAML syntax, we can continue learning how to use these files to build and share our application's container images.

Building and sharing multi-container applications

Docker Compose allows you to run multi-container applications on single nodes. These applications will not really have high availability, as you will have a single point of failure and you will probably prefer to use orchestrated clusters by using Kubernetes or Docker Swarm. But even in these situations, `docker-compose` will help you build and manage the container images for your project. In this section, we will learn how to use these features.

Your Compose YAML file will have some service definitions, and each service's container will need an image definition or a `build` directory. The `image` key will be used for either downloading this image from a registry (if it does not exist in your container runtime already) or setting the name of the service's container image to be created if a `build` folder exists. As we already mentioned in the previous section, the project's name will be used as a prefix for all your images by default, but having this `image` key overrides this. Project prefixes will help you identify all the images prepared for a project but may be confusing when a project must be executed twice (two different project instances). In such situations, it may be convenient to prepare and push your images for both projects instead of building them with default folder names.

We will now focus on the `build`-related keys:

```
services:
  lb:
    build: simplestlb
    image: myregistry/simplest-lab:simplestlb
  ...
  db:
    build: simplestdb
    image: myregistry/simplest-lab:simplestdb
  ...
  app:
    build: simplestapp
    image: myregistry/simplest-lab:simplestapp
```

As we mentioned, the `image` key defines the images to be downloaded but in this situation, the `build` key is also present with a folder string, which means that this folder will be used for building the image:

```
$ docker-compose --project-name test build \
--progress quit --quiet
$ docker image ls
REPOSITORY                TAG           IMAGE
ID        CREATED     SIZE
myregistry/simplest-lab   simplestapp   26a95450819f   3 days
ago    73.7MB
myregistry/simplest-lab   simplestdb    7d43a735f2aa   3 days
ago    243MB
myregistry/simplest-lab   simplestlb    3431155dcfd0   3 days
ago    8.51MB
```

As you may notice, a project name is included to avoid using the default directory name as a prefix, but the images created used the repository and tag strings defined.

Let's remove the `image` key lines and launch the `build` process again:

```
$ docker-compose --project-name test build \
--progress quit --quiet
$ docker image ls
REPOSITORY                TAG           IMAGE
```

```
ID          CREATED      SIZE
test-app                 latest        b1179d0492be    3 days
ago    73.7MB
myregistry/simplest-lab  simplestapp   b1179d0492be    3 days
ago    73.7MB
test-db                  latest        8afd263a1e89    3 days
ago    243MB
myregistry/simplest-lab  simplestdb    8afd263a1e89    3 days
ago    243MB
test-lb                  latest        4ac39ad7cefd    3 days
ago    8.51MB
myregistry/simplest-lab  simplestlb    4ac39ad7cefd    3 days
ago    8.51MB
```

New images were built (cached layers were used) with new names. Notice the project's name prefix and the folder name. No tags were added, so latest was used by default. This is what we expect in such situations, but we could have used any of the following keys to modify the build process:

- context: This key must be included inside the build key to identify the context used for each image. All the files included in this context directory will be passed to the container runtime for analysis. Take care to remove any unnecessary files in this path.

- dockerfile: By default, the container runtime will use any existing Dockerfile in your build folder, but we can use this key to change this filename and use our own.

- dockerfile_inline: This key may be very interesting, as it allows us to use inline definitions, as we already learned in *Chapter 2*, *Building Docker Images*. These quick definitions don't permit any COPY or ADD keys.

- args: This key is equivalent to --build-arg and it allows us to add any required arguments to our build process. Remember that you should include appropriate ARG keys in your Dockerfile.

- labels: We can include labels in our Dockerfile, and we can also add new ones or overwrite those already defined by using the labels key. We will include a list with these labels in key-value format.

- targets: This key will identify which targets should be compiled in the build process. This may be of interest when you want to separate the base image and some additional debug ones from the production-ready final container images.

- tags: We can add more than one tag at a time. This may be pretty interesting for defining a build process that creates a container image for different registries. By using this key, you will be able to push to all registries at the same time (you will need to be already logged in or you will be asked for your username and password).

- `platforms`: Remember that we learned that `buildx` allowed us to prepare images for different container runtimes architectures (we learned how to create images for ARM64, AMD64, and so on in *Chapter 2, Building Docker Images*). In our Compose file, we can write which architectures must always be included in the `build` process. This is very interesting for automating your software supply chain.

- `secrets`: Sometimes, we need to include a token, an authentication file with a username and password, or a certificate for accessing some SSL-protected site during the `build` process. In such situations, we can use a secret to introduce this information only at such a stage. You should always avoid adding sensitive information to your container images. Secrets will only be accessible to the containers created for building the image; thus they will not be present in the final image. We will need to define a `secrets` object in our Compose file, but this time, it will be used in the `build` process instead of the container runtime. Here is an example of adding a certificate required to access a server:

```
services:
  frontend:
    build:
      secrets:
        - server-certificate
secrets:
  server-certificate:
    file: ./server.cert
```

We can use the long syntax, which is always recommended because it allows us to set the destination path for the included file (by default, the secrets will be present in `/run/secrets`) and its permissions and ownership:

```
services:
  frontend:
    build:
      secrets:
        - source: server-certificate
          target: server.cert
          uid: "103"
          gid: "103"
          mode: 0440
secrets:
  server-certificate:
    file: ./server.cert
```

There are some keys that may be interesting to you if you plan on modifying the default caching behavior during the `build` processes. You can find additional information at `https://docs.docker.com/compose/compose-file/build/`.

You can push images to the registries to share them using docker-compose if you included a registry in the image key definition; otherwise, images will be local.

By default, all images will be pushed when you execute docker-compose push, but as with any other Compose action, you may need to pass a service as an argument. In this case, it is useful to use the --include-deps argument to push all the images of the services defined in the depends_on key. This will ensure that your service will not miss any required images when it is executed:

```
$ docker-compose --project-name test push --include-deps app
[+] Running 0/26bc077c4d137 Layer already exists   3.6s
⋮ Pushing lb: 0bc077c4d137 Layer already exists     3.8s
...
⋮. Pushing db: a65fdf68ac5a Layer already exists    3.7s
...
⋮. Pushing app: 7dfc1aa4c504 Layer already exists   3.7s
```

Notice, in this example, that even though we have just pushed the app service image, lb and db are also pushed to our registry because they were declared as dependencies.

In this section, we learned how to use docker-compose for building and sharing our application's components. In the next section, we will run containers defined in our Compose YAML files and review their logs and status.

Running and debugging multi-container applications

Applications executed using Docker Compose will be orchestrated but without high availability. This doesn't mean you can't use it in production, but you may need additional applications or infrastructure to keep your components always on.

Docker Compose provides an easy way of running applications using a *single point of management*. You may have more than one YAML file for defining your application's components, but the docker-compose command will merge them into a single definition. We will simply use docker-compose up to launch our complete application, although we can manage each component separately by simply adding its service's name. docker-compose will refresh the components' status and will just recreate those that have stopped or are non-existent. By default, it will attach our terminal to all application containers, which may be very useful for debugging but it isn't useful for publishing our services. We will use -d or --detach to launch all container processes in the background.

The docker-compose up execution will first verify whether all the required images are present on your system. If they aren't found, the container runtime will receive the order of creating them if a build key is found. If this key isn't present, images will be downloaded from the registry defined in your image key. This process will be followed every time you execute docker-compose up. If you are changing some code within your application's component folders, you will need to recreate them by changing your compose YAML image tags or the image's digest.

> **Important note**
>
> You can use `docker-compose up --d --build` to specifically ask your container runtime to rebuild all the images (or part of them if you specified a service). As you may expect, the runtime will check each image layer (RUN and COPY/ADD keys) and rebuild only those that have changed. This will avoid the use of an intermediate `docker-compose build` process. Remember to maintain your container runtime disk space by pruning old unnecessary images.

As we already mentioned in this section, containers will always be created if they don't exist when we execute `docker-compose up`. But in some cases, you will need to execute a fresh start of all containers (maybe some non-resolved dependencies in your code may need to force some order or reconnection). In such situations, we can use the `--force-recreate` argument to enforce the recreation of your services' containers.

We will use the `entrypoint` and `command` keys to *overwrite* the ones defined in the images used for creating each service container and we will specifically define which services will be available for the users by publishing them. All other services will use the internally defined network for their communications.

As you may expect, everything we learned about **container networking** in *Chapter 4, Running Docker Containers*, will also apply here. Therefore, an internal DNS is available for all the communications provided in the internal network, and services will be published with their defined names. This is key to understanding how your application's components will know each other. You shouldn't use their instance name (`<PROJECT_NAME>_<SERVICE_NAME>_<INSTANCE>`); we will instead use the service's name to locate a defined service. For example, we will use db in our app component connection string. Take care because your instance name will also be available but shouldn't be used. This will really break the portability and dynamism of your applications if you move them to clustered environments where instances' names may not be usable or if you use more than one replica for some services.

> **Important note**
>
> We can manage the number of container replicas for a service by using the `replicas` key. These replicas will run in isolation, and you will not need a load balancer service to redirect the service requests to each instance. Consider the following example:
>
> ```
> services:
>   app:
> ...
>     deploy:
>       replicas: 2
> ...
> ```
>
> In such a situation, two containers of our app service will be launched. Docker Swarm and Kubernetes will provide TCP load balancer capabilities. If you need to apply your own balancer rules (such as specific weights), you need to add your own load balancer service. Your container runtime will just manage OSI layer 3 communications (`https://en.wikipedia.org/wiki/OSI_model`).

Multi-container applications defined in a Compose file will run after we execute docker-compose up --detach, and to review their *state*, we will use docker-compose ps. Remember to add your project in all your commands if you need to overwrite the default project's name (current folder). We can use common --filter and --format arguments to filter and modify the output of this command. If some of the service's containers are missing, maybe they didn't start correctly; by default, docker-compose ps will only show the running containers. To review all the containers associated with our project, we will use the --all argument, which will show the running and stopped containers.

If any issues are found in our project's containers, we will see them as exited in the docker-compose ps command's output. We will use docker-compose logs to review all container logs at once, or we can choose to review only the specific service in error by adding the name of the service to this command.

We can use the --file (or -f) argument to define the complete path to our Compose YAML file. For this to work, it is very useful to first list all the Compose applications running in our system by using docker-compose ls. The full path to each application's Compose YAML file will be shown along with its project's name, as in this example:

```
$ docker-compose ls
NAME                    STATUS              CONFIG FILES
test                    running(3)          /home/frjaraur/labs/simplest-
lab/docker-compose.yaml
```

In this case, we can add the path to the Compose file to any docker-compose action:

```
$ docker-compose --project-name test \
--file /home/frjaraur/labs/simplest-lab/docker-compose.yaml ps
NAME                    IMAGE                                      CO
MMAND           SERVICE            CREATED           STATU
S               PORTS
test-app-1              docker.io/frjaraur/simplest-lab:sim-
plestapp    "node simplestapp.js…"    app                 25 hours
ago         Up 28 minutes       3000/tcp
test-db-1               docker.io/frjaraur/simplest-lab:sim-
plestdb     "docker-entrypoint.s…"    db                  25 hours
ago         Up 24 minutes       5432/tcp
test-lb-1               docker.io/frjaraur/simplest-lab:simplestlb    "/
entrypoint.sh /bin…"    lb              25 hours ago       Up 24
minutes         0.0.0.0:8080->80/tcp
```

This will work even with build actions, as the Compose YAML file location will be used as a reference for all commands. The context key may be included to modify its behavior.

We can review the port exposed for the application in the docker-compose ps output. To review our application's logs, we can use docker-compose logs, and each service will be represented in a different random color. This is very useful for following the different entries in each service. We can specify a single service by passing its name as an argument.

The following screenshot shows the output of `docker-compose logs` using `--tail 5` to only retrieve the latest five lines:

```
Ubuntu 22.04.2 LTS                                                                              —
frjaraur@sirius:~/tests/dcadeg/chapter5/simplest-lab$ docker-compose --project-name test logs --tail 5
test-app-1  | Connecting to database postgres://demo:d3m0@simplestdb:5432/demo
test-app-1  | SELECT: select serverip, count (*) as hits from hits group by serverip
test-app-1  | Request received from 172.16.0.1
test-app-1  | Connecting to database postgres://demo:d3m0@simplestdb:5432/demo
test-app-1  | SELECT: select serverip, count (*) as hits from hits group by serverip
test-db-1   | 2023-04-16 10:06:44.034 UTC [1] LOG:  listening on IPv4 address "0.0.0.0", port 5432
test-db-1   | 2023-04-16 10:06:44.034 UTC [1] LOG:  listening on IPv6 address "::", port 5432
test-db-1   | 2023-04-16 10:06:44.056 UTC [1] LOG:  listening on Unix socket "/var/run/postgresql/.s.PGSQL.5432"
test-db-1   | 2023-04-16 10:06:44.136 UTC [52] LOG:  database system was shut down at 2023-04-16 10:06:43 UTC
test-db-1   | 2023-04-16 10:06:44.162 UTC [1] LOG:  database system is ready to accept connections
frjaraur@sirius:~/tests/dcadeg/chapter5/simplest-lab$ docker-compose --project-name test logs db --tail 5
test-db-1   | 2023-04-16 10:06:44.034 UTC [1] LOG:  listening on IPv4 address "0.0.0.0", port 5432
test-db-1   | 2023-04-16 10:06:44.034 UTC [1] LOG:  listening on IPv6 address "::", port 5432
test-db-1   | 2023-04-16 10:06:44.056 UTC [1] LOG:  listening on Unix socket "/var/run/postgresql/.s.PGSQL.5432"
test-db-1   | 2023-04-16 10:06:44.136 UTC [52] LOG:  database system was shut down at 2023-04-16 10:06:43 UTC
test-db-1   | 2023-04-16 10:06:44.162 UTC [1] LOG:  database system is ready to accept connections
frjaraur@sirius:~/tests/dcadeg/chapter5/simplest-lab$
```

Figure 5.2 – Service container logs retrieved by using docker-compose logs

Notice that in this simple test, we only have two services, and colors are applied to each one. We retrieved only the latest five lines of each container by adding `--tail 5`. This argument applies to all containers (we didn't get the latest five lines of all logs merged). It is also important to mention that service names must be used as arguments when we need to use an action in a specific service. We will never use the container names; hence, we need to include the appropriate project name.

We can use the same approach to access a container's namespace by using the `exec` action. Remember that we learned in *Chapter 4*, *Running Docker Containers*, that we can execute a new process inside our container (it will share all the container's process kernel namespaces). By using `docker-compose exec <SERVICE_NAME>`, we can execute a new process inside any of our service's containers:

```
$ docker-compose --project-name test exec db ps -ef
PID    USER      TIME   COMMAND
   1 postgres   0:00 postgres
  53 postgres   0:00 postgres: checkpointer
  54 postgres   0:00 postgres: background writer
  55 postgres   0:00 postgres: walwriter
  56 postgres   0:00 postgres: autovacuum launcher
  57 postgres   0:00 postgres: stats collector
  58 postgres   0:00 postgres: logical replication launcher
  90 root       0:00 ps -ef
```

In summary, we will be able to run the same actions for containers by using `docker-compose`.

For you as a developer, Docker Compose can really help you develop applications faster. You will be able to run all application containers at once. In the development stage, you can include your code in specific containers by mounting a volume, and you can verify how your changes affect other components. For example, you can mount the code of one application component and change it while other components are running. Of course, you can do this without the `docker-compose` command line, but you will need to automate your deployments with scripts and verify the containers' state. Docker Compose orchestrates this for you, and you can focus on changing your code. If you work in a team and all other developers provide container images and you share some application information, you can run these images locally while you are still working on your component.

Now that we know how to run and interact with multi-container applications, we will end this chapter by learning how to use environment variables to deploy your applications under different circumstances.

Managing multiple environments with Docker Compose

In this section, we will learn how to prepare our Compose YAML files as templates for running our applications in different environments and under different circumstances, such as developing or debugging.

If you are familiar with the use of environment variables in different operating systems, this section will seem pretty easy. We already learned how to use variables to modify the default behavior of Dockerfiles (*Chapter 2, Building Docker Images*) and containers at runtime (*Chapter 4, Running Docker Containers*). We used variables to overwrite the default values defined and modify the `build` process or the execution of container image processes. We will use the same approach with Compose YAML files. We will now review some of the different options we have to use variables with the `docker-compose` command line.

We can define a `.env` file with all the variables we are going to use in a Compose YAML file defined as a template. Docker Compose will search for this file in our project's root folder by default, but we can use `--env-file <FULL_FILE_PATH>` or the `env_file` key in our Compose YAML file. In this case, the key must be set for each service using the environment file:

```
env_file:
  - ./debug.env
```

The environment file will overwrite the values defined in our images. Multiple environment files can be included; thus, the order is critical. The lower ones in your list will overwrite the previous values, but this also happens when we use more than one Compose YAML file. The order of the arguments passed will modify the final behavior of the execution.

You, as a developer, must prepare your Compose YAML files with variables to modify the execution passed to your container runtime. The following example shows how we can implement some variables to deploy applications in different environments:

```
services:
  lb:
    build:
      context: ./simplestlb
```

```
    args:
      alpineversion: "1.14"
    dockerfile: Dockerfile.${environment}
    labels:
      org.codegazers.description: "Test image"
  image: ${dockerhubid}/simplest-lab:simplestlb
  environment:
    - APPLICATION_ALIAS=simplestapp
    - APPLICATION_PORT=${backend_port}
  networks:
    simplestlab:
        aliases:
        - simplestlb
  ports:
    - "${loadbalancer_port}:80"
```

In this example, we can complete our variables with the following .env file:

```
environment=dev
dockerhubid=frjaraur
loadbalancer_port=8080
backend_port=3000
```

This environment file will help us define a base build and deployment. Different Dockerfiles will be included – Dockerfile.dev and Dockerfile.prod, for example.

We can then verify the actual configuration applied using docker-compose:

```
$ docker-compose --project-name test \
--file myapp-docker-compose.yaml config
name: test
services:
  lb:
...
      context: /home/frjaraur/tests/dcadeg/chapter5/simplest-lab/
simplestlb
      dockerfile: Dockerfile.dev
      args:
        alpineversion: "1.14"
      labels:
        org.codegazers.description: Test image
    environment:
      APPLICATION_ALIAS: simplestapp
      APPLICATION_PORT: "3000"
    image: frjaraur/simplest-lab:simplestlb
```

```
  . . .
    ports:
    - mode: ingress
      target: 80
      published: "8080"
      protocol: tcp
  networks:
  . . .
```

All the values have already been assigned using the `.env` file, but these can be overridden manually:

```
$ dockerhubid=myid \
  docker-compose --project-name test \
  --file myapp-docker-compose.yaml config
...
    image: myid/simplest-lab:simplestlb
. . .
```

Remember that profiles and targets can also be used to prepare specific images and then run the services completely customized.

We can now review some labs that will help us better understand some of the content of this chapter.

Labs

The following labs will help you deploy a simple demo application by using some of the commands learned in this chapter. The code for the labs is available in this book's GitHub repository at `https://github.com/PacktPublishing/Containers-for-Developers-Handbook.git`. Ensure that you have the latest revision available by simply executing `git clone https://github.com/PacktPublishing/Docker-for-Developers-Handbook.git` to download all its content or `git pull` if you have already downloaded the repository before. All commands and content used in these labs will be located inside the `Docker-for-Developers-Handbook/Chapter5` directory.

In this chapter, we learned how to deploy a complete application using `docker-compose`. Let's put this into practice by deploying a sample application.

Deploying a simple demo application

In this lab, we will learn how to deploy an application with three components: a load balancer, a frontend, and a database.

There are hundreds of good Docker Compose examples and, in fact, there are many vendors who provide their applications packaged in the Compose YAML format, even for production. We chose this pretty simple application because we are focusing on the Docker command line and not on the application itself.

If you list the content of the `Chapter5` folder, you will see a folder named `simplestapp`. There is a subfolder for each component and a Compose file that will allow us to deploy the full application.

The Compose YAML file that defines our application contains the following code:

```yaml
version: "3.7"
services:
  lb:
    build: simplestlb
    image: myregistry/simplest-lab:simplestlb
    environment:
      - APPLICATION_ALIAS=simplestapp
      - APPLICATION_PORT=3000
    networks:
      simplestlab:
          aliases:
          - simplestlb
    ports:
      - "8080:80"
  db:
    build: simplestdb
    image: myregistry/simplest-lab:simplestdb
    environment:
        - "POSTGRES_PASSWORD=changeme"
    networks:
        simplestlab:
        aliases:
          - simplestdb
    volumes:
      - pgdata:/var/lib/postgresql/data
  app:
    build: simplestapp
    image: myregistry/simplest-lab:simplestapp
    environment:
      - dbhost=simplestdb
      - dbname=demo
      - dbuser=demo
      - dbpasswd=d3m0
    networks:
        simplestlab:
        aliases:
            - simplestapp
    depends_on:
      - lb
```

```
      - db
volumes:
  pgdata:
networks:
  simplestlab:
    ipam:
      driver: default
      config:
        - subnet: 172.16.0.0/16
```

This application is a very simplified demo for showing how various components could be deployed. Never use environment variables for your sensitive data. We already learned how to use `configs` and `secrets` objects in this chapter. It is good to also notice that we didn't use a non-root user for the database and load balancer components. You, as a developer, should always try to keep security at the maximum on your application components. It is also important to notice the lack of health checks at the Dockerfile and Compose levels. We will learn more about application health checks in Kubernetes later in this book because it may not always be a good idea to include some **Transmission Control Protocol** (**TCP**) check tools in your images. In *Chapter 8*, *Deploying Applications with the Kubernetes Orchestrator*, we will learn how this orchestration platform provides internal mechanisms for such tasks and how we can enforce better security options.

In this lab, only one volume will be used for the database component, and the only service published is the load balancer. We included this service just to let you understand how we can integrate a multilayer application and only share one visible component. All images will be created locally (you may want to upload to your own registry or Docker Hub account). Follow the next steps to deploy the `simplestapp` application described in the `compose` file:

1. To build all the images for this project, we will use `docker-compose build`:

    ```
    $ docker-compose --file \
    simplestlab/docker-compose.yaml \
    --project-name chapter5 build
    [+] Building 0.0s (0/0)
    ...
     => => writing image sha256:2d88460e20ca557f-
    cd25907b5f026926b0e61d93fde58a8e0b854c-
    fa0864c3bd                         0.0s
     => => naming to docker.io/myregistry/simplest-lab:simplest
    app                                0.0s
    ```

    We could have directly used `docker-compose run` to build or pull the images and run all containers, but this way, we can review the process step by step.

Important note

If you reset your Docker Desktop before starting the labs, you may find some errors regarding an old Docker container runtime integration on your WSL environment:

```
$ docker-compose --file simplestlab/docker-compose.yaml --pro-
ject-name chapter5 build

docker endpoint for "default" not found
```

The solution is very easy: simply remove your old Docker integration by removing your .docker directory, located in your home directory: `$ rm -rf ~/.docker`.

2. We can take a look at the images created locally:

```
$ docker image ls
REPOSITORY                     TAG            IMAGE
ID         CREATED             SIZE
myregistry/simplest-lab        simplestapp    2d88460e20ca    8 minutes
ago    73.5MB
myregistry/simplest-lab        simplestdb     e872ee4e9593    8 minutes
ago    243MB
myregistry/simplest-lab        simplestlb     bab86a191910    8 minutes
ago    8.51MB
```

As you may notice, all the images created follow the names defined in the Compose YAML file. Because the `build` key exists, the build process is executed instead of pulling images directly.

3. Let's now create the container for the database service:

```
$ docker-compose --file \
simplestlab/docker-compose.yaml \
--project-name chapter5 create db
[+] Running 3/3
 :: Network chapter5_simplest-
lab  Created                                                    0.8s
 :: Volume "chapter5_pgdata"        Create
d                                                               0.0s
 :: Container chap-
ter5-db-1          Created                         0.2s
```

All the objects required for the database service are created. It is not running yet, but it is ready for that.

4. We run this service alone and review its status:

```
$ docker-compose --file \
simplestlab/docker-compose.yaml \
--project-name chapter5 up -d db
[+] Running 1/1
 :: Container chapter5-db-1  Started
```

If you omit the Compose filename and the project's name, we will get neither the services nor the containers:

```
$ docker-compose ps
no configuration file provided: not found
$ docker-compose --file simplestlab/docker-compose.yaml ps
NAME              IMAGE           COMMAND           SERV
ICE          CREATED         STATUS            PORTS
```

Always ensure you use the appropriate name and Compose file for all the commands related to a project:

```
$ docker-compose --file \
simplestlab/docker-compose.yaml \
--project-name chapter5 ps
NAME              IMAGE           COMMAN
D                 SERVICE         CREATED           STATU
S                 PORTS
chapter5-db-1     myregistry/simplest-lab:simplestdb   "dock-
er-entrypoint.s…"  db              4 minutes ago     Up 3
minutes           5432/tcp
```

5. We will now run the `lb` and `app` services by using `docker-compose up -d`:

```
$ docker-compose --file \
simplestlab/docker-compose.yaml \
--project-name chapter5 up -d
[+] Running 3/3
 :: Container chap-
ter5-lb-1    Started                                   2.0s
 :: Container chap-
ter5-db-1    Running                            0.0s
 :: Container chap-
ter5-app-1   Started                                   2.9s
```

Your services will quickly change their status from `Created` to `Started`.

6. We can now review the status of all our application components and the ports exposed:

```
$ docker-compose --file \
simplestlab/docker-compose.yaml \
--project-name chapter5 ps
NAME              IMAGE           COMMAN
D                 SERVICE         CREATED           STATU
S                 PORTS
chapter5-app-1    myregistry/simplest-lab:simplestapp   "node
simplestapp.js…"  app             9 minutes ago     Up 9
minutes           3000/tcp
chapter5-db-1     myregistry/simplest-lab:simplestdb   "dock-
er-entrypoint.s…"  db              16 minutes ago    Up
```

```
15 minutes          5432/tcp
chapter5-lb-1       myregistry/simplest-lab:simplestlb     "/
entrypoint.sh /bin..."    lb                 9 minutes
ago        Up 9 minutes       0.0.0.0:8080->80/tcp
```

Once running, you can access the `simplestlab` application by connecting with your browser to `http://127.0.0.1:8080`:

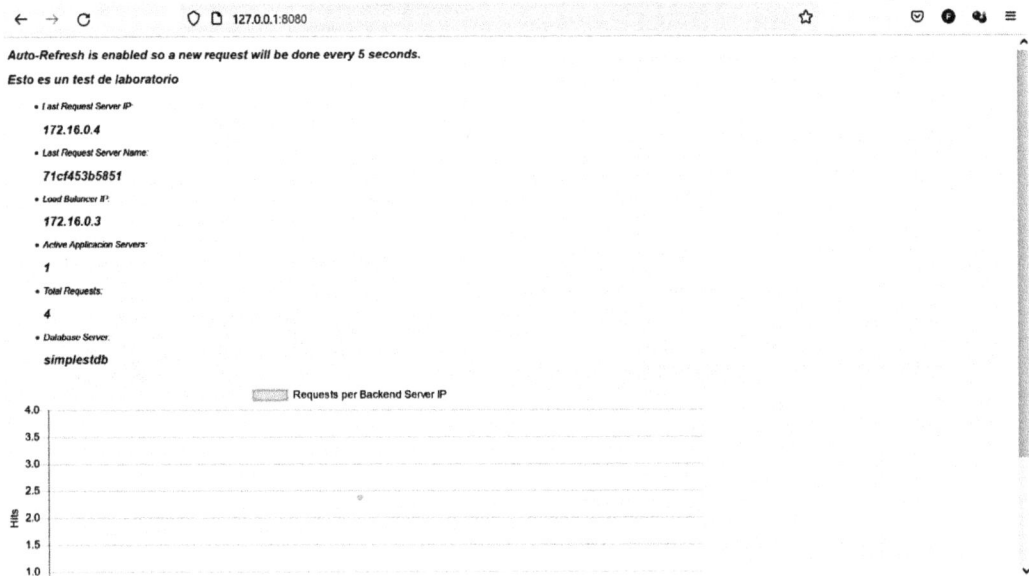

Figure 5.3 – The simplestlab application

This allows us to graphically review how requests are distributed when multiple backends are available.

7. We can scale our `app` component in this example. This option may be complicated or impossible in other deployments, as it really depends on your own application code and logic. For example, you should scale a database component without appropriate database internal scale logic (you should review the database server vendor's documentation):

```
$ docker-compose --file \
simplestlab/docker-compose.yaml \
--project-name chapter5 up --scale app=2 -d
[+] Running 4/4
 :: Container
chapter5-db-1    Running                          0.0s
 :: Container
chapter5-lb-1    Running                          0.0s
 :: Container chapter5-
app-2  Created                         0.2s
```

```
   :: Container chapter5-
app-1  Recreated                              10.8s
```

8. We can now review the app service's logs. We will retrieve both containers' logs:

```
$ docker-compose --file \
simplestlab/docker-compose.yaml \
--project-name chapter5 logs app
chapter5-app-1  | dbuser: demo dbpasswd: d3m0
...
chapter5-app-1  | dbuser: demo dbpasswd: d3m0
...
chapter5-app-2  | Can use environment variables to avoid '/APP/
dbconfig.js' file configurations.
```

Finally, your application is up and running, and we can move on to the next lab, in which we will use the same Compose file to deploy a second project with the same application.

Deploying another project using the same Compose YAML file

In this simple example, we will review and discuss the problems we may encounter by running two projects using the same Compose YAML file. To do this, we will follow these instructions:

1. Let's create a new project by using a new project name:

```
$ docker-compose --file \
simplestlab/docker-compose.yaml \
--project-name newdemo create
[+] Running 0/0
 :: Network newdemo_simplestlab  Error     0.0s
failed to create network newdemo_simplestlab: Error response
from daemon: Pool overlaps with other one on this address space
```

As you may notice, we defined a specific network in the **classless inter-domain routing** (CIDR) format for our project network. The Docker container runtime assigns IP ranges by using its own **IP address management** (IPAM); thus, it manages any IP overlap automatically for us. By using a specific range, we broke the dynamism of the platform.

2. Let's remove the IP address range from our docker-compose definition:

```
networks:
  simplestlab:
    ipam:
      driver: default
```

And now we try to deploy the application again:

```
$ docker-compose --file \
simplestlab/docker-compose.yaml \
--project-name newdemo create
[+] Running 5/5
 :: Network newdemo_simplest-
lab  Created                              0.9s
 :: Volume "newdemo_pgdata"     Create
d                         0.0s
 :: Container newde-
mo-db-1         Created                           0.2s
 :: Container newdemo-lb-1      Created                        0.2s
 :: Container newde-
mo-app-1        Created                    0.2s
```

We didn't have any problems this time. The `volume` and `network` objects were created with the project prefix. We will not be able to reuse the project name because object names must be unique.

3. Let's run all the application components now:

```
$ docker-compose --file simplestlab/docker-compose.yaml
--project-name newdemo start
[+] Running 1/2
 :: Container newde-
mo-db-1  Started                              1.4s
 :: Container newdemo-lb-1  Start-
ing                         1.4s
Error response from daemon: driver failed programming external
connectivity on endpoint newdemo-lb-1 (bb03c1b0a14a90a3022a-
ca3c3a9a9d506b3e312cc864f0dcda6a5360d58ef3d0): Bind for
0.0.0.0:8080 failed: port is already allocated
```

You may notice that we also defined a specific port for our `lb` service. This seems fine for production, but defining a specific port in development, where multiple copies of an application can be expected, also breaks the dynamism of container-based components. For this to work, we could just simply change this port number, allow the system to choose a random one for us, or define a variable that will allow us to define a port for each project.

4. We change our Compose YAML and add the `LB_PORT` variable as the port for exposing our application:

```
services:
  lb:
...
    ports:
      - "${LB_PORT}:80"
```

Then, we test it again:

```
$ LB_PORT=8081 docker-compose --file \
simplestlab/docker-compose.yaml \
--project-name newdemo up lb
[+] Running 1/1
 :: Container newdemo-lb-1  Recreated
```

Let's review the component status:

```
$ docker-compose --file simplestlab/docker-compose.yaml
--project-name newdemo ps
WARN[0000] The "LB_PORT" variable is not set. Defaulting to a
blank string.
NAME                    IMAGE                                COMMAN
D                       SERVICE            CREATED           STATU
S                       PORTS
newdemo-db-1            myregistry/simplest-lab:simplestdb   "dock-
er-entrypoint.s..."    db                 11 minutes ago    Up 8
minutes                 5432/tcp
newdemo-lb-1            myregistry/simplest-lab:simplestlb   "/
entrypoint.sh /bin..."  lb                 46 seconds        Up 34 seconds        0.0.0.0:8081->80/tcp
ago     Up 34 seconds       0.0.0.0:8081->80/tcp

$ docker-compose --file simplestlab/docker-compose.yaml
--project-name newdemo up app -d
[+] Running 3/3
 :: Container
newdemo-db-1    Running                                     0.0s
 :: Container newdemo-lb-1    Running                        0.0s
 :: Container newdemo-
app-1  Started                     1.4s
```

5. Once we changed the unique and fixed values in the docker-compose.yaml file, we were able to deploy a second project using a unique Compose YAML file. We can list the projects deployed in our host along with their number of components:

```
$ docker-compose ls
NAME                    STATUS             CONFIG FILES
chapter5                running(4)         /home/frjaraur/labs/
Chapter5/simplestlab/docker-compose.yaml
newdemo                 running(3)         /home/frjaraur/labs/
Chapter5/simplestlab/docker-compose.yaml
```

With this simple example of the usual problems you may find while preparing your applications, we will end this chapter's labs by removing all the objects created.

Removing all projects

To remove all the created projects, we will perform the following steps:

1. We will remove all the deployed projects by using `docker-compose down`:

    ```
    $ docker-compose -f  /home/frjaraur/labs/Chapter5/simplestlab/
    docker-compose.yaml --project-name chapter5 down
    WARN[0000] The "LB_PORT" variable is not set. Defaulting to a
    blank string.
    [+] Running 5/5
    :: Container chapter5-
    app-2       Removed                          0.1s
    :: Container chapter5-
    app-1       Removed                          0.1s
    :: Container
    chapter5-db-1        Removed                          0.1s
    :: Container
    chapter5-lb-1        Removed                          0.1s
    :: Network chapter5_
    simplestlab  Removed                    0.6s
    ```

    You may notice that volumes are not listed as removed. We can review the current volumes on your system:

    ```
    $ docker volume ls
    DRIVER     VOLUME NAME
    local      chapter5_pgdata
    local      newdemo_pgdata
    ```

2. We will manually remove the volume present for the `chapter5` project, but we will use the `--volumes` argument to remove all the volumes associated with a project:

    ```
    $ docker-compose -f  /home/frjaraur/labs/Chapter5/simplestlab/
    docker-compose.yaml --project-name newdemo down --volumes
    WARN[0000] The "LB_PORT" variable is not set. Defaulting to a
    blank string.
    [+] Running 5/5
    :: Container newde-
    mo-app-1       Removed                          0.0s
    :: Container newde-
    mo-lb-1       Removed                          0.1s
    :: Container newde-
    mo-db-1       Removed                          0.1s
    :: Volume newdemo_pgdata        Remove
    d                              0.1s
    :: Network newdemo_simplest-
    lab  Removed                    0.6s
    ```

The volume from the `newdemo` project was removed, as we can verify now, but the volume from the `Chapter5` project is still present:

```
$ docker volume ls
DRIVER      VOLUME NAME
local       chapter5_pgdata
```

3. We remove the remaining volume manually:

```
$ docker volume rm  chapter5_pgdata
chapter5_pgdata
```

Additional labs are included in this book's GitHub repository.

Summary

In this chapter, we covered the basic usage of Docker Compose and discussed how it will help you develop and run your multi-component applications locally. You will be able to run and review the status of all your application's components using a single command line. We learned about the syntax of Compose YAML files and how to prepare template-like files for developing your applications using different customizations. You will probably run your applications in production using a clustered orchestrated environment such as Docker Swarm or Kubernetes, but `docker-compose` does also provide a basic orchestration for running your multi-container applications in production using a single server. Understanding the basic orchestration and networking features of Docker Compose will help us introduce more sophisticated orchestration methods, which we will learn about in the next chapters.

In the next chapter, we will briefly review orchestration platforms that will help us run our applications cluster-wide.

Part 2: Container Orchestration

In this part of the book, we will cover the **orchestration** of containers in cluster-wide environments. We will learn how to deploy distributed component applications on different hosts in a cluster, allowing users to interact with components and publish services.

This part has the following chapters:

- *Chapter 6, Fundamentals of Orchestration*
- *Chapter 7, Orchestrating with Swarm*
- *Chapter 8, Deploying Applications with the Kubernetes Orchestrator*

6

Fundamentals of
Container Orchestration

So far, we have learned what software containers are, how they work, and how to create them. We focused on using them, as developers, to create our applications and distribute functionalities into different components running in containers. This chapter will introduce you to a whole new perspective. We will learn how our applications run in production using containers. We will also introduce the concept of container orchestrators and cover what they can deliver and the key improvements we need to include in our applications to run them in a distributed cluster-wide fashion.

In this chapter, we will cover the following topics:

- Introducing the key concepts of orchestration
- Understanding stateless and stateful applications
- Exploring container orchestrators

We will then go on to study how to leverage Docker Swarm and Kubernetes orchestrators' features in *Chapter 7, Orchestrating with Swarm*, and *Chapter 8, Deploying Applications with the Kubernetes Orchestrator*. This chapter does not include any labs as it is intended to teach you the theory behind Docker Swarm and Kubernetes.

Introducing the key concepts of orchestration

Running an application on a host may be complicated, but executing this same application on a distributed environment composed of multiple hosts would be very tedious. In this section, we will review some of the key concepts regarding the orchestration of application components, regardless of whether they are run using containers or as different virtual machines.

Orchestrators are special software components that help us manage the different interactions and dependencies between our application components. As you can imagine, if you divide your application into its many different functionalities, each with its own entity, orchestrating them together is key. We have to say here that some special functionalities, such as dependency management, may not be available in your orchestrator and therefore you will need to manage them by yourself. This gives rise to an important question: what do we need to know about orchestrators to prepare our applications for them? Orchestrators keep our processes up and running, manage the communications between our applications' components, and attach the storage required for these processes.

Focusing specifically on container-based applications, it is easy to understand that container runtimes will be part of the orchestration infrastructure as they are required to run containers. The biggest challenge when working with containers is the intrinsic dynamism associated with their networking features. This will probably not be a problem in virtual machine environments, but containers commonly use different IP addresses on each execution (and although we can manually assign IP addresses to containers, it is not good practice). Note that in the previous chapters' labs (*Chapter 4, Running Docker Containers*, and *Chapter 5, Creating Multi-Container Applications*), we used service names instead of container IP addresses.

Let's review the key concepts to bear in mind when designing an application to run its components in a distributed cluster-wide fashion:

- **Dependencies resolution**: Some dependencies may exist in your application. This means that some services are required *before* others to enable your application's functionality. Some orchestration solutions such as Docker Compose (standalone orchestration) include this feature, but most others usually don't. It is up to you, as a developer, to resolve any issues arising from these dependencies in your code. A simple example is a database connection. It is up to you to determine what to do if the connection from some components is lost due to a database failure. Some components may function correctly while others may need to reconnect. In such situations, you should include a verification of the connectivity before any transaction and prepare to manage queued transactions that may accrue before your component realizes that the database isn't working.

- **Status**: Knowing the current status of each component is critical at any time. Some orchestrators have their own features to check and verify the status of each component, but you, as the developer of your application, know best which paths, processes, ports, and so on are required and how to test whether they are alive. In *Chapter 2, Building Docker Images*, we discussed how the simpler the content on your container images, the better the security. If your application requires some additional software for testing its health, you should include it; however, it may be a better idea to include some test endpoints or testing functions that can be called when required to verify the health of the application's components.

- **Circuit breakers**: If we have already successfully managed our application's dependencies, circuit breakers will allow us to identify any problems and make appropriate decisions when some components are down. Orchestrators don't provide any circuit breakers natively; you will need additional software or infrastructure components to implement such solutions. Depending on the complexity of your application, it can be beneficial to integrate some circuit breakers. For example, we can stop all the components that require a healthy database whenever this is not available, while other components can continue running and providing their functionality as usual.

- **Scalability**: Perhaps some of your application's components run with more than one replica. Scalability should be built in by design. Orchestrators allow you to execute more than one replica of any component, but it is up to you to manage their co-existence. A database component running with more than one replica will corrupt your data if it is not prepared for such a situation. In this example, you would need to use a master-slave or distributed database architecture. Other issues regarding data transactions may appear, such as in your frontends, if you don't manage user sessions. In such situations, session integrity may require additional components to ensure that all transactions follow a coordinated workflow. Orchestrators do not know about any of your application's components' functionality and as such, it is up to you to determine whether some components can scale up or down. Some orchestrators will provide you with rules for deciding when this should occur, but they just trigger the actions for managing your component's replicas. All replicas will be treated in the same way and you will need to add additional components if you require weight distribution functionality to ensure some replicas receive more data than others.

- **High availability**: The concept of high availability may vary depending on whether you ask someone from the infrastructure team or the application side, but both will agree that it should be transparent for the end user. In terms of infrastructure, we can think of high availability at various levels:

  - **Duplicity of infrastructure**: When physical or virtual machines are used to provide high availability for our applications, we need to have all our infrastructure duplicated, including the underlying communications layers and storage. Most devices know how to manage the special configurations that must be changed when a replicated device acts as a master. Servers should maintain quorum or master-slave relations to decide which of them will handle user requests. This will be required even if we plan to have active-passive (where only one instance handles all requests and the others take over only if any error is found) or active-active responses (all instances serve at the same time).

  - **Routes to healthy components**: Duplicated infrastructure requires a layer of extra components to route the user's request. Load balancers help present the different applications' endpoints in a completely transparent way for the users.

  - **Storage backends**: As we would expect, storage not only has to maintain the data safely and securely, but also attach it to any running and serving instance. In active-passive environments, storage will switch from a damaged instance to a healthy one. You, as the developer, may need to ensure data integrity after a switchover.

High availability means that services will never be interrupted, even when maintenance tasks require one instance to be stopped. Container orchestrators don't provide high availability by themselves; it must be factored in as part of your application design.

- **Service state definition**: We usually define how an application is to be deployed, including the number of replicas that must be available for the service to be considered healthy. If any replica fails, the orchestrator will create or start a new one for you. We don't need to trigger any event, just define the monitors to allow the orchestrator to review the status of each application component.

- **Resilience**: Applications may fail, and the orchestrator will try to start them again. This is the concept of *resilience*, which orchestrators provide by default. When your application is run in containers, the container runtimes keep the application alive by starting them again when something goes wrong. Orchestrators interact with container runtimes to manage the start and stop processes for your containers cluster-wide, trying to mitigate the impact of the failures in your applications. Therefore, you must design your applications to stop and start fast. Usually, applications running in containers don't take more than a few seconds to get running, so your users may not even notice an outage. So, to avoid significant issues for users, we must provide high availability for our application's components.

- **Distributed data**: When your applications run in a distributed cluster-wide fashion, different hosts will run your application's components, hence the data required needs to be made available when needed. The orchestrator will interact with container runtimes to mount container volumes. You can mount your data in all the possible hosts in anticipation that the containers will use it, and this may seem a good idea at first. However, managing an application's data permissions may leave you with unexpected consequences. For example, you can misconfigure the permissions for a root directory where different applications are storing their data, allowing some applications to read others' data files. It is usually better to entrust the volume management to the orchestrator itself, using the features it makes available for this.

- **Interoperability**: Communications between your application components can be complicated, but orchestration provides you with a simplified network layer. In *Chapter 7*, *Orchestrating with Swarm*, and *Chapter 8*, *Deploying Applications with the Kubernetes Orchestrator*, we will see how both of these orchestrators provide a different layer of communications for your applications. It is very important to learn and understand how they deploy your application's communications and design your application to avoid any network lock-in from the beginning. This will allow your applications to work on any of the available orchestrators.

- **Dynamic addressing**: Container environments are based on dynamic objects managed by the container runtime. Some of these objects can have static properties such as IP addresses within containers. Managing static properties in very dynamic infrastructures can be difficult and is not recommended. Orchestrators will manage dynamism for you if you follow their rules.

- **Service discoverability**: Services published within orchestrators will be announced internally and can be reached from any other application component. By default, such services only work internally, which improves the security of your full application as you only externally publish those frontend services that must be reached by users.

- **Multisite**: Having multiple data centers for follow-the-sun or disaster recovery architectures is common in big enterprises. You must always ensure that your applications can run anywhere. We can go further than this because some companies may have cloud-provisioned infrastructure as well as on-premises services – in such situations, your applications may run some components on cloud infrastructure while others run locally. Apart from the infrastructure synchronization challenges that this scenario may create, if you design your applications with these advanced environments in mind and you understand the breakpoints to avoid (for example, quorum between components if communications are lost), you will still be able to run your applications in extreme situations.

One of the key aspects of deploying applications cluster-wide is the management of component status. In the next section, we will review the importance of setting a component's status outside the container's life cycle when designing our applications.

Understanding stateless and stateful applications

Previously, we provided a brief description of the key points and concepts regarding orchestrating your applications and running them distributed across a cluster, in which you may have noticed that providing a service to your users without outages can be complex. We reviewed how orchestrators help us deliver resilient processes and saw how high availability has to be designed into applications. One of the key aspects of such a design concerns how your application manages the processes' state over time.

Applications fall into two different categories, depending on how they manage their processes' state: **stateful** and **stateless**.

Before learning about each one, it is important to understand what the state of an application or process means. The state of a system is the condition in which it is at a specific time. This system can be running or stopped, or even in between both when it is either starting or stopping. It is important to be able to identify and manage the status of the application's components. To manage the state of a system, such as triggering an action to start or stop the system, we must know how to retrieve the state. In many cases, this is not simple, and particularly complex situations may require managing some dependencies or interactions with other external systems.

Now that we have a definition for the state of a system, let's describe stateful and stateless applications with some examples.

Stateful applications

Imagine a situation where a process needs some data to be loaded before it can start. If the process reads all its configurations when this starts, we will need to restart it for any change to take effect. But if this process reads part of the configuration (or the full content) whenever it is needed for some functions or actions, a restart may not be needed. That's why we will need to know whether the process has already started or not to load the required data. In some cases, we can design a full process that loads the data whenever the process starts without reviewing whether it was started before. However, other times, this can't be done because we can't replace the data or load it more than once. A simple file can be used as a flag to specify whether the load process was already executed or if we need to load the data again.

This can be managed locally quite easily if our application is running on a host as a simple process but isn't easy when working with containers. When a container runs in a host, it uses its own storage layers unless we specify a volume to store some data. A new container running the same process can reuse a previous volume if your application stores this flag file outside the container's life cycle by design. This appears easy enough in a standalone host where all the processes run in the same host.

The volume can be either a bind mount (a directory from the host's filesystem) or a named volume (a volume with a known and reusable name). When you run your application cluster-wide, this approach may not work correctly. This is because a bind mount is attached to a host, and that directory will not exist on other hosts. Remote filesystems can be used to persist the flag and make it available in other hosts. In this case, we use a volume and the orchestrator will manage the mounting of the required filesystem.

However, managing an application's state is more difficult when more than one process is involved. In such situations, it is recommended to take this requirement into account at the very beginning of your application's design process. If we were designing a web application, for example, we would need to store some user data to identify who made a given request. In this scenario, we don't just have to manage the process state – we also need to manage users' data, so more than one file will be required and we will have to use a database to store this data. We usually say that such an application is **stateful** and requires **persistent data**.

Stateless applications

Stateless applications, on the other hand, do not require any data persistence. We can restart these components whenever needed without persisting any data. The application itself contains all the required information. Imagine a service that receives some data, and if the service doesn't respond, we send the data again until it gives a response. This service can do some operations with the received data and send a response without needing to save any data. In such a situation, this service is **stateless**. It may require some external data to do operations, but if something goes wrong and we need to restart the service, we aren't concerned with the status of any pending operations. We will simply send the data again until we get a valid response. When some operations are still pending, we have both a stateful process, which requires us to load some pending requests, and a stateless process, because it doesn't store the requests by itself. The service sending the request may need to store it while the one processing the operation doesn't.

As you may imagine, stateless applications are easier to manage in distributed environments. We don't need to manage process states and their associated data in different locations across a cluster.

In the next section, we will review some of the most popular container orchestrators.

Exploring container orchestrators

Now that we know what to expect from any container orchestrator, let's review some of the most important and technically relevant ones available. We will also take a quick look at the strengths and weaknesses of each option presented.

We will start with the currently most popular and widely extended container orchestrator, Kubernetes.

Kubernetes

Kubernetes is an open source container orchestration platform, fast becoming the de facto standard for running microservices on cloud providers and local data centers. It started as a Google project for managing the company's internal applications back in 2003. This project was initially called Borg and was created for deploying workloads distributed across different nodes and clusters. This project evolved into a more complex orchestration platform called Omega, which focused on bigger clusters running thousands of workloads for very different applications. In 2014, Google published Borg's code to the open source community and it finally became Kubernetes that same year. In 2015, the first release, Kubernetes 1.0, was published after Red Hat, IBM, Microsoft, and Docker joined the community project. The Kubernetes community is now huge and a very important part of why this orchestrator has become so popular nowadays.

The most important feature of Kubernetes is that its core is focused on executing a few tasks and delegating more complicated tasks to external plugins or controllers. It is so extensible that many contributors add new features daily. Nowadays, as Kubernetes is the most popular and extended container orchestrator, it is quite common for software vendors to provide their own Kubernetes definitions for their applications when you ask them for high availability. Kubernetes does not provide container runtimes, cluster network capabilities, or clustered storage by default. It is up to us to decide which container runtime to use for running the containers to be deployed and maintained by the orchestrator.

On the network side, Kubernetes defines a list of rules that must be followed by any **Container Network Interface** (**CNI**) we want to be included in our platform to deliver inter-container communications cluster-wide, as we will learn in *Chapter 8, Deploying Applications with the Kubernetes Orchestrator*. The Kubernetes network model differs from other orchestration solutions in the way it presents a plain or flat network (no routing is required between containers), where all containers are reachable by default. Many open source and proprietary options for deploying the Kubernetes network are also available, including Flannel, Weave, Cilium, and Calico. These network providers define the overlay networks and IPAM configurations for our Kubernetes cluster and even encrypt communications between nodes.

Kubernetes provides many cloud provider integrations because it was designed to be cloud-ready. A cloud controller is available for managing integrations with publishing applications or using some special cloud-provided storage backends. As mentioned earlier in this section, Kubernetes does not provide a solution for deploying any cluster-wide storage backend, but you can integrate NFS and some AWS, Google, and Azure storage backends for your applications. To extend your storage possibilities, you can use **Container Storage Interfaces** (**CSI**), which are different vendor or community-driven storage backends that can easily be integrated into Kubernetes to provide different storage solutions for our orchestrated containers.

Many cloud providers and software vendors package and share or sell their own Kubernetes flavors. For example, Red Hat/IBM provide their own Kubernetes platform inside their OpenShift product. Microsoft, Amazon, Google, and indeed almost all cloud service providers have their own Kubernetes implementations ready for use. In these Kubernetes platforms, you don't even have to manage any of the control plane features – the platforms are offered as Kubernetes-managed solutions for you, as the developer, to use for delivering your applications. These solutions are known as **Kubernetes-as-a-Service** platforms and are where you pay for your workloads and the bandwidth used in your applications.

The Kubernetes project publishes a release approximately every 4 months and maintains three minor releases at a time (thus providing almost a year of patches and support for each release). Some changes and deprecations are always expected between releases, so it is very important to review the change notes for each release.

Kubernetes clusters have nodes with different roles: **master** nodes create the control plane for delivering containers, while **worker** nodes execute the workloads assigned to them. This model allows us to deploy a Kubernetes cluster with high availability by replicating some of the master node's services. An open source key-value database, called etcd, is used for managing all objects' (known in Kubernetes as **resources**) references and states.

Kubernetes has evolved so fast that nowadays, we can even manage and integrate virtual machines into Kubernetes clusters by using operators such as KubeVirt. An additional great aspect of Kubernetes is that you can create your own resources (**Kubernetes Custom Resource Definitions**) for your application when some special requirements for your application are not met by the Kubernetes core resources.

Let's quickly summarize the pros of using Kubernetes:

- Available for customers of many cloud providers and from many on-premises software solutions for container providers

- Very thoroughly documented with lots of examples and guides for learning the basics

- Highly extensible via standardized interfaces such as the CNI and CSI

- Lots of objects or resources that will meet most of your application's requirements for running cluster-wide

- Comes with many security features included, such as role-based access control, service accounts, security contexts, and network policies

- Offers different methods for publishing your applications

- Used as a standard deployment method for many software vendors, you may easily find your applications packaged in Kubernetes manifest format

However, it does have some disadvantages as well. These include the following:

- It is not easy to master due to its continuous evolution and many resource types. The learning curve may seem higher compared to other orchestration solutions.

- The many releases per year may necessitate a lot of effort to maintain the platform.

- Having many flavors can be a problem when each vendor introduces their own particularities on their platforms.

- You will never use Kubernetes for just one application or a few small ones because it takes a lot of maintenance and deploying efforts. Indeed, Kubernetes-as-a-Service providers such as Microsoft's Azure Kubernetes Service will help you with minimal maintenance efforts.

We will learn about all the Kubernetes features and how we will prepare and deploy our applications in this orchestrator in more detail in *Chapter 8, Deploying Applications with the Kubernetes Orchestrator*.

Docker Swarm

Docker Swarm is a container orchestration solution created by Docker Inc. It is intended to provide a simple orchestration platform that includes everything needed to run our containerized applications cluster-wide by default. This includes overlay networking (which can be encrypted) and isolation by creating different networks for each project if required.

Different objects can be used to deploy our applications, such as **global** or **replicated** services, each with its own properties for managing how the containers are to be spread cluster-wide. As we have seen with Kubernetes, a master-worker or (master-slave) model is also used. The master node creates the full control plane and the worker nodes execute your containers.

It is important to mention a big difference in the way changes are managed within the cluster. While Kubernetes uses etcd as its key-value database, Docker Swarm manages its own object database solution using the Raft Consensus Algorithm with a complete command-line interface for this. Docker Engine installation is enough to get working with Docker Swarm as the container runtime binaries also include SwarmKit features. Moby is the open source project behind Docker Inc., having created kits for delivering and improving container communications (VPNKit) and improving the default `docker build` features (BuildKit), some of which we covered in *Chapter 2, Building Docker Images*, with the `buildx` extended build command-line. SwarmKit is the Moby project behind Docker Swarm and provides the cluster functionality, security, and simplicity of its model. Docker Swarm is quite simple, but this doesn't mean it isn't production-ready. It provides the minimum features required to deploy your applications with high availability.

It is important to mention that Compose YAML files allow us to deploy our applications using a set of manifests for creating and managing all our application objects in Docker Swarm. Some of the keys we learned about in *Chapter 5, Creating Multi-Container Applications*, will not work here, such as depends_on, hence application dependency management must be covered in your code itself.

Here are some of the advantages of Docker Swarm:

- Easier to learn than other container orchestrators
- Integrated inside Docker Engine and managed using the Docker command line
- Single-binary deployment
- Compatible with Compose YAML files

Some of the disadvantages are as follows:

- There are fewer objects or resources for deploying our applications, which may affect the logic applied in our applications. It is important to understand that you, as a developer, can implement the logic for your application inside your code and avoid any possible issues associated with the orchestration.
- It only works with the Docker container runtime, so vendor lock-in is present and the security improvements offered by other container runtimes can't be used here.
- Although Docker Swarm provides some plugins associated with the container runtime, it isn't as open and extensible as Kubernetes.
- Publishing applications is easier, but this means that we can't apply any advanced features without external tools.

We will learn more about Docker Swarm in *Chapter 7, Orchestrating with Swarm*.

Nomad

HashiCorp **Nomad** is a platform that allows us to run containers, virtual machines (using **QEMU**, which is a well-known open source virtualization engine), and Java applications. It focuses on scheduling application workloads and checking which services, such as discovery, health check monitoring, DNS, and secrets, are delivered by other HashiCorp tools, including **Consul** and **Vault**. Nomad bases its security on **Access Control Lists (ACLs)**, including tokens, policies, roles, and capabilities. In terms of networking, it uses CNI plugins for the bridge working mode. Its multi-site features allow us to run applications in different regions from a single-orchestration perspective using **federation**.

Nomad adopts some of the architecture features mentioned with Kubernetes and Docker Swarm, where some nodes act as the control plane (servers) while others (clients) execute all the workloads. Servers accept jobs from users, manage clients, and determine the workload placements.

HashiCorp provides a Community Edition and **Software-as-a-Service (SaaS)** platform in its cloud. It can be integrated via an API with some CI/CD environments and scripts for infrastructure automation can be included. Some of the advantages of this are as follows:

- Simplicity in usage and maintenance

- Single-binary deployment

- Flexibility to deploy and manage virtual machines, along with containerized and non-containerized applications

Here are some of its limitations:

- Although HashiCorp provides good documentation, they don't have many users and thus less of a community behind the project.

- Nomad appeared at the same time as Docker Swarm (a legacy platform) and Kubernetes were starting out, but Nomad initially focused on virtual machines and applications. Container orchestration became associated with Docker Swarm and Kubernetes, both of which gained popularity in this field as a result.

- With fewer associated projects, Nomad was left in the hands of a single company, which may have been more mature. This makes the evolution of the product or the addition of new features slower than community-driven projects.

Apache Mesos

Mesos is a project created by the Apache organization in 2009 for running cluster-wide workloads. This happened before containers became widely used and as a result, containers were only integrated into the project when most of the architecture's logic had already been designed. This means that Mesos can run containers and normal application workloads cluster-wide, as Nomad can.

Hadoop and other big data workload managers are the main frameworks that are managed by Apache Mesos, while its usage for containers is quite limited or at least less popular than the other solutions listed in this section. The benefit of Mesos is that integrating Apache projects' workloads, such as those for Spark, Hadoop, and Kafka, is easy because it was designed for them.

However, the disadvantages include the following:

- Special packages for workloads or frameworks may require manual configuration and thus are not as standardized as Kubernetes or Docker Compose YAML files

- This orchestrator is not very popular and thus has a smaller community, with only a few tutorial examples available compared to Kubernetes or Docker Swarm

Cloud vendor-specific orchestration platforms

Now that we have had a quick overview of the most important on-premises orchestrators (some of which are also available as cloud solutions), let's examine some of the orchestration solutions created specifically by the cloud vendors for their own platforms. You can expect a degree of vendor lock-in when using their specific features. Let's look at the most important ones:

- **Amazon Elastic Container Service (ECS)** and **Fargate**: Amazon ECS is an Amazon-managed container orchestration service. ECS relies on your EC2 contracted resources (storage, virtual networks, and load balancers) and as such, you can increase or decrease the hardware available for the platform by adding more resources or nodes. Amazon **Elastic Kubernetes Service (EKS)** is completely different from this simplified option as it deploys a complete Kubernetes cluster for you. On the other hand, AWS Fargate is a simpler technology that allows you to run containers without having to manage servers or clusters on the Amazon compute platform. You simply package your application in containers and specify the base operating system and the CPU and memory requirements. You will just need to configure a few networking settings and the **Identity and Access Management (IAM)** to secure your access. These are all the requirements so that you can finally run your application.

- **Google Anthos** and **Google Cloud Run**: Although Google Cloud Platform offers its own Kubernetes-as-a-Service platform, **Google Kubernetes Engine (GKE)**, it also provides Anthos and Cloud Run. Anthos is a hybrid and cloud-agnostic container management platform that allows the integration of applications running in containers in Google cloud and on your data center. It focuses on preparing and running applications in containers used as virtual machines. On the other hand, Google Cloud Run is more flexible, offering the ability to scale workloads on demand and integrate CI/CD tools and different container runtimes.

- **Azure Service Fabric** and **Azure Container Instances** (**ACI**): Microsoft provides different solutions for running simple containers while **Azure Kubernetes Service (AKS)** is also available. Azure Service Fabric provides a complete microservice-ready platform for your applications, while ACI is a simplified version in which you run containers as if they were small virtual machines. You simply code and build your container images for your applications instead of managing the infrastructure that runs them.

All these cloud platforms allow you to test and even run your applications in production without having to manage any of the underlying orchestration. Just using a simple web UI, you determine what should be done when your application's components fail and can add as many resources as needed. The cloud storage services available from each provider can be used by your application and you can monitor the complete cost of your application thanks to the reports available for your cloud platform. Here are some of the pros of these cloud-vendor platforms:

- Easier to use than any other container orchestration

- A full Kubernetes cluster may not be necessary for testing or even executing your applications, making cloud-vendor solutions a reasonable choice

Their disadvantages include the following:

- Vendor lock-in is always present with these platforms as you will use many cloud vendor-embedded services

- They can be used effectively for testing or even publishing some simple applications, but when using microservices, some issues may arise, depending on the complexity of your application's components

Summary

In this chapter, we reviewed the common orchestration concepts we will employ in our applications and the different platforms available. We learned about the most important features on offer to help us decide which options are more suited for our application. Learning how orchestrators work will be a big boon to you, as a developer, in designing your applications to run cluster-wide and with high availability in production, thanks to the unique features that orchestration offers. In the following chapters, we will delve deeper into Kubernetes, which is the most popular and widely extended container orchestration platform, and Docker Swarm, both of which are available in the cloud and on-premises.

7

Orchestrating with Swarm

As a developer, you can create your applications based on microservices. Using containers to distribute your applications into different components will allow you to provide them with different functionalities and capabilities, such as **scalability** or **resilience**. Working with a standalone environment is simple with tools such as Docker Compose, but things get difficult when containers can run cluster-wide on different hosts. In this chapter, we are going to learn how **Docker Swarm** will allow us to orchestrate our application containers with a full set of features for managing scalability, networking, and resilience. We will review how orchestration requirements are included in the Docker container engine and how to implement each of our application's specific needs.

This chapter will cover the following topics:

- Deploying a Docker Swarm cluster
- Providing high availability with Docker Swarm
- Creating tasks and services for your applications
- A review of stacks and other Docker Swarm resources
- Networking and exposing applications with Docker Swarm
- Updating your application's services

Technical requirements

We will use open source tools to build, share, and run a simple but functional Docker Swarm environment. The labs included in this chapter will help you to understand the content presented, and they are published at `https://github.com/PacktPublishing/Containers-for-Developers-Handbook/tree/main/Chapter7`. The *Code In Action* video for this chapter can be found at `https://packt.link/JdOIY`.

Deploying a Docker Swarm cluster

Docker Swarm is the orchestration platform developed by Docker Inc. It is probably the simplest orchestration solution for beginning to deploy your containerized applications. It is included inside the Docker container runtime and no additional software is required to deploy, manage, and provide a complete and secure Docker Swarm cluster solution. However, before we learn how to do this, let's explore the architecture of Docker Swarm.

Understanding Docker Swarm's architecture

Docker Swarm's architecture is based on the concepts of a **control plane**, **management plane**, and **data plane** or **workload plane**. The control plane supervises the status of the cluster, the management plane provides all the platform management features, and finally, the data plane executes the user-defined tasks. These planes can be isolated from each other using multiple network interfaces (but this should be completely transparent to you as a developer). This model is also present in other orchestrators, such as Kubernetes (simplified into **role nodes**). Different roles will be used to define the work associated with the nodes within the cluster. The main difference between Docker Swarm and other orchestrators is that these roles are easily interchangeable within nodes; hence, a control/management plane node can be converted into a workload-ready node with just a command. Docker Swarm manages all the control plane communications securely by using **Transport Layer Security** (**TLS**)-encrypted networks. The internal **certificate authority** (**CA**) and its certificates will be completely managed by Docker Swarm.

> **Important note**
>
> Most container orchestration platforms will define **master nodes** as the nodes used to manage the platform, while **worker nodes** will finally execute all the workloads. These roles can also be shared and master nodes can execute some specific tasks. Docker Swarm allows us to completely change a node's role with the command line without having to reinstall or recreate the node in the cluster.

We will use the concept of a **service** to define the workloads in our cluster. The services have different properties to modify and manage how traffic will be delivered to our applications' workloads. We will define the number of **replicas** for a service to be considered alive. The orchestrator will be in charge of making sure this number of replicas is always running. With this in mind, to scale a service up or down, we will just modify the number of replicas required to be considered healthy. When a cluster node goes down, Docker Swarm will schedule its tasks on other available hosts.

Applications that use many different container-based components will require multiple services to run, which makes it important to define communications between them. Docker Swarm will manage both the workload states and the networking layer (**overlay networks**). Encryption can also be enabled for service communications. To ensure that all service replicas are reachable, Docker Swarm manages an internal DNS and load-balances service requests among all healthy replicas.

The cluster will also manage the application's services' rolling upgrades any time a change is made to any of the components. Docker Swarm provides different types of deployments for specific needs. This feature allows us to execute maintenance tasks such as updates by simply replacing or degrading a service, avoiding any possible outages.

We can summarize all of Docker Swarm features in the following points:

- Docker Swarm is a built-in Docker container runtime and no additional software is required to deploy a container orchestrator cluster.

- A control plane, a management plane, and a data plane are deployed to supervise cluster states, manage all tasks, and execute our applications' processes, respectively.

- Cluster nodes can be part of the **control and management planes** (with a manager role), simply execute assigned workloads (with a worker or compute role), or have both roles. We can easily change a node's role from manager to worker with the Docker client command line.

- An application's workloads are defined as services, represented by a number of healthy replicas. The orchestrator will automatically oversee the execution of a reconciliation process when any replica fails to meet the service's requirements.

- All the cluster control plane and service communications (overlay networks) are managed by Docker Swarm, providing security by default with TLS in the control plane and with encryption features for services.

- Docker Swarm provides internal service discovery and homogeneous load balancing between all service replicas. We define how service replicas will be updated when we change any content or workload features, and Docker Swarm manages these updates.

Now we can advance in this section and learn how to deploy a Docker Swarm cluster and its architecture. You as a developer can apply your own knowledge to decide which cluster features and resources will help you run your applications with supervision from this orchestrator.

Docker Swarm manager nodes

As mentioned earlier in this section, the control plane is provided by a set of hosts. These hosts are known as **manager nodes** in Swarm. These nodes in the cluster are critical for delivering all the control plane's features. They all manage the Docker Swarm cluster environment. An internal key-value database is used to maintain the metadata of all the objects created and managed in the cluster.

To provide high availability to your cluster, we deploy more than one manager node and we will share this key-value store to prevent a failure if a manager goes down. One of the manager nodes acts as the leader and writes all the objects' changes to its data store. The other managers will replicate this data into their own databases. The good thing here is that all this is managed internally by Docker Swarm. It implements the **Raft consensus algorithm** to manage and store all the cluster states. This ensures information distribution across multiple managers equally.

The first node created in a Docker Swarm cluster automatically becomes the cluster leader and an election process is always triggered when the leader fails. All healthy manager nodes vote for a new leader internally; hence, a consensus must be reached before electing a new one. This means that we need at least $N/2+1$ healthy managers to elect a new leader. We need to deploy an odd number of manager nodes and they all maintain the cluster's health, serve the Docker Swarm HTTP API, and schedule workloads on healthy, available compute nodes.

All the communications between manager and worker nodes are encrypted by using TLS (mutual TLS) by default. We don't need to manage any of this encryption; an internal CA is deployed and servers' certificates are rotated automatically.

Now that we understand how the cluster is managed, let's review how applications are executed in the compute nodes.

Docker Swarm worker nodes

The leader of the manager nodes reviews the status of the platform and decides where to run a new task. All nodes report their statuses and loads to help the leader decide what the best location for executing the service replicas is. Worker nodes talk with manager nodes to inform them about the status of their running containers, and this information reaches the leader node.

Worker nodes will just execute containers; they never participate in any scheduling decisions and they are part of the data plane, where all services' internal communications are managed. These communications (overlay networks) are based on UDP VXLAN tunneling and they can be encrypted, although this isn't enabled by default since some overhead is expected.

> **Important note**
> It is important to know that Docker Swarm manager nodes also have the worker role. This means that by default, any workload can run either on a manager node or a worker node. We will use additional mechanisms, such as workload locations, to avoid the execution of an application's containers on manager nodes.

We can continue now and learn how to create a simple cluster.

Creating a Docker Swarm cluster

Docker Swarm's features are completely embedded into the Docker container runtime; hence, we don't need any additional binaries to create a cluster.

To create a Docker Swarm cluster, we will start by initializing it. We can choose any host interface for creating the cluster, and by default, the first one available will be used if none is selected. We will execute `docker swarm init` in a cluster node and this will become the leader. It is important to understand that we can have a fully functional cluster with just one node (leader), although we will not be able to provide high availability to our applications in this case. By default, any manager node, including the leader, will be able to run any application's workloads.

Once a Docker Swarm cluster is created, the Docker container runtime starts to work in **swarm mode**. At this point, some new Docker objects become available, which may make it interesting for you as a developer to deploy your own cluster:

- **Swarm**: This object represents the cluster itself, with its own properties.
- **Nodes**: Each node within the cluster is represented by a **node object**, no matter whether it's a leader, manager, or worker node. It can be very useful to add some labels to each node to help the internal scheduler allocate workloads to specific nodes (remember that all nodes can run any workload).
- **Services**: The service represents the minimal workload scheduling unit. We will create a service for each application's component, even if it just runs a single container. We will never run standalone containers in a Docker Swarm cluster, as these containers will not be managed by the orchestrator.
- **Secrets**: These objects allow us to securely store all kinds of sensitive data (up to a maximum of 500 KB). Secrets will be mounted and used inside service containers and the cluster will manage and store their content.
- **Configs**: Config objects will work like secrets, but they are stored in clear text. It is important to understand that configs and secrets are spread cluster-wide, which is critical, as containers will run in different hosts.
- **Stacks**: These are a new type of object used to deploy applications in Docker Swarm. We will use the Compose YAML file syntax to describe all the application's components and their storage and networking configurations.

> **Important note**
>
> The Docker Swarm cluster platform does not require as many resources as Kubernetes; hence, it is possible to deploy a three-node cluster on your laptop for testing. You will be able to verify how your applications work and maintain the service level when some of the application's components fail or a cluster node goes completely offline. We use a standalone Docker Swarm cluster in order to use special objects such as **secrets** or **configs**.

As we mentioned before in this section, we can just create a Docker Swarm cluster by executing `docker swarm init`, but many arguments can modify the default behavior. We will review a few of the most important ones just to let you know how isolated and secure a cluster can be:

- `--advertise-addr`: We can define the interface that will be used to initiate the cluster with this. All other nodes will use this IP address to join the recently created cluster. By default, the first interface will be used. This option will allow us to set which interface will be used to announce the control plane.

- `--data-path-addr` and `--data-path-port`: We can isolate the data plane for the applications by setting a host's specific interface IP address and port using these arguments. Traffic can be encrypted, and this will be completely transparent to your applications. Docker Swarm will manage this communication; some overhead may be expected due to the encryption/decryption processes.

- `--listen-addr`: By default, the Docker Swarm API will be listening on all host interfaces, but we can secure the API by answering on a defined interface.

- `--autolock`: Docker Swarm will store all its data under `/var/lib/docker/swarm` (by default, depending on your runtime root data path). This directory contains the CA, used for creating all the nodes' certificates, and the snapshots automatically created by Docker Swarm to preserve the data in case of failure. This information must be secure, and the `--autolock` option allows us to lock the content until a passphrase is provided.

> **Important note**
>
> Locking Docker Swarm content may affect your cluster's high availability. This is because every time the Docker runtime is restarted, you must use an unlock action to retrieve the directory's content, and you will be asked for the autolock passphrase. Hence, an automatic restart of components is broken since manual intervention is required.

When a Docker swarm is initialized, a couple of cluster tokens are created. These tokens should be used to join additional nodes to the cluster. One token should be used to join new manager nodes and the other one should only be used to integrate worker nodes. Remember that the node's roles can be changed later if a manager node fails, for example. The following code shows how the tokens are presented:

```
$ docker swarm init
Swarm initialized: current node (s3jekhby2s0vn1qmbhm3ulxzh) is now a
manager.
```

To add a worker to this swarm, you can run the following command:

```
    docker swarm join --token SWMTKN-1-17o42n70mys1l3qklmew87n82lrgym
trr65exmaga9jp57831g-4g2kzh4eoec297317hc561bte 192.168.65.4:2377
```

To add a manager to this swarm, run `docker swarm join-token manager` and follow the instructions.

We use `docker swarm join` followed by `--token` and the appropriate token for a new manager or worker node. This token will be shown at cluster initialization, but it can be retrieved whenever needed by simply using `docker swarm join-token`. This action can also be used to rotate the current token (automatic rotation will be triggered by default every 90 days).

Docker Swarm nodes can leave the cluster whenever it is needed by executing `docker swarm leave`. It is important to understand that losing one manager may leave your cluster in danger. Be careful with the manager nodes, especially when you change their role to a worker node or when you remove them from a cluster.

> **Important note**
>
> Some Swarm object properties, such as autolock or certificate expiration, can be modified by using `docker swarm update`.

In the next section, we will learn what is required to provide high availability to a Docker Swarm cluster and the requirements for our applications.

Providing high availability with Docker Swarm

The Docker Swarm orchestrator will provide out-of-the-box **high availability** if we use an odd number of manager nodes. The **Raft protocol** used to manage the internal database requires an odd number of nodes to maintain it healthily. Having said that, the minimum number of healthy managers for the cluster to be fully functional is *N/2+1*, as discussed earlier in this chapter. However, no matter how many managers are working, your application's functionality may not be impacted. Worker nodes will continue working even when you don't have the minimum number of required manager nodes. An application's services will continue running unless a container fails. In this situation, if the managers aren't functional, your containers will not be managed by the cluster and thus your application will be impacted. It is important to understand this because it is the key to preparing your clusters for these situations.

Although your cluster runs with fully high availability, you must prepare your applications. By default, resilience is provided. This means that if a running container fails, a new one will be created to replace it, but this will probably impact your application even if you run a fully stateless service.

Services integrate **tasks** or **instances**, which finally represent a container. Hence, we must set the number of replicas (or tasks) required for a service to be considered healthy. The Docker container runtime running the workload will check whether the container is healthy by executing the health checks included within the container image or the ones defined at execution time (written using the Compose YAML file format in which the service is defined).

Definitely, the number of **replicas** will impact the outage of your service when things go wrong. Therefore, you should prepare your applications for this situation by executing, for example, more than one replica for your services. Of course, this requires that you think of your application's components' logic from the very beginning. For example, even if your application is completely stateless and uses a stateful service, such as a database, you will probably have to think about how to provide high availability or at least fault tolerance to this component. Databases can run inside containers but their logic may need some tweaks. Sometimes, you can just replace your SQL database with a NoSQL database distributed solution.

In the previous application example, with a database component, we didn't take into account the problem of managing the stateful data using volumes (even if using a distributed solution), but every stateful application should be able to move from one cluster node to another. This also affects the associated volumes that must be attached to containers wherever they run, no matter which node in the cluster receives a task. We can use remote storage filesystem solutions, such as NFS, or sync filesystems or folders between nodes. You as a developer don't have to manage the infrastructure, but you must prepare your applications, for example, by verifying the existence of certain files. You should also ask yourself what will happen if more than one replica tries to access your data. This situation will definitely corrupt a database, for example. Other orchestrators, such as Kubernetes, provide more interesting solutions for these situations, as we will learn in *Chapter 8, Deploying Applications with the Kubernetes Orchestrator*.

Docker Swarm nodes can be **promoted** from the worker role to the manager role, and vice versa, a manager can be **demoted** to a worker. We can also **drain** and **pause** nodes, which allows us to completely move all containers from a node to another available worker, and disable scheduling in the nodes defined, respectively. All these actions are part of infrastructure management. You should at least verify how your application will behave when a drain action is triggered and your containers stop on one node and start on another. How will your application's components manage such circumstances? How will this affect component containers that consumed some of the affected services? This is something you have to solve in your application's logic and code as a developer.

Next, let's learn how to schedule our applications in Docker Swarm.

Creating tasks and services for your applications

The first thing you should know is that we will never schedule containers on a Docker Swarm cluster. We will always run **services**, which are the minimal deployment units in a Docker Swarm cluster.

Each service is defined by a number of replicas, known in Docker Swarm as **tasks**. And finally, each task will run one container.

> **Important note**
> Docker Swarm is based on Moby's **SwarmKit** project, which was designed to run any kind of task cluster-wide (virtual machines, for example). Docker created Docker Swarm by implementing SwarmKit in the orchestrator, but specifically for running containers.

We will use a **declarative model** to schedule services in our Docker Swarm cluster by setting the desired state for our services. Docker Swarm will take care of their state continuously and take corrective measures in case of any failure to reconcile its state. For example, if the number of running replicas is not correct because one container has failed, Docker Swarm will create a new one to correct the service's state.

Let's continue by creating a simple `webserver` service using an `nginx:alpine` container image:

```
$ docker service create --name webserver nginx:alpine
k40w64wkmr5v582m4whfplca5
overall progress: 1 out of 1 tasks
1/1: running   [=====================================================>]
verify: Service converged
```

We have just defined a service's name and the image to be used for the associated containers. By default, services will be created with one replica.

We can verify the service's state by simply executing `docker service ls` to list all the Docker Swarm services:

```
$ docker service ls
ID                NAME        MODE         REPLICAS    IMAGE
PORTS
k40w64wkmr5v      webserver   replicated   1/1         nginx:alpine
```

As you may have noticed, a service ID is created (object ID) and we can use any of the actions learned about in *Chapter 2* and *Chapter 4* for listing, inspecting, and removing Docker objects.

We can verify in which node the service's containers are running by using `docker node ps`. This will list all the containers running in the cluster that are associated with services:

```
$ docker node ps
ID                NAME          IMAGE          NODE         DESIRED
STATE    CURRENT STATE           ERROR     PORTS
og3zh7h7ht9q      webserver.1   nginx:alpine   docker-
desktop    Running             Running 7 minutes ago
```

In this example, one container is running in the `docker-desktop` host (Docker Desktop environment). We didn't specify a port for publishing our web server; hence, it will work completely internally and will be unreachable to users. Only one replica was deployed because the default value was used when we didn't set anything else. Therefore, to create a real service, we will usually need to specify the following information:

- The repository from which the image should be downloaded
- The number of healthy replicas/containers required by our service to be considered alive
- The published port, if the service must be reachable externally

It is also important to mention that Docker Swarm services can be either replicated (by default, as we have seen in the previous example) or global (run on all cluster nodes).

A **replicated service** will create a number of replicas, known as **tasks**, and each will create one container. You as a developer can prepare your application's logic to run more than one replica per service to provide simple but useful high availability (this will help you lose half of your service in case of failure). This will reduce the impact in case of failure and really help with the upgrade processes when changes need to be introduced.

On the other hand, a **global service** will run one replica of your service in each cluster node. This is very powerful but may reduce the overall performance of your cluster if you can distribute your application's load into different processes. This type of service is used to deploy monitoring and logging applications, and they work as agents, automatically distributed in all nodes at once. It is important to notice that Docker Swarm will schedule a task for each service on any node joined to the cluster. You may use global services when you need to run agent-like applications on your cluster.

You as a developer should think about which service type suits your application best and use the --mode argument to create an appropriate Docker Swarm service.

> **Important note**
>
> You may think that it's a good idea to run a distributed database (MongoDB or any simpler key-value store) or a queue management solution (such as RabbitMQ or Apache Kafka) as a global service to ensure its availability, but you have to take care of the final number of running containers. Global services do not guarantee an odd number of containers/processes and may break your application if you join new nodes to the cluster. Every time you join a new node, a new container is created as part of the global service.

We can use labels to define locations for certain services. They will affect all the replicas at the same time. For example, we can create a global service that should only run on nodes labeled as web:

```
$ docker service create --detach \
--name global-webserver --mode global \
--constraint node.labels.web=="true" nginx:alpine
n9e24dh4s5731q37oboo8i7ig
```

The Docker Desktop environment works like a one-node Docker Swarm cluster; hence, the global service should be running if the appropriate label, web, is present:

```
$ docker service ls
ID               NAME               MODE         REPLICAS    IMAGE
PORTS
n9e24dh4s573     global-webserver   global       0/0         nginx:alpine
k40w64wkmr5v     webserver          replicated   1/1         nginx:alpine
```

Let's add the label to the only cluster node we have:

```
$ docker node update --label-add web="true" docker-desktop
docker-desktop
$ docker node inspect docker-desktop \
--format="{{.Spec.Labels}}"
map[web:true]
```

Automatically, Docker Swarm detected the node label change and scheduled the global service container for us:

```
$ docker service ls
ID                NAME                  MODE         REPLICAS    IMAGE
PORTS
n9e24dh4s573      global-webserver      global       1/1         nginx:alpine
k40w64wkmr5v      webserver             replicated   1/1         nginx:alpine
```

As you can see, Docker Swarm allows us to change the default location of any service. Let's review some of the options available to place our application's tasks in specific nodes or pools of nodes:

- --constraint: This option fixes where to run our service's containers. It uses labels, as we saw in the previous example. We can verify the placement requirements of a service by using docker service inspect:

  ```
  $ docker service inspect global-webserver \
  --format="{{.Spec.TaskTemplate.Placement}}"
  {[node.labels.web==true] [] 0 [{amd64 linux} { linux} { linux}
  {arm64 linux} {386 linux} {ppc64le linux} {s390x linux}]}
  ```

- --placement-pref: Sometimes, we are looking for a preferred location, but we need to ensure that the application will execute even if this doesn't exist. We will use a placement preference in such situations.

- --replicas-max-per-node: Another way of setting the location under certain circumstances will be to avoid more than a specific number of replicas per cluster node. This will ensure, for example, that replicas will not compete for resources in the same host.

By using placement constraints or preferred locations, you can ensure, for example, that your application will run in certain nodes with GPUs or faster disks.

> **Important note**
>
> You as a developer should design your application to run almost anywhere. You will need to ask your Docker Swarm administrators for any location labeling or preferences and use them in your deployments. These kinds of infrastructure features may impact how your applications run and you must be aware of them.

We can also execute **jobs** in Docker Swarm. A job may be considered a type of service that should only run once. In these cases, the service's tasks run a container that exists. If this execution is correct, the task is marked as `Completed` and no other container will be executed. Docker Swarm allows the execution of both global or replicated jobs, the `global-job` and `replicated-job` service types, respectively.

Docker Swarm services can be updated at any time, for example, to change the container image or other properties, such as their scheduling location within the cluster nodes.

To update any available service's property, we will use `docker service update`. In the following example, we will just update the number of replicas of the service:

```
$ docker service update webserver --replicas=2
webserver
overall progress: 2 out of 2 tasks
1/2: running   [==========================================================>]
2/2: running   [==========================================================>]
verify: Service converged
$ docker service ls
ID              NAME               MODE         REPLICAS    IMAGE
PORTS
n9e24dh4s573    global-webserver   global       1/1         nginx:alpine
k40w64wkmr5v    webserver          replicated   2/2         nginx:alpine
```

Now that we have two replicas or instances running for the `webservice` service, we can verify how Docker Swarm will check and manage any failure:

```
$ docker node ps
ID              NAME                                               IMAGE
NODE              DESIRED STATE    CURRENT STATE        ERROR
PORTS
g72t1n2myffy    global-webserver.s3jekhby2s0vn1qmbhm3ulxzh    nginx:
alpine    docker-desktop    Running          Running 3 hours ago
og3zh7h7ht9q    webserver.1                                        nginx:
alpine    docker-desktop    Running          Running 16 hours ago
x2u85bbcrxip    webserver.2                                        nginx:
alpine    docker-desktop    Running          Running 2 minutes ago
```

Using the `docker` runtime client, we can list all the containers running (this works because we are using just one node cluster, the `docker-desktop` host):

```
$ docker ps
CONTAINER ID    IMAGE          COMMAND              CREATED
STATUS          PORTS     NAMES
cfd1b37cca93    nginx:alpine   "/docker-entrypoint...."    9 minutes ago
Up 9 minutes    80/tcp    webserver.2.x2u85bbcrxip1svtmj08x69z5
```

```
9bc7a5df593e    nginx:alpine    "/docker-entrypoint...."    3 hours ago
Up 3 hours      80/tcp    global-webserver.s3jekhby2s0vn1qmbhm3ulxzh.
g72t1n2myffy0abl1bj57m6es
3d784315a0af    nginx:alpine    "/docker-entrypoint...."    16 hours ago
Up 16 hours     80/tcp    webserver.1.og3zh7h7ht9qpv1xkip84a9gb
```

We can kill one of the `webserver` service's containers and verify that Docker Swarm will create a new container to reconcile the service's status:

```
$ docker kill webserver.2.x2u85bbcrxiplsvtmj08x69z5
webserver.2.x2u85bbcrxiplsvtmj08x69z5
```

A second after the failure is detected, a new container runs:

```
$ docker service ls
ID                NAME               MODE          REPLICAS    IMAGE
PORTS
n9e24dh4s573      global-webserver   global        1/1         nginx:alpine
k40w64wkmr5v      webserver          replicated    2/2         nginx:alpine
```

We can verify that Docker Swarm managed the container issue:

```
$ docker service ps webserver
ID                NAME             IMAGE          NODE
DESIRED STATE     CURRENT STATE            ERROR
PORTS
og3zh7h7ht9q      webserver.1      nginx:alpine   docker-desktop
Running           Running 16 hours ago
x02zcj86krq7      webserver.2      nginx:alpine   docker-desktop
Running           Running 4 minutes ago
x2u85bbcrxip      \_ webserver.2   nginx:alpine   docker-desktop
Shutdown          Failed 4 minutes ago     "task: non-zero exit (137)"
```

The preceding snippet shows a short history with the failed container ID and the new one created to maintain the health of the service.

As you may have noticed, the containers created within Docker Swarm have the prefix of the service associated, followed by the instance number. These help us identify which services may be impacted when we have to execute maintenance tasks on a node. We can list current containers to view how services' tasks are executed:

```
$ docker container ls
CONTAINER ID    IMAGE          COMMAND            CREATED
STATUS          PORTS     NAMES
52779dbe389e    nginx:alpine    "/docker-entrypoint...."    13 minutes ago
Up 13 minutes   80/tcp    webserver.2.x02zcj86krq7im2irwivwzvww
```

```
9bc7a5df593e    nginx:alpine    "/docker-entrypoint.…"    4 hours ago
Up 4 hours      80/tcp      global-webserver.s3jekhby2s0vn1qmbhm3ulxzh.
g72t1n2myffy0abl1bj57m6es
3d784315a0af    nginx:alpine    "/docker-entrypoint.…"    16 hours ago
Up 16 hours     80/tcp      webserver.1.og3zh7h7ht9qpv1xkip84a9gb
```

You must remember that we are running containers; hence, services can inherit all the arguments we used with containers (see *Chapter 4, Running Docker Containers*). One interesting point is that we can include some Docker Swarm internal keys using the Go template format as variables for our application deployments:

- **Service**: This object provides the `.Service.ID`, `.Service.Name`, and `.Service.Labels` keys. Using these service labels may be interesting for identifying or including some information in your application.

- **Node**: The node object allows us to use its hostname within containers, which may be very interesting when you need to identify a node within your application (for example, for monitoring). If you use a host network, this is not required, as the hostname will be included as part of the network namespace, but for security reasons, your environment may not allow you to use the host's namespaces. This object provides `.Node.ID` and `.Node.Hostname`.

- **Task**: This object provides `.Task.ID`, `.Task.Name`, and `.Task.Slot`, which may be interesting if you want to manage the behavior of your container within some application's components.

Let's see a quick example of how we can use such variables:

```
$ docker exec webserver.2.x02zcj86krq7im2irwivwzvww hostname
52779dbe389e
$ docker exec webserver.1.og3zh7h7ht9qpv1xkip84a9gb hostname
3d784315a0af
```

Now we update our service with a new hostname:

```
$ docker service update --hostname="{{.Node.Hostname}}" webserver
webserver
overall progress: 2 out of 2 tasks
1/2: running    [==================================================>]
2/2: running    [==================================================>]
verify: Service converged
```

We can now verify that the container's hostname has changed.

Let's continue with the definition of a complete application, as we already did with Compose in standalone environments, but this time running cluster-wide.

A review of stacks and other Docker Swarm resources

Docker Swarm allows us to deploy applications with multiple services by running stacks. This new object defines, in a Compose YAML file, the structure, components, communications, and interactions with external resources of your applications. Therefore, we will use **infrastructure as code (IaC)** to deploy our applications on top of the Docker Swarm orchestrator.

> **Important note**
> Although we use a Compose YAML file, not all the docker-compose keys are available. For example, the depends_on key is not available for stacks because they don't include any dependency declarations. That's why it is so important to prepare your application's logic in your code. Health checks will let you decide how to break some circuits when some components fail, but you should include status verifications on dependent components when, for example, they need some time to start. Docker Compose runs applications' containers on standalone servers while Docker Swarm stacks deploy applications' services (containers) cluster-wide.

In a stack YAML file, we will declare our application's network, volumes, and configurations. We can use any Compose file with a few modifications. In fact, the Docker container runtime in swarm mode will inform you and fail if you use a forbidden key. Other keys, such as depends_on, are simply omitted when we use a docker-compose file with Docker Swarm. Here is an example using the Compose YAML file found in *Chapter 5, Creating Multi-Container Applications*. We will use docker stack deploy:

```
$ docker stack deploy --compose-file docker-compose.yaml test
Ignoring unsupported options: build
Creating network test_simplestlab
Creating service test_lb
Creating service test_db
Creating service test_app
```

As you can see, the container runtime informed us of an unsupported key, Ignoring unsupported options: build. You can use a Compose YAML file to build, push, and then use the container images within your application, but you must use a registry for your images. By using a registry, we can ensure that all container runtimes will get the image. You can download the images, save them as files, and copy them to all nodes, but this is not a reproducible process and it may cost some time and effort to synchronize all changes. It seems quite logical to use a registry to maintain all the images available to your clusters.

We can now review the deployed stack and its services:

```
$ docker stack ls
NAME        SERVICES    ORCHESTRATOR
test        3           Swarm
```

```
$ docker service ls
ID                 NAME                    MODE              REPLICAS    IMAGE
                                    PORTS
4qmfnmibqqxf    test_app                  replicated       0/1         myregistry/
simplest-lab:simplestapp       *:3000->3000/tcp
asr3kwfd6b8u    test_db                   replicated       0/1         myregistry/
simplest-lab:simplestdb        *:5432->5432/tcp
jk603a8gmny6    test_lb                   replicated       0/1         myregistry/
simplest-lab:simplestlb        *:8080->80/tcp
```

Notice that the REPLICAS column shows 0/1 for all the services; this is because we are using a fake registry and repository. The container runtime will not pull images in this example because we are using an internal registry that doesn't exist, but this still shows how to deploy a complete application. Working with registries may require the use of --with-registry-auth to apply certain authentications to our services. Credentials should be used to pull the images associated with each of your services if you are using private registries.

You will probably have also realized that all services have the stack's name as a prefix, as we already learned about for projects and their services in *Chapter 5, Creating Multi-Container Applications*.

Let's now quickly review how configurations are managed cluster-wide. As we'd expect, running applications cluster-wide may require a lot of synchronization effort. Every time we create a service with some configuration or persistent data, we will need to ensure its availability on any host. Docker Swarm helps us by managing the synchronization of all configurations within the cluster. Docker Swarm provides us with two types of objects for managing configurations: secrets and configs. As we have already learned how secrets and configs work with Compose, we will just have a quick review since we will use them in this chapter's labs.

Secrets

Secrets allow us to store and manage sensitive data such as tokens, passwords, or certificates inside a Docker Swarm cluster and containers deployed in a standalone environment. To use secrets and configs, we need to enable swarm mode in our container runtime because it is needed to provide a store in which information is encrypted. This data is stored in the Raft log database managed by the cluster (an embedded key-value store). When we run a container that needs to use a stored secret, the host who runs that container will ask the managers via an API for the information and it will be mounted inside the container using a temporal filesystem (in-memory tmpfs on Linux hosts). This ensures that the information will be removed when the container dies. It may be considered volatile and therefore it is only available on running containers. By default, secrets are mounted as files in the form /run/secrets/<SECRET_NAME>, and they include the secret object's content. This path can be changed, as well as the file permissions and ownership.

We can use secrets inside environment variables, which is fine because they are only visible inside containers. However, you can also use secrets to store a complete configuration file, even if not all its content must be secured. Secrets can only contain 500 KB of data; thus, you may need to split your configuration into different secrets if you think it may not be enough.

Secrets can be created, listed, removed, inspected, and so on like any other Docker container object, but they can't be updated.

> **Important note**
>
> As secrets are encrypted, `docker secret inspect` will show you their labels and other relevant information, but the data itself will not be visible. It is important to also understand that secrets can't be updated, so they should be recreated if need be (removed and created again).

Configs

Configs are similar to secret objects, but they are not encrypted and can be updated. This makes them the perfect combination for easily reconfiguring your applications, but remember to always remove any sensitive information, such as connection strings where passwords are visible, tokens, and so on, that could be used by an attacker to exploit your application. Config objects are also stored in the Docker Swarm Raft Log database in clear text; therefore, an attacker with access to the Docker Swarm information can view them (remember that this information can be locked with a passphrase). These files can contain a maximum of 500 KB, but you can include even binary files.

Config objects will be mounted inside containers as if they were bind-mounted files, owned by the main process user and with read-all permissions. As with secrets, config-mounted file permissions and ownership can be changed depending on your own needs.

In both cases, Docker Swarm takes care of syncing these objects cluster-wide without any additional action on our end.

> **Important note**
>
> In both cases, you can decide the path at which secret or config files will be mounted and its owner and permissions. Please take care of the permissions you give to your files and ensure that only the minimum necessary permissions for reading the file are granted.

We will now learn how our applications will be published internally and externally and how the application's services will be announced cluster-wide.

Networking and exposing applications with Docker Swarm

We already learned how container runtimes provide network capabilities to our containers by setting network namespaces and virtual interfaces attached to the host's bridge network interfaces. All these features and processes will also work with Docker Swarm but communication between hosts is also required, and this is where overlay networks come in.

Understanding the Docker Swarm overlay network

To manage all communications cluster-wide, a new network driver, **overlay**, will be available. The overlay network works by setting UDP VXLAN tunnels between all the cluster's hosts. These communications can be encrypted with some overhead and Docker Swarm sets the routing layer for all containers. Docker Swarm only takes care of overlay networks while the container runtime will manage all other local scope networks.

Once we have initialized a Docker Swarm cluster, two new networks will appear, `docker_gwbridge` (bridge) and `ingress` (overlay), with two different functions. The first one is used to interconnect all container runtimes, while the second one is used to manage all service traffic. By default, all services will be attached to the `ingress` network if no additional network is provided during their creation.

> **Important note**
>
> If you find issues with Docker Swarm, check whether your firewall blocks your overlay networking; `2377/TCP` (cluster management traffic), `7946/TCP-UDP` (node intercommunication), and `4789/UDP` (overlay networking) traffic should be permitted.

All services attached to the same overlay network will be reachable by other services also connected to the same overlay network. We can also run containers attached to these networks, but remember that standalone containers will not be managed by Docker Swarm. By default, all overlay networks will be unencrypted and non-attachable (standalone containers can't connect); hence, we need to pass `--opt encrypted --attachable` arguments along with `--driver overlay` (required to create overlay networks) to encrypt them and make them attachable.

We can create different overlay networks to isolate our applications, as containers attached to one network will not see those attached to a different one. It is recommended to isolate your applications in production and define any allowed communication by connecting your services to more than one network if required. Configurations such as the subnet or IP address range within a subnet can be used to create your custom network, but remember to specify the `--driver` argument to ensure you create an overlay network.

Let's see now how we can access our services and publish them.

Using service discovery and internal load balancing

Docker Swarm provides its own internal IPAM and DNS. Each service receives an IP address from the attached network range and a DNS entry will be created for it. An internal load-balancing feature is also available to distribute requests across service replicas. Therefore, when we access our service's name, available replicas will receive our traffic. However, you as a developer don't have to manage anything – Docker Swarm will do it all for you – but you must ensure that your application's components are attached to appropriate networks and that you use the appropriate service's names. The internal load balancer receives the traffic and routes your requests to your service's task containers. Never use a container's IP address in your applications as it will probably change (containers die and new ones are created), but a service's IP addresses will stay as they are unless you recreate your service (as in, remove and create a new one again). A service's IP addresses are assigned by an internal IPAM from a specific set.

Publishing your applications

You may be asking yourself, what about the overlay `ingress` network created by default? Well, this network will be used to publish our applications. As we already learned for standalone environments, containers can run attached to a network and expose their processes' ports internally, but we can also expose them externally by using –`publish` options. In Docker Swarm, we have the same behavior. If no port is exposed, the ports declared in the image will be published internally (you can override the ports' definitions, but your application may not be reached). However, we can also publish our service's containers externally, exposing its processes either in a random port within the 30000-32767 range or in a specifically defined port (as usual, more than one port can be published per container).

All nodes participate in the overlay `ingress` network, and the published container ports will be attached using the port NAT, in all available hosts. Docker Swarm provides internal OSI Layer 3 routing using a mesh to guide requests to all available services' tasks. Therefore, we can access our services on the defined published port on any cluster host, even if they don't have a running service container.

An external load balancer can be used to assign an IP address and forward the clients' requests to certain cluster hosts (enough to provide high availability to our service).

Let's see a quick example by creating a new service and publishing the container port, 80, on the host port, 1080:

```
$ docker service create --name webserver-published \
--publish published=1080,target=80,protocol=tcp \
--quiet nginx:alpine
gws9iqphhswujnbqpvncimj5f
```

Now, we can verify its status:

```
$ docker service ls
ID                NAME                  MODE         REPLICAS    IMAGE
                              PORTS
gws9iqphhswu     webserver-published    replicated   1/1
nginx:alpine                            *:1080->80/tcp
```

We can test port 1080 on any cluster host (we only have one host on Docker Desktop):

```
$ curl 0.0.0.0:1080 -I
HTTP/1.1 200 OK
Server: nginx/1.23.4
Date: Fri, 12 May 2023 19:05:43 GMT
Content-Type: text/html
Content-Length: 615
Last-Modified: Tue, 28 Mar 2023 17:09:24 GMT
Connection: keep-alive
ETag: "64231f44-267"
Accept-Ranges: bytes
```

As we have seen, containers are available on a host's port. In fact, in a cluster with multiple hosts, this port is available on all hosts, and this is the default mechanism for publishing applications in Docker Swarm. However, there are other methods for publishing applications in this orchestrator:

- The host mode allows us to set the port only on those nodes actually running a service's container. Using this mode, we can specify a set of cluster hosts where service instances will run by setting labels and then forward the clients' traffic to these hosts using an external load balancer.

- The dnsrr mode allows us to avoid a service's virtual IP address; hence, no IP address from the IPAM will be set and a service's name will be associated directly with a container's IP address. We can use the --endpoint-mode argument to manage the publishing mode when we create a service. In dnsrr mode, the internal DNS will use **round-robin resolution**. Cluster-internal client processes will resolve a different container IP address every time they ask the DNS for the service's name.

Now that we have learned how to publish applications running inside a Docker Swarm cluster to be consumed by applications inside and outside the cluster itself, let's move on to review how service containers and other properties can be updated automatically.

Updating your application's services

In this section, we are going to review how Docker Swarm will help our applications' stability and availability when we push changes to them. It is important to understand that whatever platform we are using to run our containers, we need to be able to modify our application content to fix issues or add new functionality. In production, this will probably be more restricted but automation should be able to do this too, ensuring a secure supply chain.

Docker Swarm provides a rolling update feature that deploys new changes without interrupting current replicas and automatically switches to an older configuration when the update goes wrong (rolls back).

You as a developer must think about which update method fits your application best. Remember to deploy multiple replicas if you want to avoid any outages. This way, by setting the update parallelism (`--update-parallelism`), the delay in seconds between container updates (`--update-delay`), and the order in which to deploy the change (`--update-order`) – which allows us to stop the previous container before starting a new one (default), or do the reverse – we can ensure our service health when we apply changes. It is very important to understand that your application must allow you to run more than one container replica at a time because this may be needed to access a volume at the same time. Remember, this may break your application data if your processes don't allow it (for example, a database may get corrupted).

When our service deploys many replicas, for example, a stateless frontend service, it is very important to decide what to do when issues arise during the upgrade process.

By default, Docker Swarm will wait five seconds to start monitoring the status of each task update. If your application requires more time to be considered healthy, you may need to set up an appropriate value by using the `--update-monitor` argument.

The update process works by default as follows:

1. Docker Swarm stops the first service's container (the first replica/task; the container's suffix shows the task number).

2. Then, the update is triggered for this stopped task.

3. A new container starts to update the task.

4. Then, two situations may occur:

    - If the process goes fine, the update of a task returns `RUNNING`. Then, Docker Swarm waits for the defined delay time between updates and triggers the update process again for the next service task.

    - If the process fails, for example, the container doesn't start correctly, the updated task returns `FAILED` and the current service update process is paused.

5. When the service update process is paused, we have to decide whether we have to manually roll back to a previous version (configurations, container images, and so on – in fact, any change deployed since the latest correct update) or execute a new update again.

6. We will use the `--update-failure-action` argument to automate the process when something goes wrong during the updates. This option allows us to either *continue* with the update, even if some containers fail, *pause* the update process (default), or automatically trigger a *rollback* in case of any error.

It is really recommended to test your deployments and updates to have a clear idea of how your application can be compromised in case of failure.

All the options described to define the update process are also available for the rollback procedure; hence, we have a lot of options for managing our application stability even when we trigger service changes.

In the following section, we will prepare an application for Docker Swarm and review some of the features learned about in this chapter.

Labs

The following labs will help you deploy a simple demo application on top of a Docker Swarm cluster to review the most important features provided by this container orchestrator. The code for the labs is available in this book's GitHub repository at `https://github.com/PacktPublishing/Containers-for-Developers-Handbook.git`. Ensure you have the latest revision available by simply executing `git clone https://github.com/PacktPublishing/Containers-for-Developers-Handbook.git` to download all its content or `git pull` if you have already downloaded the repository before. Additional labs are included in GitHub. All commands and content used in these labs will be located inside the `Containers-for-Developers-Handbook/Chapter7` directory.

We will start by deploying our own Docker Swarm cluster.

Deploying a single-node Docker Swarm cluster

In this lab, we will create a one-node Docker Swarm cluster using the Docker Desktop environment.

> **Important note**
> Deploying a single-node cluster is enough to review the most important features learned about in this chapter but, of course, we wouldn't be able to move service tasks to another node. If you are interested in a situation like that and want to review advanced container scheduling scenarios, you can deploy multiple-node clusters following any of the methods described in the specific `multiple-nodes-cluster.md` Markdown file located in this chapter's folder.

To create a single-node Docker Swarm cluster, we will follow these steps:

1. Use the `docker` CLI with the `swarm` object. In this example, we will use default IP address values to initialize a Docker Swarm cluster:

```
$ docker swarm init
Swarm initialized: current node (pyczfubvyyih2kmeth8xz9yd7) is
now a manager.
```

To add a worker to this swarm, run the following command:

```
docker swarm join --token SWMTKN-1-3dtlnakr275se91p7b5
gj8rpk97n66jdm7o1tn5cwsrf3g55yu-5ky1xrr61mdx1gr2bywi5v0o8
192.168.65.4:2377
```

To add a manager to this swarm, run `docker swarm join-token manager` and follow the instructions.

2. We can now verify the current Docker Swarm nodes:

```
$ docker node ls
ID                          HOSTNAME         STATUS
AVAILABILITY    MANAGER STATUS    ENGINE VERSION
pyczfubvyyih2kmeth8xz9yd7 *   docker-
desktop    Ready      Active          Leader         20.10.22
```

3. Overlay and specific bridge networks were created, as we can easily verify by listing the available networks:

```
$ docker network ls
NETWORK ID      NAME             DRIVER     SCOPE
75aa7ed7603b    bridge           bridge     local
9f1d6d85cb3c    docker_gwbridge  bridge     local
07ed8a3c602e    host             host       local
7977xslkr9ps    ingress          overlay    swarm
cc46fa305d96    none             null       local
```

4. This cluster has one node; hence, this node is the manager (leader) and also acts as a worker (by default):

```
$ docker node inspect docker-desktop \
--format="{{ .Status }}"
{ready  192.168.65.4}

$ docker node inspect docker-desktop \
--format="{{ .ManagerStatus }}"
{true reachable 192.168.65.4:2377}
```

This cluster is now ready to run Docker Swarm services.

Reviewing the main features of Docker Swarm services

In this lab, we are going to review some of the features of the most important services by running a replicated and global service:

1. We will start by creating a simple `webserver` service using Docker Hub's `nginx:alpine` container image:

    ```
    $ docker service create --name webserver nginx:alpine
    m93gsvuin5vly5bn4ikmi69sq
    overall progress: 1 out of 1 tasks
    1/1:
    running    [==================================================>]
    verify: Service converged
    ```

2. After a few seconds, the service's task is running and we can list the services and the tasks associated with them using `docker service ls`:

    ```
    $ docker service ls
    ID                NAME          MODE          REPLICAS     IMAGE
    PORTS
    m93gsvuin5vl     webserver     replicated     1/1           nginx:alpine
    $ docker service ps webserver
    ID                NAME           IMAGE            NODE
    DESIRED STATE    CURRENT STATE              ERROR        PORTS
    138u6vpyq5zo     webserver.1    nginx:alpine    docker-
    desktop    Running          Running about a minute ago
    ```

    Notice that by default, the service runs in replicated mode and deploys just one replica. The task is identified as `webserver.1` and it runs on the `docker-desktop` node; we can verify the associated container by listing the containers on that node:

    ```
    $ docker container ls
    CONTAINER ID     IMAGE         COMMAND              CREATED
    STATUS           PORTS         NAMES
    63f1dfa649d8     nginx:alpine    "/docker-
    entrypoint...."    3 minutes ago     Up 3 minutes     80/
    tcp     webserver.1.138u6vpyq5zo9qfyc70g2411x
    ```

    It is easy to track the containers associated with services. We can still run containers directly using the container runtime, but those will not be managed by Docker Swarm.

3. Let's now replicate this service by adding a new task:

    ```
    $ docker service update --replicas 3 webserver
    webserver
    overall progress: 3 out of 3 tasks
    1/3:
    running    [==================================================>]
    ```

```
2/3:
running   [========================================================>]
3/3:
running   [========================================================>]
verify: Service converged
```

4. We can verify its status by using `docker service ps webserver` again:

```
$ docker service ps webserver
ID               NAME           IMAGE          NODE            
DESIRED STATE    CURRENT STATE                 ERROR      PORTS
138u6vpyq5zo     webserver.1    nginx:alpine   docker-desktop
Running          Running about an hour ago
j0at9tnwc3tx     webserver.2    nginx:alpine   docker-desktop
Running          Running 4 minutes ago
vj6k8cuf0rix     webserver.3    nginx:alpine   docker-desktop
Running          Running 4 minutes ago
```

5. Each container gets its own IP address and we will reach each one when we publish the service. We verify that all containers started correctly by reviewing the service's logs:

```
$ docker service logs webserver --tail 2
webserver.1.138u6vpyq5zo@docker-desktop     | 2023/05/14 09:06:44
[notice] 1#1: start worker process 31
webserver.1.138u6vpyq5zo@docker-desktop     | 2023/05/14 09:06:44
[notice] 1#1: start worker process 32
webserver.2.j0at9tnwc3tx@docker-desktop     | 2023/05/14 09:28:02
[notice] 1#1: start worker process 33
webserver.2.j0at9tnwc3tx@docker-desktop     | 2023/05/14 09:28:02
[notice] 1#1: start worker process 34
webserver.3.vj6k8cuf0rix@docker-desktop     | 2023/05/14 09:28:02
[notice] 1#1: start worker process 32
webserver.3.vj6k8cuf0rix@docker-desktop     | 2023/05/14 09:28:02
[notice] 1#1: start worker process 33
```

6. Let's publish the service and verify how clients will reach the `webserver` service:

```
$ docker service update \
--publish-add published=8080,target=80 webserver
webserver
overall progress: 3 out of 3 tasks
1/3:
running   [========================================================>]
2/3:
running   [========================================================>]
3/3:
running   [========================================================>]
```

We can review the service's status again and see that the instances were recreated:

```
$ docker service ps webserver
ID              NAME             IMAGE         NODE
DESIRED STATE   CURRENT STATE          ERROR       PORTS
u7i2t7u60wzt    webserver.1      nginx:alpine  docker-
desktop    Running           Running 26 seconds ago
138u6vpyq5zo    \_ webserver.1   nginx:alpine  docker-
desktop    Shutdown          Shutdown 29 seconds ago
i9ia5qjtgz96    webserver.2      nginx:alpine  docker-
desktop    Running           Running 31 seconds ago
j0at9tnwc3tx    \_ webserver.2   nginx:alpine  docker-
desktop    Shutdown          Shutdown 33 seconds ago
9duwbwjt6oow    webserver.3      nginx:alpine  docker-
desktop    Running           Running 35 seconds ago
vj6k8cuf0rix    \_ webserver.3   nginx:alpine  docker-
desktop    Shutdown          Shutdown 38 seconds ago
```

7. We list which port was chosen (we didn't specify any port for the service; hence, port 80 was assigned to a random host port):

```
$ docker service ls
ID              NAME          MODE        REPLICAS   IMAGE
PORTS
m93gsvuin5vl    webserver     replicated  3/3        nginx:alpine
*:8080->80/tcp
```

8. And now we can test the service with `curl`:

```
$ curl localhost:8080 -I
HTTP/1.1 200 OK
Server: nginx/1.23.4
...
Accept-Ranges: bytes
```

Repeat this `curl` command a few times to access more than one service's replica.

9. Now we can check the logs again:

```
$ docker service logs webserver --tail 2
...
webserver.2.afp6z72y7y1p@docker-desktop    | 10.0.0.2 - -
[14/May/2023:10:36:11 +0000] "HEAD / HTTP/1.1" 200 0 "-"
"curl/7.81.0" "-"

...
webserver.3.ub28rsqbo8zq@docker-desktop    | 10.0.0.2 - -
[14/May/2023:10:38:11 +0000] "HEAD / HTTP/1.1" 200 0 "-"
"curl/7.81.0" "-"

...
```

As you may have noticed, multiple replicas were reached; hence, internal load balancing worked as expected.

10. We end this lab by removing the created service:

```
$ docker service rm webserver
webserver
```

This lab showed how to deploy and modify a simple replicated service. It may be interesting for you to deploy your own global service and review the differences between them.

We will now run a simple application using a Compose YAML file.

Deploying a complete application with Docker

In this lab, we will run a complete application using a stack object. Take a good look at the YAML file that we will use to deploy our application:

1. We first create a couple of secret objects that we will use in the stack:

```
$ echo demo|docker secret create dbpasswd.env -
2tmqj06igbkjt4cot95enyj53

$ docker secret create dbconfig.json dbconfig.json
xx0pesu1t19bvexk6dfs8xspx
```

We used `echo` to create the secret and included only the `demo` string inside the `dbpasswd.env` secret, while we included a complete JSON file in the `dbconfig.json` secret. These secret names can be changed because we will use the full format to reference them in the Compose file.

2. To create an initial database with our own data structure, we add a new config, `init-demo.sh`, to overwrite the file included in the image:

```
$ docker config create init-demo.sh init-demo.sh
zevf3fg2x1a6syze2i54r2ovd
```

3. Let's take a look at our final Compose YAML file before deploying the demo stack. There is a header defining the version to use and then the `services` section will contain all the services to create. First, we have the `lb` service:

```
version: "3.9"
services:
  lb:
    image: frjaraur/simplestlab:simplestlb
    environment: # This environment definitions are in clear-
text as they don't manange any sensitive data
```

```
        - APPLICATION_ALIAS=app # We use the service's names
        - APPLICATION_PORT=3000
      networks:
        simplestlab:
      ports:
      - target: 80
        published: 8080
        protocol: tcp
```

After the lb service, we have the definition for the database. Each service's section includes the container image, the environment variables, and the networking features of the service:

```
   db:
      image: frjaraur/simplestlab:simplestdb
      environment: # Postgres images allows the use of a password
file.
        - POSTGRES_PASSWORD_FILE=/run/secrets/dbpasswd.env
      networks:
        simplestlab:
      secrets:
      - dbpasswd.env
      configs: # We load a initdb script to initialize our demo
database.
      - source: init-demo.sh
        target: /docker-entrypoint-initdb.d/init-demo.sh
        mode: 0770
      volumes:
      - pgdata:/var/lib/postgresql/data
```

Notice that this component includes secrets, configs, and volumes sections. They allow us to include data inside the application's containers. Let's continue with the app service:

```
   app:
      image: frjaraur/simplestlab:simplestapp
      secrets: # A secret is used to integrate de database
connection into our application.
      - source: dbconfig.json
        target: /APP/dbconfig.json
        mode: 0555
      networks:
        simplestlab:volumes:
   pgdata: # This volume should be mounted from a network
resource available to other hosts or the content should be
synced between nodes
```

At the end of the file, we have the definitions for `networks`, `configs`, and `secrets` included in each service definition:

```
networks:
  simplestlab:
configs:
  init-demo.sh:
    external: true
secrets:
  dbpasswd.env:
    external: true
  dbconfig.json:
    external: true
```

You may notice that all secrets and configs are defined as external resources. This allows us to create them outside of the stack. It is not a good idea to include the sensitive content of secrets in cleartext in your Compose YAML files.

> **Important note**
>
> We haven't used a network volume because we are using a single-node cluster, so it isn't needed. But if you plan to deploy more nodes in your cluster, you must prepare either a network storage or a cluster-wide synchronization solution to ensure the data is available wherever the database component is running. Otherwise, your database server won't be able to start correctly.

4. Now we can deploy the Compose YAML file as a Docker stack:

```
$ docker stack deploy -c docker-compose.yaml chapter7
Creating network chapter7_simplestlab
Creating service chapter7_db
Creating service chapter7_app
Creating service chapter7_lb
```

5. We verify the status of the deployed stack:

```
$ docker stack ps chapter7
ID                NAME              IMAGE
NODE              DESIRED STATE     CURRENT STATE              ERROR
PORTS
zaxo9aprs42w      chapter7_app.1    frjaraur/simplestlab:simplestapp
docker-desktop    Running             Running 2 minutes ago
gvjyiqrudi5h      chapter7_db.1     frjaraur/simplestlab:simplestdb
docker-desktop    Running             Running 2 minutes ago
tyixkplpfy6x      chapter7_lb.1     frjaraur/simplestlab:simplestlb
docker-desktop    Running             Running 2 minutes ago
```

6. We now review which ports are available for accessing our application:

```
$ docker stack services chapter7
ID              NAME            MODE            REPLICAS
IMAGE                           PORTS
dmub9x0tis1w    chapter7_app    replicated      1/1            frjaraur/
simplestlab:simplestapp
g0gha8n57i7n    chapter7_db     replicated      1/1            frjaraur/
simplestlab:simplestdb
y2f8iw6vcr5w    chapter7_lb     replicated      1/1            frjaraur/
simplestlab:simplestlb    *:8080->80/tcp
```

7. We can test the chapter7 application stack using our browser (http://localhost:8080):

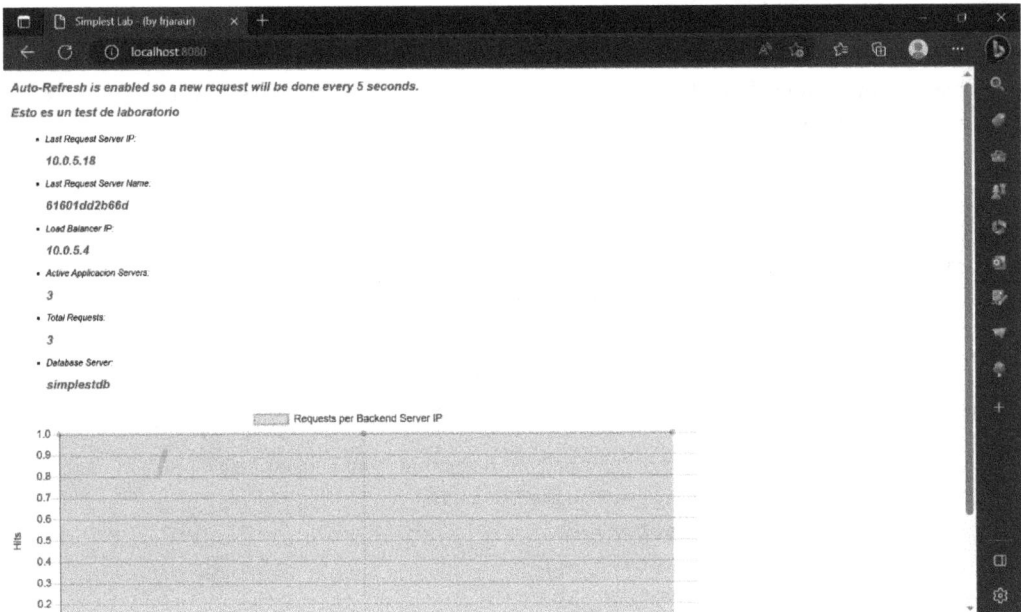

Figure 7.1 – Application is accessible at http://localhost:8080

Before removing the application using docker stack rm chapter7, it may be interesting for you to experiment with scaling up and down the app component and changing some content (you have the code, configurations, and secrets deployed). This will help you experiment with how rolling updates and rollbacks are managed by Docker Swarm.

This lab helped you understand how you can parametrize a Docker Compose file to deploy a complete application into a Docker Swarm cluster.

Summary

In this chapter, we covered the basic usage of Docker Swarm. We learned how to deploy a simple cluster and how to run our applications by taking advantage of Docker Swarm's features. We learned how to use Compose YAML files to deploy stacks and define an application completely using services and tasks to finally execute its containers. Docker Swarm manages complicated networking communication cluster-wide, helping us to publish our applications for users or other applications to access. It also provides mechanisms to ensure the availability of our applications even when we trigger component updates, such as a change to a container image.

In the next chapter, we will learn the basics of Kubernetes, the most popular and advanced container orchestrator currently available.

8

Deploying Applications with the Kubernetes Orchestrator

Developing containers for your applications on your workstation or laptop really improves your development process by running other applications' components while you focus on your own code. This simple standalone architecture works perfectly in your development stage, but it does not provide **high availability (HA)** for your applications. Deploying container orchestrators cluster-wide will help you to constantly keep your applications running healthy. In the previous chapter, we briefly reviewed Docker Swarm, which is simpler and can be a good introductory platform before moving on to more complex orchestrators. In this chapter, we will learn how to prepare and run our applications on top of **Kubernetes**, which is considered a standard nowadays for running containers cluster-wide.

In this chapter, we will cover the following topics:

- Introducing the main features of Kubernetes
- Understanding Kubernetes' HA
- Interacting with Kubernetes using `kubectl`
- Deploying a functional Kubernetes cluster
- Creating Pods and Services
- Deploying orchestrated resources
- Improving your applications' security with Kubernetes

Technical requirements

You can find the labs for this chapter at `https://github.com/PacktPublishing/Containers-for-Developers-Handbook/tree/main/Chapter8`, where you will find some extended explanations, omitted in the chapter's content to make it easier to follow. The *Code In Action* video for this chapter can be found at `https://packt.link/JdOIY`.

Now, let's start this chapter by learning about the main features of Kubernetes and why this orchestrator has become so popular.

Introducing the main features of Kubernetes

We can say without any doubt that Kubernetes is the new standard for deploying applications based on containers. However, its success didn't happen overnight; Kubernetes started in 2015 as a community project based on Google's own workload orchestrator, **Borg**. The first commit in Kubernetes' GitHub repository occurred in 2014, and a year later, the first release was published. Two years later, Kubernetes went mainstream thanks to its great community. I have to say that you will probably not use Kubernetes alone; you will deploy multiple components to achieve a fully functional platform, but this isn't a bad thing, as you can customize a Kubernetes platform to your specific needs. Also, Kubernetes by default has a lot of integrations with cloud platforms, as it was designed from the very beginning with them in mind. For example, cloud storage solutions can be used without additional components.

Let's take a moment to briefly compare Kubernetes' features with those of Docker Swarm.

Comparing Kubernetes and Docker Swarm

I have to declare that, personally, my first impressions of Kubernetes weren't good. For me, it provided a lot of features for many simple tasks that I was able to solve with Docker Swarm at that time. However, the more complex your applications are, the more features you require, and Docker Swarm eventually became too simple as it hasn't evolved too much. Docker Swarm works well with simple projects, but microservices architecture usually requires complex interactions and a lot of portability features. Kubernetes has a really steep learning curve, and it's continuously evolving, which means you need to follow the project almost every day. Kubernetes' core features are usually improved upon in each new release, and lots of pluggable features and side projects also continuously appear, which makes the platform's ecosystem grow daily.

We will see some differences between the Docker Swarm orchestration model and the Kubernetes model. We can start with the definition of the workloads within a cluster. We mentioned in *Chapter 7, Orchestrating with Swarm*, that Docker Swarm doesn't schedule containers; in fact, it schedules **Services**. In Kubernetes, we schedule **Pods**, which is the minimum scheduling unit in this orchestrator. A Pod can contain multiple Pods, although most of them will just run one. We will explore and learn more about Pods later in this chapter.

Exploring the control plane

Container orchestrators should provide a **control plane** for all management tasks, providing us with scheduling capabilities to execute our application workloads on a data plane and cluster-wide networking features. The Kubernetes control plane components are designed to manage every cluster component, schedule workloads, and review events that emerge in the platform. It also manages the node components, which really execute containers for us thanks to their container runtimes. Kubernetes follows the manager-worker model, as with Docker Swarm, in which two different node roles are defined. Manager nodes will manage the control plane components, while worker nodes will execute the tasks assigned by the control plane nodes.

Now, let's review some key processes.

Understanding the key processes

The following is a list of the key processes that run in the Kubernetes control plane:

- **kube-apiserver**: The API server is a component that interacts with all other components and the user. There isn't any direct communication between components; hence, kube-apiserver is essential in every Kubernetes cluster. All the cluster management is provided by exposing this component's API, and we can use different clients to interact with the cluster. Different endpoints allow us to retrieve and set Kubernetes resources.

- **etcd**: This is a component that provides data storage for all cluster components. It is a key-value store that can be consumed via its HTTP REST API. This reliable key-value store contains sensitive data, but, as mentioned before, only the kube-apiserver component can access it.

- **kube-scheduler**: This is in charge of allocating workloads to the container runtimes deployed in the nodes. To decide which nodes will run the different containers, kube-scheduler will ask kube-apiserver for the hardware resources and availability of all nodes included in the cluster.

- **kube-controller-manager**: Different controller processes run inside the Kubernetes cluster to maintain the status of the platform and the applications running inside. The kube-controller-manager is responsible for managing these controllers, and different tasks are delegated to each controller:

  - The **node controller** manages nodes' statuses

  - The **job controller** is responsible for managing workloads' tasks and creating Pods to run them

  - The **endpoint controller** creates endpoint resources to expose Pods

  - The **service account controller** and **token controller** manage accounts and API access token authorizations

- **cloud-controller-manager**: This is a separate component that manages different controllers that talk with underlying cloud providers' APIs:

  - The **node controller** manages the state and health of nodes deployed in your cloud provider.

  - The **route controller** creates routes in the cloud provider, using its specific API to access your deployed workloads.

  - The **service controller** manages cloud providers' load balancer resources. You will never deploy this component in your own local data center because it is designed for cloud integrations.

Next, let's review **node components**, which are in charge of executing and giving visibility to the workload processes.

Understanding node components

Node components run on worker nodes (as in Docker Swarm, manager nodes can also have the worker role). Let's take a closer look at them:

- **Container runtime**: The runtime for running containers is key, as it will execute all the workloads for us. The Kubernetes orchestrator schedules Pods on each host, and they run the containers for us.

- **kubelet**: We can consider kubelet as the Kubernetes integration agent. All nodes with the worker role have to run kubelet in order to communicate with the control plane. In fact, the control plane will manage communications to receive the health of each worker node and the status of their running workloads. kubelet will only manage containers deployed in the Kubernetes cluster; in other words, you can still execute containers in the workers' container runtime, but those containers will not be managed by Kubernetes.

- **kube-proxy**: This component is responsible for Kubernetes communications. It is important to mention here that Kubernetes does not really provide full networking capabilities by itself, and this component will only manage the integration of Kubernetes Service resources within a cluster. Additional communications components will be required to have a fully functional cluster. It is fair to say that kube-proxy works at the worker-node level, publishing the applications within the cluster, but more components will be needed in order to reach other Services deployed in other cluster nodes.

> **Important note**
>
> A **Service resource** (or simply **Service**) is designed to make your applications' Pods accessible. Different options are available to publish our applications either internally or externally for users. Service resources will get their own IP address to access the associated Pods' endpoints. We can consider Service resources as logical components. We will use Services to access our applications because Pods can die and be recreated, acquiring new IP addresses, but Services will remain visible with their specified IP address.

Worker nodes can be replaced when necessary; we can perform maintenance tasks whenever it's required, moving workloads from one node to another. However, control plane components can't be replaced. To achieve Kubernetes' HA, we need to execute more than one replica of control plane components. In the case of etcd, we must have an odd number of replicas, which means that at least three are required for HA. This requirement leads us to a minimum of three manager nodes (or master nodes, in Kubernetes nomenclature) to deploy a Kubernetes cluster with HA, although other components will provide HA with only two replicas. Conversely, the number of worker nodes may vary. This really depends on your application's HA, although a minimum of two workers is always recommended.

Networking in Kubernetes

It is important to remember that Kubernetes networking differs from the Docker Swarm model. By itself, Kubernetes does not provide cluster-wide communications, but a standardized interface, the **Container Network Interface** (**CNI**), is provided. Kubernetes defines a set of rules that any project integrating a communications interface must follow:

- **Container-to-container communications**: This communication works at the Pod level. Containers running inside a Pod share the same network namespace; hence, they receive the same IP address from the container runtime **IP Address Management** (**IPAM**). They can also use `localhost` to resolve their internal communications. This really simplifies communications when multiple containers need to work together.

- **Host-to-container communications**: Each host can communicate with Pods running locally using its container runtime.

- **Pod-to-Pod communications**: These communications will work locally but not cluster-wide, and Kubernetes imposes that communications must be provided without any **network address translation** (**NAT**). This is something the CNI must resolve.

- **Pod-to-Service interactions**: Pods will never consume other Pods as their IP address can change over time. We will use Services to expose Pods, and Kubernetes will manage their IP addresses, but the CNI must manage them cluster-wide.

- **Publishing Services**: Different approaches exist to publish our applications, but they are resolved by Service types and Ingress resources, and cluster-wide communications must be included in the CNI.

Because NAT isn't allowed, this model declares a flat network, where Pods can see each other when the CNI is included in the Kubernetes deployment. This is completely different from Docker Swarm, where applications or projects can run in isolated networks. In Kubernetes, we need to implement additional mechanisms to isolate our applications.

There are a lot of CNI plugins available to implement these cluster-wide communications. You can use any of them, but some are more popular than others; the following list shows recommended ones with some of their key features:

- **Flannel** is a simple overlay network provider that works very well out of the box. It creates a VXLAN between nodes to propagate the Pods' IP address cluster-wide, but it doesn't provide network policies. These are Kubernetes resources that can drop or allow Pods' connectivity.

- **Calico** is a network plugin that supports different network configurations, including non-overlay and overlay networks, with or without **Border Gateway Protocol** (**BGP**). This plugin provides network policies, and it's adequate for almost all small environments.

- **Canal** is used by default in SUSE's Rancher environments. It combines Flannel's simplicity and Calico's policy features.

- **Cilium** is a very interesting network plugin because it integrates **extended Berkeley Packet Filter (eBPF)** Linux kernel features in Kubernetes. This network provider is intended for multi-cluster environments or when you want to integrate network observability into your platform.

- **Multus** can be used to deploy multiple CNI plugins in your cluster.

- Cloud providers offer their own cloud-specific CNIs that allow us to implement different network scenarios and manage Pods' IP addresses within our own private cloud infrastructure.

The CNI plugin should always be deployed once the Kubernetes control plane has started because some components, such as the internal DNS or kube-apiserver, need to be reachable cluster-wide.

Namespace scope isolation

Kubernetes provides project or application isolation by using **namespaces**, which allow us to group resources. Kubernetes provides both cluster-scoped and namespace-scoped resources:

- **Cluster-scoped** resources are resources available cluster-wide, and we can consider most of them as cluster management resources, owned by the cluster administrators.

- **Namespace-scoped** resources are those confined at the namespace level. Services and Pods, for example, are defined at the namespace level, while node resources are available cluster-wide.

Namespace resources are key to isolating applications and restricting users from accessing resources. Kubernetes provides different authentication and authorization methods, although we can integrate and combine additional components such as an external **Lightweight Directory Access Protocol (LDAP)** or Microsoft Active Directory.

Internal resolution

The internal DNS is based on the **CoreDNS** project and provides autodiscovery of Services by default (additional configurations can also publish Pods' IP addresses). This means that every time a new Service resource is created, the DNS adds an entry with its IP address, making it accessible to all Pods cluster-wide with the following **Fully Qualified Domain Name (FQDN)** by default: `SERVICE_NAME.NAMESPACE.svc.cluster.local`.

Attaching data to containers

Kubernetes includes different resource types to attach storage to our workloads:

- **Volumes** are Kubernetes-provided resources that include multiple cloud providers' storage APIs, temporal storage (`emptyDir`), host storage, and **Network File System (NFS)**, among other remote storage solutions. Other very important volume-like resources are Secrets and ConfigMaps, which can be used to manage sensitive data and configurations cluster-wide, respectively.

- **Persistent volumes** are the preferred solution when you work on production in a local data center. Storage vendors provide their own drivers to integrate **network-attached storage** (**NAS**) and **storage area network** (**SAN**) solutions in our applications.

- **Projected volumes** are used to map several volumes inside a unique Pod container's directory.

Providing persistent storage to our applications is key in container orchestrators, and Kubernetes integrates very well with different dynamic provisioning solutions.

Publishing applications

Finally, we will introduce the concept of **Ingress** resources. These resources simplify and secure the publishing of applications running in Kubernetes by linking Service resources with specific applications' URLs. An Ingress controller is required to manage these resources, and we can integrate into this component many different options, such as NGINX, Traefik, or even more complex solutions, such as Istio. It is also remarkable that many network device vendors have also prepared their own integrations with Kubernetes platforms, improving performance and security.

Now that we have been quickly introduced to Kubernetes, we can take a deep dive into the platform's components and features.

Understanding Kubernetes' HA

Deploying our applications with HA requires a Kubernetes environment with HA. At least three replicas of etcd are required and two replicas of other control plane components. Some production architectures deploy etcd externally in dedicated hosts, while other components are deployed in additional master nodes. This isolates completely the key-value store from the rest of the control plane components, improving security, but it adds additional complexity to the environment. You will usually find three master nodes and enough worker nodes to deploy your production applications.

A Kubernetes installation configures and manages its own internal **certificate authority** (**CA**) and then deploys certificates for the different control plane and kubelet components. This ensures TLS communications between kube-apiserver and other components. The following architecture diagram shows the different Kubernetes components in a single-master node scenario:

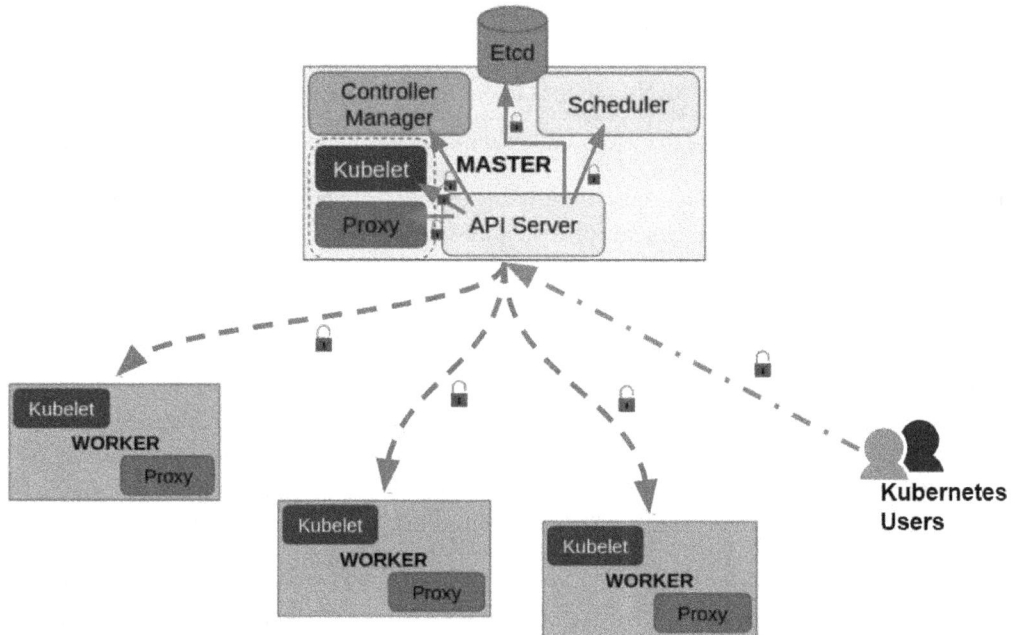

Figure 8.1 – Kubernetes cluster architecture with HA

Worker nodes are those designated to run workloads. Depending on the Kubernetes installation method, you will be able to run specific workloads on master nodes if they also run the kubelet and kube-proxy components. We can use different affinity and anti-affinity rules to identify which nodes should finally execute a container in your cluster.

However, simply replicating the control plane does not provide HA or resilience to your applications. You will need a CNI to manage communications between your containers cluster-wide. Internal load balancing will route requests to deployed Pods within your Kubernetes cluster.

Running your applications on different hosts requires appropriate storage solutions. Whenever a container starts with a container runtime, the required volumes should be attached. If you work on-premises, you will probably use a **Container Storage Interface** (**CSI**) in your infrastructure. However, as a developer, you should consider your storage requirements, and your infrastructure administrators will provide you with the best solution. Different providers will present filesystems, blocks, or object storage, and you can choose which best fits your application. All of them will work cluster-wide and help you provide HA.

Finally, you have to think about how your application's components work with multiple replicas. Your infrastructure provides resilience to your containers, but your application's logic must support replication.

Running a production cluster can be hard, but deploying our own cluster to learn how Kubernetes works is a task that I really recommend to anyone who wants to deploy applications on these container architectures. Using **kubeadm** is recommended to create Kubernetes clusters for the first time.

Kubeadm Kubernetes deployment

Kubeadm is a tool that can be used to easily deploy a fully functional Kubernetes cluster. In fact, we can even use it to deploy production-ready clusters.

We will initialize a cluster from the first deployment node, executing `kubeadm init`. This will create and trigger the bootstrapping processes to deploy the cluster. The node in which we execute this action will become the cluster leader, and we will join new master and worker nodes by simply executing `kubeadm join`. This really simplifies the deployment process; every step required for creating the cluster is automated. First, we will create the control plane components; hence, `kubeadm join` will be executed in the rest of the designated master nodes. And once the master nodes are installed, we will join the worker nodes.

Kubeadm is installed as a binary in your operating system. It is important to note here that the Kubernetes master role is only available on Linux operating systems. Therefore, we can't install a Kubernetes cluster only with Microsoft Windows or macOS nodes.

This tool does not only install a new cluster. It can be used to modify current kubeadm-deployed cluster configurations or upgrade them to a newer release. It is a very powerful tool, and it is good to know how to use it, but unfortunately, it is out of the scope of this book. Suffice it to say that there are many command-line arguments that will help us to fully customize Kubernetes deployment, such as the IP address to be used to manage communications within the control plane, the Pods' IP address range, and the authentication and authorization model to use. If you have time and hardware resources, it is recommended to create at least a two-node cluster with kubeadm to understand the deployment process and the components deployed by default on a Kubernetes cluster.

Here is the link for a Kubernetes deployment process with the kubeadm tool: `https://kubernetes.io/docs/setup/production-environment/tools/kubeadm`. We will not use kubeadm to deploy Kubernetes in this book. We will use Docker Desktop, Rancher Desktop, or Minikube tools, which provide fully automated deployments that work out of the box on our laptop or desktop computers.

Docker or any other container runtime just cares about containers. We learned in *Chapter 7, Orchestrating with Swarm*, how Docker provides the command line to manage Docker Swarm clusters, but this doesn't work with Kubernetes, as it is a completely different platform. The kube-apiserver component is the only component accessible to administrators and end users. The Kubernetes community project provides its own tool to manage Kubernetes clusters and the resources deployed on them. Thus, in the next subsection, we will learn the basics of `kubectl`, the tool that we will use in a lot of examples in this book to manage configurations, content, and workloads on clusters.

Interacting with Kubernetes using kubectl

In this section, we will learn the basics of the `kubectl` command line. It is the official Kubernetes client, and its features can be extended by adding plugins.

The installation process for this tool is quite simple, as it is a single binary written in the Go language; therefore, we can download it from the official Kubernetes binaries repository. To download from this repository, you must include the release to use in the URL. For example, `https://dl.k8s.io/release/v1.27.4/bin/linux/amd64/kubectl` will link you to the `kubectl` Linux binary for Kubernetes 1.27.4. You will be able to manage Kubernetes clusters using binaries from a different release, although it is recommended to maintain alignment with your client and Kubernetes server releases. How to install the tool for each platform is described in `https://kubernetes.io/docs/tasks/tools/#kubectl`. As we will use Microsoft Windows in the *Labs* section, we will use the following link to install the tool's binary: `https://kubernetes.io/docs/tasks/tools/install-kubectl-windows`.

Let's start with `kubectl` by learning the syntax for executing commands with this tool:

```
kubectl [command] [TYPE] [NAME] [flags]
```

`TYPE` indicates that `kubectl` can be used with many different Kubernetes resources. We can use the singular, plural, or abbreviated forms, and we will use them in case-insensitive format.

The first thing to know before learning some of the uses of `kubectl` is how to configure access to any cluster. The `kubectl` command uses, by default, a `config` configuration file in the home of each user, under the `.kube` directory. We can change which configuration file to use by adding the `--kubeconfig <FILE_PATH_AND_NAME>` argument or setting the `KUBECONFIG` variable with the configuration file location. By changing the content of the `kubeconfig` file, we can easily have different cluster configurations. However, this path change is not really needed because the configuration file structure allows different contexts. Each context is used to uniquely configure a set of user and server values, allowing us to configure a context with our authentication and a Kubernetes cluster endpoint.

> **Important note**
>
> You will usually access a Kubernetes cluster using an FQDN (or its resolved IP address). This name or its IP address will be load-balanced to all Kubernetes clusters' available instances of kube-apiserver; hence, a load balancer will be set in front of your cluster servers. In our local environments, we will use a simple IP address associated with our cluster.

Let's see what a configuration file looks like:

```
apiVersion: v1
kind: Config
clusters:
```

```
- cluster:
    certificate-authority-data: BASE64_CLUSTER_CA or the CA file path
    server: https://cluster1_URL:6443
  name: cluster1
contexts:
- context:
    cluster: cluster1
    user: user1
  name: user1@cluster1
current-context: user1@cluster1
users:
- name: user1
  user:
    client-certificate-data: BASE64_USER_CERTIFICATE or the cert file
path
    client-key-data: BASE64_USER_KEY or the key file path
```

We can add multiple servers and users and link them in multiple contexts. We can switch between defined contexts by using `kubectl config use-context CONTEXT_NAME`.

We can use `kubectl api-resources` to retrieve the type of resources available in the defined cluster. This is important because the `kubectl` command line retrieves data from a Kubernetes cluster; hence, its behavior changes depending on the endpoint. The following screenshot shows the API resources available in a sample Kubernetes cluster:

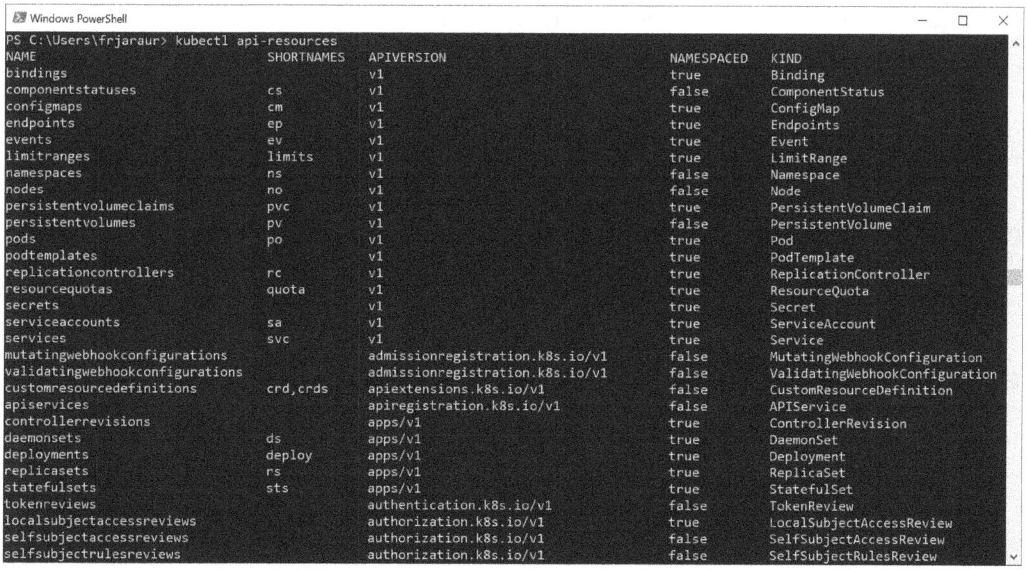
Windows PowerShell

PS C:\Users\frjaraur> kubectl api-resources

NAME	SHORTNAMES	APIVERSION	NAMESPACED	KIND
bindings		v1	true	Binding
componentstatuses	cs	v1	false	ComponentStatus
configmaps	cm	v1	true	ConfigMap
endpoints	ep	v1	true	Endpoints
events	ev	v1	true	Event
limitranges	limits	v1	true	LimitRange
namespaces	ns	v1	false	Namespace
nodes	no	v1	false	Node
persistentvolumeclaims	pvc	v1	true	PersistentVolumeClaim
persistentvolumes	pv	v1	false	PersistentVolume
pods	po	v1	true	Pod
podtemplates		v1	true	PodTemplate
replicationcontrollers	rc	v1	true	ReplicationController
resourcequotas	quota	v1	true	ResourceQuota
secrets		v1	true	Secret
serviceaccounts	sa	v1	true	ServiceAccount
services	svc	v1	true	Service
mutatingwebhookconfigurations		admissionregistration.k8s.io/v1	false	MutatingWebhookConfiguration
validatingwebhookconfigurations		admissionregistration.k8s.io/v1	false	ValidatingWebhookConfiguration
customresourcedefinitions	crd,crds	apiextensions.k8s.io/v1	false	CustomResourceDefinition
apiservices		apiregistration.k8s.io/v1	false	APIService
controllerrevisions		apps/v1	true	ControllerRevision
daemonsets	ds	apps/v1	true	DaemonSet
deployments	deploy	apps/v1	true	Deployment
replicasets	rs	apps/v1	true	ReplicaSet
statefulsets	sts	apps/v1	true	StatefulSet
tokenreviews		authentication.k8s.io/v1	false	TokenReview
localsubjectaccessreviews		authorization.k8s.io/v1	true	LocalSubjectAccessReview
selfsubjectaccessreviews		authorization.k8s.io/v1	false	SelfSubjectAccessReview
selfsubjectrulesreviews		authorization.k8s.io/v1	false	SelfSubjectRulesReview

Figure 8.2 – Kubernetes API resources in a sample cluster

As you can see, there is a column that indicates whether a Kubernetes resource is namespaced or not. This shows the scope where the resource must be defined. Resources can have a cluster scope, be defined and used at the cluster level, or be namespace-scoped, in which case they exist grouped inside a Kubernetes namespace. Kubernetes namespaces are resources that allow us to isolate and group resources within a cluster. The resources defined inside a namespace are unique within the namespace, as we will use the namespaces to identify them.

There are many commands available for kubectl, but we will focus in this section on just a few of them:

- create: This action allows us to create Kubernetes resources from a file or our terminal stdin.
- apply: This creates and updates resources in Kubernetes.
- delete: We can remove already created resources using kubectl delete.
- run: This action can be used to quickly deploy a simple workload and define a container image.
- get: We can retrieve any Kubernetes resource definition by using kubectl get. A valid authorization is required to either create or retrieve any Kubernetes object. We can also use kubectl describe, which gives a detailed description of the cluster resource retrieved.
- edit: We can modify some resources' properties in order to change them within the cluster. This will also modify our applications' behavior.

We can configure Kubernetes resources by using an **imperative** or **declarative** method:

- In an imperative configuration, we describe the configuration of the Kubernetes resource in the command line, using our terminal.
- By using a declarative configuration, we will create a file describing the configuration of a resource, and then we create or apply the content of the file to the Kubernetes cluster. This method is reproducible.

Now that we have a basic idea of Kubernetes components, the installation process, how to interact with a cluster, and the requirements to run a functional Kubernetes platform, let's see how to easily deploy our own environment.

Deploying a functional Kubernetes cluster

In this section, we will review different methods to deploy Kubernetes for different purposes. As a developer, you do not need to deploy a production environment, but it's important to understand the process and be able to create a minimal environment to test your applications. If you are really interested in the full process, it's recommended to take a look at Kelsey Hightower's GitHub repository, *Kubernetes the Hard Way* (https://github.com/kelseyhightower/kubernetes-the-hard-way). In this repository, you will find a step-by-step complete process to deploy manually a Kubernetes cluster. Understanding how a cluster is created really helps solve problems, although it's out of the scope of this book. Here, we will review automated Kubernetes solutions in which you can focus on your code and not on the platform itself. We will start this section with the most popular container desktop solution.

Docker Desktop

We have used Docker Desktop in this book to create and run containers using a **Windows Subsystem for Linux** (**WSL**) terminal. Docker Desktop also includes a one-node Kubernetes environment. Let's start using this by following these steps:

1. Click **Settings** | **Enable Kubernetes**. The following screenshot shows how Kubernetes can be set up in your Docker Desktop environment:

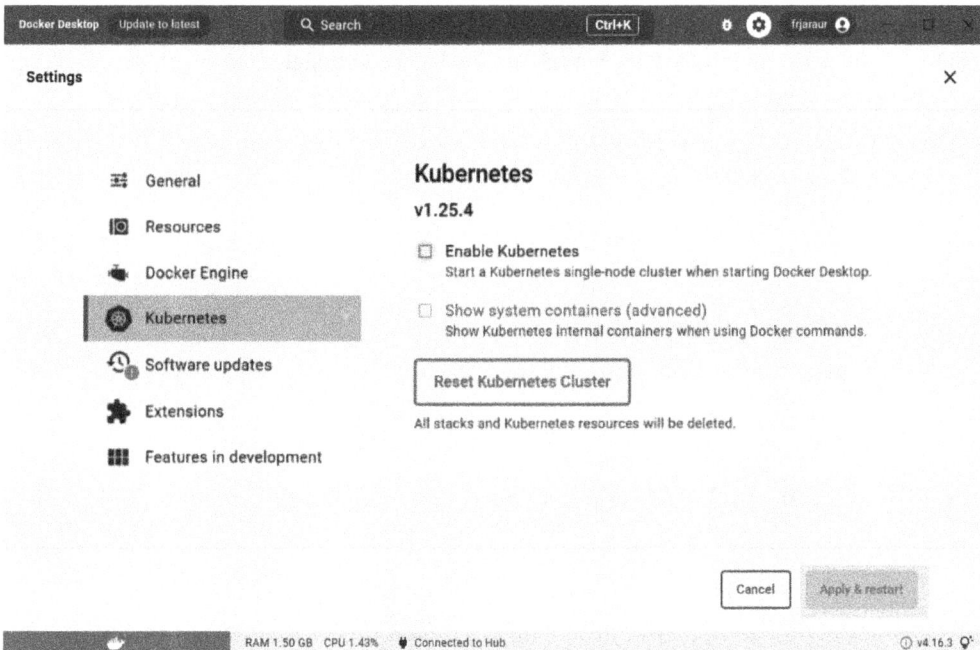

Figure 8.3 – The Docker Desktop Settings area where a Kubernetes cluster can be enabled

After the Kubernetes cluster is enabled, Docker Desktop starts the environment. The following screenshot shows the moment when Kubernetes starts:

Kubernetes

v1.25.4

☑ Enable Kubernetes
Starting ...   ━━━━━━━━━━━━━━━━━━━━━

☐ Show system containers (advanced)
Show Kubernetes internal containers when using Docker commands.

Reset Kubernetes Cluster

All stacks and Kubernetes resources will be deleted.

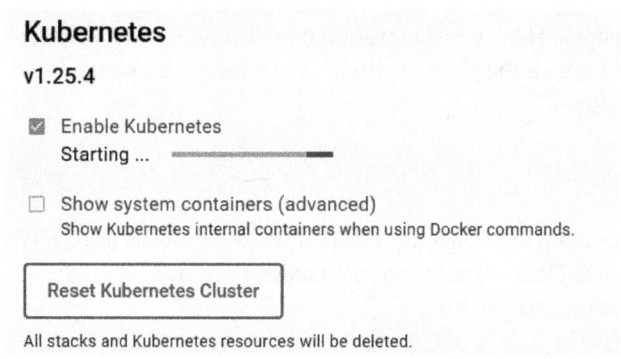

Figure 8.4 – Kubernetes starting in the Docker Desktop environment

2. Once started, we can access the cluster from our WSL terminal by using the `kubectl` command line. As you may have noticed, we haven't installed any additional software. Docker Desktop integrates the commands for us by attaching the required files to our WSL environment.

The status of the Kubernetes cluster is shown in the lower-left side of the Docker Desktop GUI, as we can see in the following screenshot:

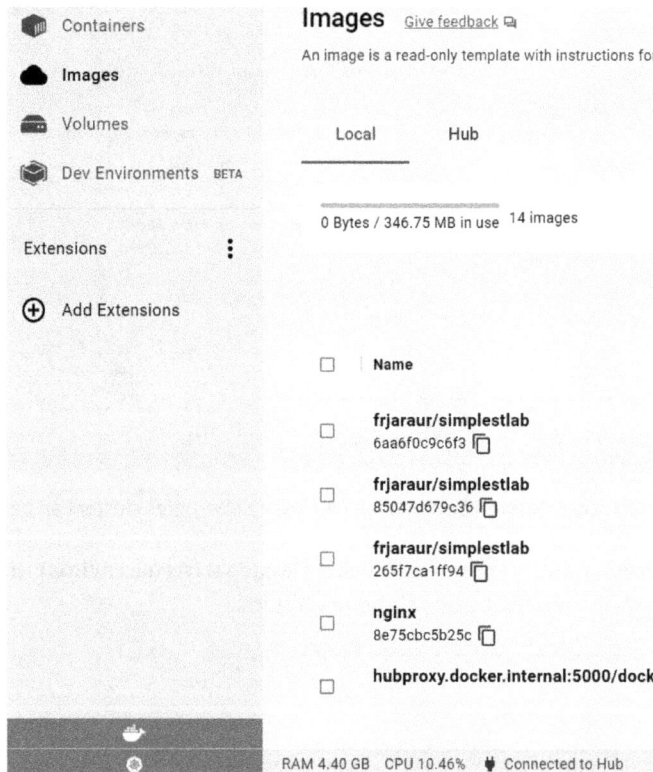

Figure 8.5 – Kubernetes' status shown in Docker Desktop

3. We can verify the number of nodes included in the deployed Kubernetes cluster by executing `kubectl get nodes`:

```
$ kubectl get nodes
NAME             STATUS    ROLES           AGE    VERSION
docker-desktop   Ready     control-plane   45m    v1.25.4
```

Notice that a Kubernetes configuration file was also added for this environment:

```
$ ls -lart ~/.kube/config
-rw-r--r-- 1 frjaraur frjaraur 5632 May 27 11:21 /home/
frjaraur/.kube/config
```

That was quite easy, and now we have a fully functional Kubernetes cluster. This cluster isn't configurable, but it is all you need to prepare your applications' deployments and test them. This solution doesn't allow us to decide which Kubernetes version to deploy, but we can reset the environment at any time from the Docker Desktop Kubernetes **Settings** page, which is very useful when we need to start the environment afresh. It can be deployed on either Microsoft Windows (using a **virtual machine** (**VM**) with Hyper-V or WSL, which is recommended as it consumes fewer resources), macOS (on Intel and Apple silicon architectures), or Linux (using a VM with **Kernel-Based Virtual Machine** (**KVM**)).

Docker prepares some images with all Kubernetes components and deploys them for us when we enable Kubernetes in Docker Desktop. By default, all containers created for such a purpose are hidden in the Docker Desktop GUI, but we can review them from the Docker command line. This Kubernetes solution is really suitable if you use the Docker command line to create your applications because everything necessary, from building to orchestrated execution, is provided. We can use different Kubernetes volume types because a **storage class** (dynamic storage provider integration) that binds the local Docker Desktop environment is provided. Therefore, we can prepare our applications as if they were to run in a production cluster. We will learn about `storageClass` resources in *Chapter 10, Leveraging Application Data Management in Kubernetes*. However, a few things are omitted in this Kubernetes deployment, which may impact your work, so it's good to understand its limitations:

* Environment internal IP addresses can't be changed.

* You cannot ping containers; this is due to the network settings of Docker Desktop, and it also affects the container runtime.

* No CNI is provided; hence, no network policies can be applied.

* No Ingress resource is provided by default. This is not really an issue because other desktop Kubernetes environments wouldn't provide it either, but you may need to deploy your own and modify your `/etc/hosts` file (or Microsoft Windows's equivalent `C:\Windows\system32\drivers\etc\hosts` file) to access your applications. We will learn about Ingress resources and controllers in *Chapter 11, Publishing Applications*.

These are the more important issues you will find by using Docker Desktop for Kubernetes deployment. It is important to understand that performance will be impacted when you enable Kubernetes, and you will need at least 4 GB of RAM free and four **virtual cores (vCores)**.

As specified in the official documentation, Docker Desktop is not an open source project, and it is licensed under the *Docker Subscription Service Agreement*. This means that it is free for small businesses (that is, fewer than 250 employees and less than $10 million in annual revenue), personal use, education, and non-commercial open source projects; otherwise, it requires a paid subscription for professional use.

We will now review another desktop solution to deploy a simplified Kubernetes environment.

Rancher Desktop

This solution comes from SUSE, and it really provides you with the Kubernetes Rancher deployment experience on your laptop or desktop computer. Rancher Desktop can be installed on Windows systems, using WSL or VMs, or macOS and Linux, using only VMs. It is an open source project and includes Moby components, `containerd`, and other components that leverage the experience of Rancher, allowing the development of new and different projects such as **RancherOS** (a container-oriented operating system) or **K3s** (a lightweight certified Kubernetes distribution). There are some interesting features on Rancher Desktop:

- We can choose which container runtime to use for the environment. This is a key difference and makes it important to test your applications using `containerd` directly.

- We can set the Kubernetes version to deploy.

- It is possible to define the resources used for the VM (on Mac and Linux).

- It provides the Rancher Dashboard, which combines perfectly with your infrastructure when your server's environments also run Kubernetes and Rancher.

The following screenshot shows how can we set up a Kubernetes release from the Rancher Desktop GUI, in the **Preferences** area. This way, we will be able to test our applications using different API releases, which may be very interesting before moving our applications to the staging or production stages. Each Kubernetes release provides its own set of API resources; you should read each release note to find out changes in the API versions and resources that may affect your project – for example, if some *beta* resources are now included in the release, or some are deprecated. The following screenshot shows the Kubernetes releases available for deploying in Rancher Desktop:

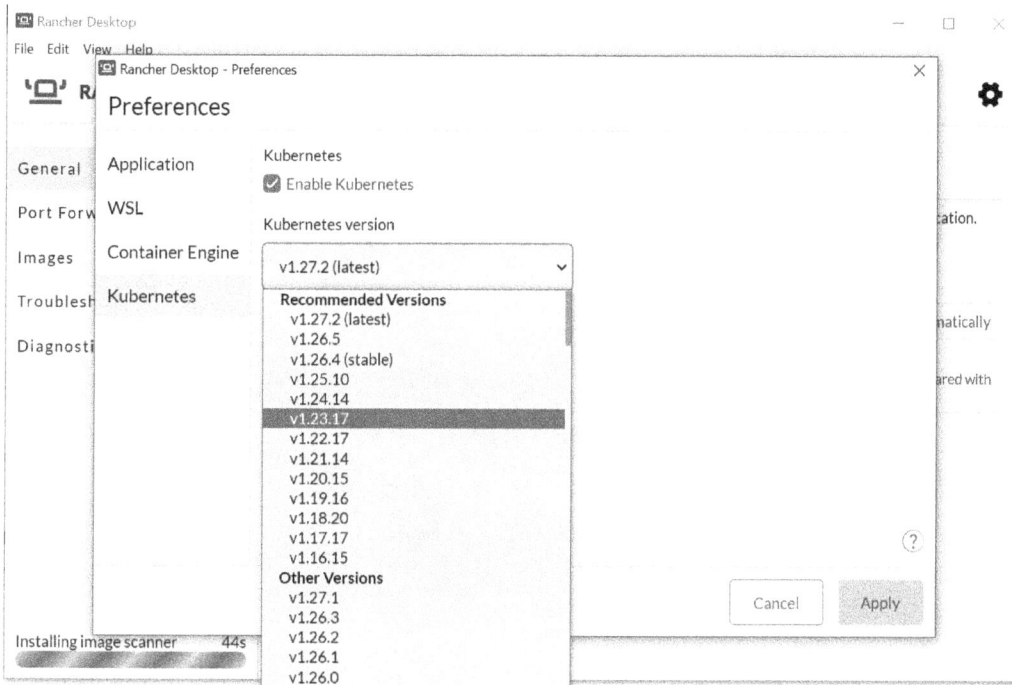

Figure 8.6 – Different Kubernetes releases can be chosen

As we have seen in Docker Desktop, Rancher Desktop also provides a simple button to completely reset the Kubernetes cluster. The following screenshot shows the **Troubleshooting** area, where we can reset the cluster:

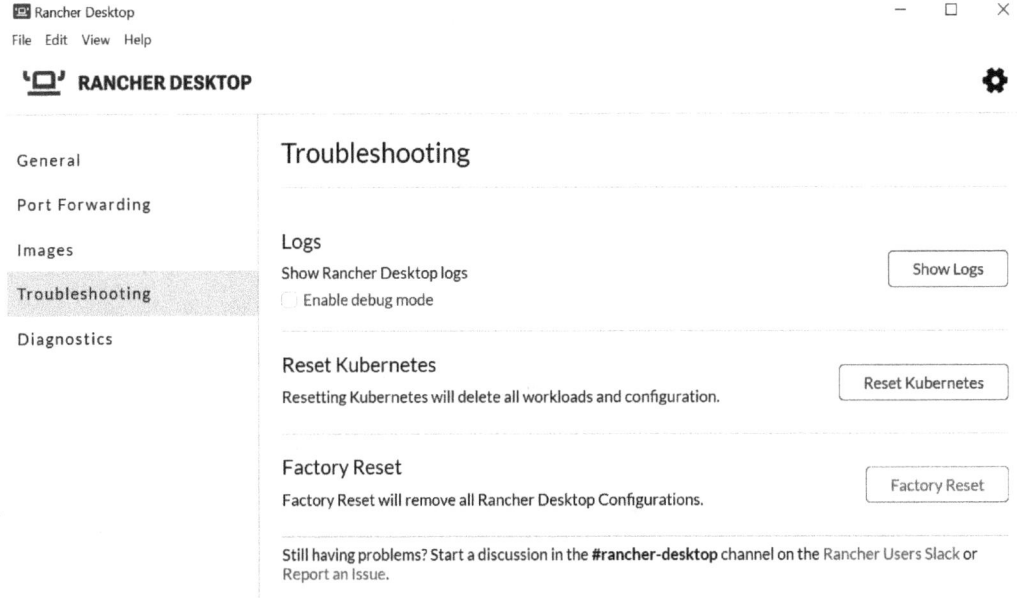

Figure 8.7 – The Troubleshooting area from the Rancher Desktop GUI

Rancher Desktop also deploys its own Ingress controller based on Traefik. This controller will help us to publish our applications, as we will learn in *Chapter 11, Publishing Applications*. We can remove this component and deploy our own Ingress controller by unselecting the **Traefik** option in the **Kubernetes Preferences** section, but it is quite interesting to have one by default.

The Rancher Dashboard is accessible by clicking on the Rancher Desktop notification icon and selecting **Open cluster dashboard**. Rancher Dashboard provides access to many Kubernetes resources graphically, which can be very useful for beginners. The following screenshot shows the Rancher Dashboard main page, where you can review and modify different deployed Kubernetes resources:

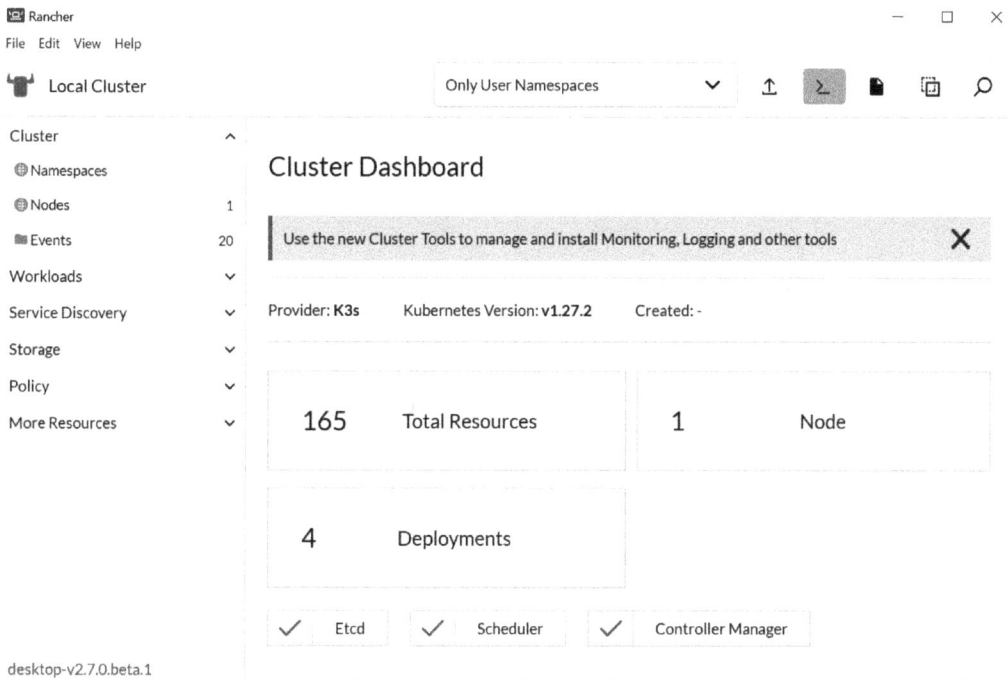

Figure 8.8 – The Rancher Dashboard main page

We can verify the Kubernetes environment from a WSL terminal by checking its version:

```
$ kubectl version --short
Flag --short has been deprecated, and will be removed in the future.
The --short output will become the default.
Client Version: v1.27.2
Kustomize Version: v5.0.1
Server Version: v1.27.2+k3s1
```

> **Important note**
>
> We can build images for the Kubernetes environment without using an external registry by using `nerdctl` with the `--namespace k8s.io` argument. This way, images will be available directly for our deployments.

It is interesting that this Kubernetes implementation is aligned with the expected network features from Kubernetes; hence, we can ping Pods and Services, or even access the Service ports from the WSL environment. It also makes our applications accessible by adding a `.localhost` suffix to our host definitions (we will deep dive into this option in *Chapter 11, Publishing Applications*). However, this cluster is still a standalone node, and we can't test the behavior of our applications under certain failures or movements between nodes. If you really need to test these features, we need to go further and deploy additional nodes with other solutions.

Both Docker Desktop and Rancher Desktop provide GUI-based Kubernetes deployments, but usually, if you don't need any GUI, we can even deploy more lightweight solutions.

We will now review Minikube, which may be the most complete and pluggable solution.

Minikube

The Minikube Kubernetes environment is very configurable and consumes considerably fewer hardware resources than other solutions, allowing us to deploy more than one node per cluster, or even multiple clusters on one single computer host. We can create a Kubernetes cluster by using either Docker, QEMU, Hyperkit, Hyper-V, KVM, Parallels, Podman, VirtualBox, or VMware Fusion/Workstation. We can use many different virtualization solutions or even container runtimes, and Minikube can be deployed on Microsoft Windows, macOS, or Linux operating systems.

These are some of the features of Minikube:

- It supports different Kubernetes releases
- Different container runtimes can be used
- A direct API endpoint improves image management
- Advanced Kubernetes customization such as the addition of **feature gates**
- It is a pluggable solution, so we can include add-ons such as Ingress for extended Kubernetes features
- It supports integration with common CI environments

Kubernetes deployment is easy, and you just require a single binary for Linux systems. Different arguments can be used to set up the environment. Let's review some of the most important:

- `start`: This action creates and starts a Kubernetes cluster. We can use the argument `--nodes` to define the number of nodes to deploy and `--driver` to specify which method to use to create a cluster. Virtual hardware resources can also be defined by using `--cpu` and `--memory`; by default, 2 CPUs and 2 GB of memory will be used. We can even choose a specific CNI to deploy with the `--cni` argument (`auto`, `bridge`, `calico`, `cilium`, `flannel`, and `kindnet` are available, but we can add our own path to a CNI manifest).
- `status`: This action shows the status of the Minikube cluster.
- `stop`: This stops a running Minikube cluster.
- `delete`: This action deletes a previously created cluster.
- `dashboard`: An open source Kubernetes Dashboard can be deployed as an add-on, which can be accessed by using `minikube dashboard`.

- `service`: This option can be very interesting to expose a deployed application Service. It returns the Service URL that can be used to access it.

- `mount`: We can mount host directories into the Minikube nodes with this option.

- `ssh`: We can access Kubernetes deployed hosts by using `minikube ssh <NODE>`.

- `node`: This action allows us to manage cluster nodes.

- `kubectl`: This runs a `kubectl` binary matching the cluster version.

- `addons`: One of the best features of Minikube is that we can extend its functionality with plugins to manage additional storage options for the cluster (for example, to define a specific a `csi-hostpath-driver`, or specify the default storage class to use, `default-storageclass`, or a dynamic `storage-provisioner`, among other options), Ingress controllers (`ingress`, `ingress-dns`, `istio`, and `kong`), and security (`pod-security-policy`). We can even deploy the Kubernetes Dashboard or the metrics server automatically, which recover metrics from all running workloads.

To create a cluster with two nodes (a master and a worker) we can simply execute `minikube start --nodes 2`. Let's see this in action:

```
PS > minikube start --nodes 2 `
--kubernetes-version=stable `
--driver=hyperv
* minikube v1.30.1 on Microsoft Windows 10 Pro 10.0.19045.2965 Build
19045.2965
* Using the hyperv driver based on user configuration
* Starting control plane node minikube in cluster minikube
...
* Done! kubectl is now configured to use "minikube" cluster and
"default" namespace by default
```

Once deployed, we can review the cluster state using `kubectl`:

```
PS > kubectl get nodes
NAME            STATUS     ROLES           AGE    VERSION
minikube        Ready      control-plane   25m    v1.26.3
minikube-m02    Ready      <none>          22m    v1.26.3
```

Minikube is a very configurable solution that provides common Kubernetes features. In my opinion, it is the best in terms of performance and features.

> **Important note**
>
> Minikube Kubernetes deployments in Microsoft Windows require administrator privileges if you use Hyper-V. Therefore, you need to open PowerShell or Command Prompt as administrator, but this may not be enough. PowerShell Hyper-V must also be included, and we will need to execute `Enable-WindowsOptionalFeature -Online -FeatureName Microsoft-Hyper-V-Tools-All -All` on a PowerShell console to enable it.

Alternative Kubernetes desktop deployments

Nowadays, there are other interesting options that use even fewer hardware resources, but they don't provide as many features as Minikube. Let's discuss some good candidates to try:

- **kind**: This solution takes advantage of a Docker runtime installation to deploy Kubernetes using containers and their own custom images. It is based on kubeadm deployment, and it really works very nicely on Linux desktop systems in which you don't usually install Docker Desktop to run containers.

- **K3s**: This Kubernetes deployment is the basis for the Rancher Desktop Kubernetes feature. It deploys a lightweight environment using less memory with customized binaries. This may impact your application's deployment if you use bleeding-edge features, as they will probably not be available. This solution comes from Rancher-SUSE, which also provides K3D to deploy Kubernetes using containers.

- **Podman Desktop**: This new solution appeared in mid-2022 and provides an open source GUI-based environment to run Kubernetes with Podman and `containerd`. It is in its early stages, but it currently seems a good solution if you don't use any Docker tool in your application development.

We can now continue and deep dive into the concepts of Pods and Services.

Creating Pods and Services

In this section, we will learn about the resources we will use in the Kubernetes orchestration platform to deploy our applications. We will start by learning how containers are implemented inside Pods.

Pods

A **Pod** is the smallest scheduling unit in Kubernetes. We can include in a Pod multiple containers, and they will always share a uniquely defined IP address within the cluster. All containers inside a Pod will resolve `localhost`; hence, we can only use ports once. Volumes associated with a Pod are also shared between containers. We can consider a Pod as a small VM in which different processes (containers) run together. Pods are considered in a healthy state when all their containers run correctly.

As Pods can contain many containers, we can think of using a Pod to deploy applications with multiple components. All containers associated with a Pod run on the same cluster host. This way, we can ensure that all application components run together, and their intercommunications will definitely be faster. Because Pods are the smallest unit in Kubernetes, we are only able to scale up or down complete Pods; hence, all the containers included will also be executed multiple times, which probably isn't what we need. Not all applications' components should follow the same scaling rules; hence, it is better to deploy multiple Pods for an application.

The following diagram shows a Pod. Two containers are included, and thus, they share the IP address and volumes of the same Pod:

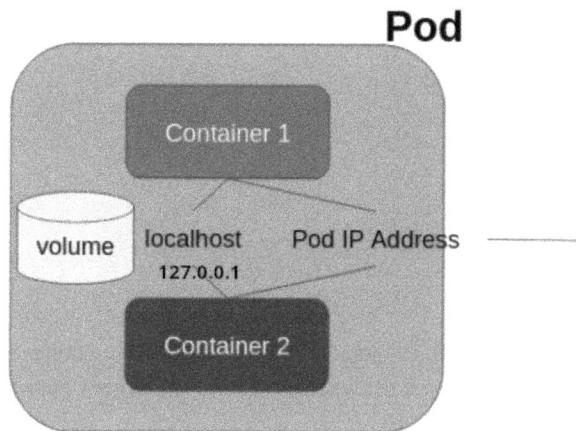

Figure 8.9 – A schema of a Pod with two containers included

We can run Pods that share the host's resources by using the host's namespaces (the host's network, **inter-process communications** or **IPCs**, processes, and so on). We should limit this type of Pod because they have direct access to a host's processes, interfaces, and so on.

The following example shows a declarative file to execute an example Pod:

```
apiVersion: v1
kind: Pod
metadata:
    name: examplepod
  labels:
    example: singlecontainer
  spec:
      containers:
      - name: examplecontainer
        image: nginx:alpine
```

We can use JSON or YAML files to define Kubernetes resources, but YAML files are more popular; hence, you have to take care of indentation when preparing your deployment files. To deploy this Pod on our Kubernetes cluster, we will simply execute `kubectl create -f <PATH_TO_THE_FILE>`.

> **Important note**
>
> When you access a Kubernetes cluster, a namespace is associated with your profile or context. By default, the namespace associated is `default`; therefore, if we don't specify any Kubernetes namespace as an argument to create a resource, the `default` namespace will be used. We can change the namespace for the current context to be applied to all commands hereafter, by executing `kubectl config use-context --current --namespace <NEW_NAMESPACE>`. The namespace for each resource can be included under the `metadata` key in the resource's YAML manifest:
>
> ```
> . . .
> metadata:
>     name: podname
>     namespace: my-namespace
> . . .
> ```

We can modify any aspect used for the container by modifying the container image's behavior, as we learned with the Docker command line in *Chapter 4, Running Docker Containers*.

If you are not sure of the available keys or are learning how to use a new Kubernetes resource, you can use `kubectl explain <RESOURCE>` to retrieve an accurate description of the keys available and the expected values:

```
PS > kubectl explain pod
KIND:      Pod
VERSION:   v1
DESCRIPTION:
      Pod is a collection of containers that can run on a host. This
resource is
...
 FIELDS:
   apiVersion   <string>
      APIVersion defines the versioned schema of this representation of
an
...
   spec <Object>
      ...
   status       <Object>
      ...
```

We can continue adding keys to obtain more specific definitions – for example, we can retrieve the keys under `pod.spec.containers.resources`:

```
PS > kubectl explain pod.spec.containers.resources
KIND:       Pod
...
RESOURCE: resources <Object>
  ...
DESCRIPTION:
    Compute Resources required by this container. Cannot be updated.
More info:
    https://kubernetes.io/docs/concepts/configuration/manage-
resources-containers/
  ...
FIELDS:
   claims          <[]Object>
   Claims lists the names of resources, defined in spec.
resourceClaims, that
   are used by this container.
  ...
    limits          <map[string]string>
    Limits describes the maximum amount of compute resources allowed.
More
  ...
    requests        <map[string]string>
    Requests describes the minimum amount of compute resources
required. If
  ...
```

Each description shows extended information with links to the Kubernetes documentation.

> **Important note**
> We can retrieve all available keys at once for a specific resource by using `kubectl explain pod --recursive`. This option really helps us to fully customize the resources.

We can test a real Pod deployment using an NGINX web server. To do this, follow these steps:

1. We will use the imperative mode with `kubectl run`:

    ```
    PS > kubectl run webserver --image=docker.io/nginx:alpine `
    --port 80
    pod/webserver created
    ```

2. We can now list the Pods to verify whether the webserver is running:

```
PS > kubectl get pods
NAME          READY   STATUS             RESTARTS   AGE
webserver     0/1     ContainerCreating  0          5s
```

3. As we can see, it is starting. After a few seconds, we can verify that our web server is running:

```
PS > kubectl get pods -o wide
NAME          READY          STATUS    RESTARTS   AGE        IP
NODE          NOMINATED NODE READINESS GATES
webserver     1/1            Running   0          2m40s      10.244.0.3
minikube      <none>                   <none>
```

4. We use `-o wide` to modify the command's output. As you can see, we now have the associated IP address.

5. We can get into the `minikube` node and verify the Pod's connectivity. The following screenshot shows the interaction with the cluster node:

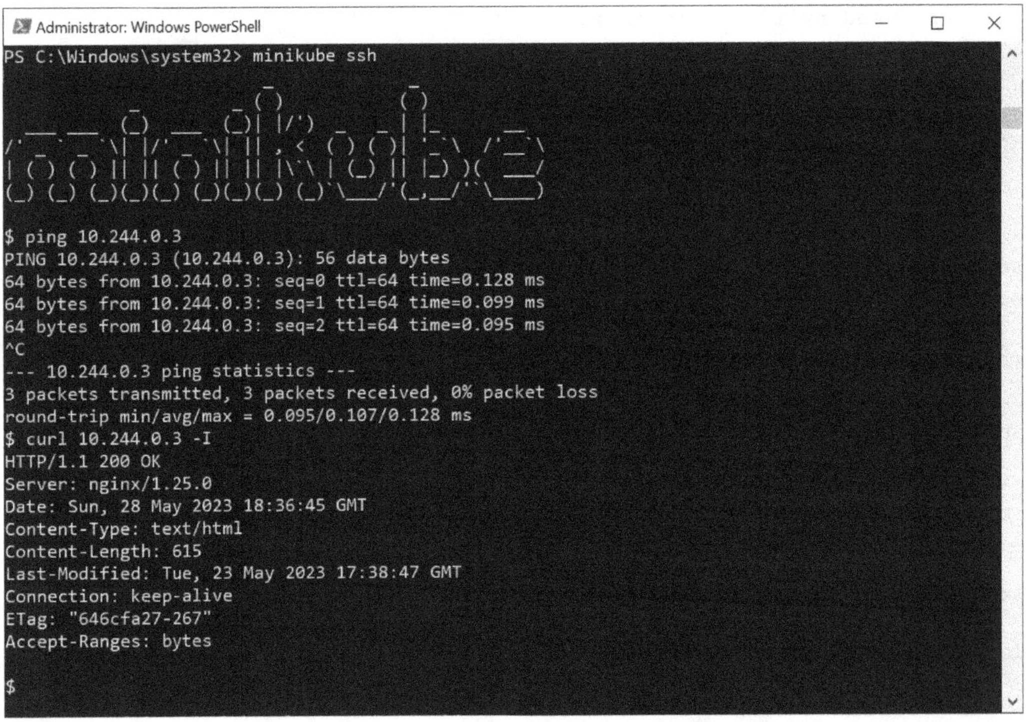

Figure 8.10 – Testing connectivity from the minikube node

6. If we now delete the Pod and create a new one, we can easily see that that new Pods can receive a new IP address, and thus, our application may need to change the IP addresses continuously:

```
PS > kubectl delete pod webserver
pod "webserver" deleted
PS > kubectl run webserver `
--image=docker.io/nginx:alpine --port 80
pod/webserver created
PS > kubectl get pods -o wide
NAME          READY    STATUS      RESTARTS     AGE    IP
NODE          NOMINATED NODE    READINESS GATES
webserver     1/1      Running     0            4s     10.244.0.4
minikube      <none>            <none>
```

This is a real problem, and that's why we never use Pod IP addresses. Let's talk now about Services and how they can help us to solve this problem.

Services

Services are abstract objects in Kubernetes; they are used to expose a set of Pods and, thus, they serve an application component. They will get an IP address from the internal Kubernetes IPAM system, and we will use this to access the associated Pods. We can also associate Services with external resources to make them accessible to users. Kubernetes offers different types of Services to be published either internally or outside a cluster. Let's quickly review the different Service types:

- `ClusterIP`: This is the default Service type. Kubernetes associates an IP address from the defined Service's IP address range, and containers will be able to access this Service by either its IP address or its name. Containers running in the same namespace will be able to simply use the Service's name, while other containers will need to use its Kubernetes FQDN (`SERVICE_NAME.SERVICE_NAMESPACE.svc.cluster.local`, where `cluster.local` is the FQDN of the Kubernetes cluster itself). This is due to Kubernetes' internal **service discovery** (**SD**), which creates DNS entries for all Services cluster-wide.

- `Headless`: These Services are a variant of the `ClusterIP` type. They don't receive an IP address, and the Service's name will resolve all the associated Pods' IP addresses. We commonly use headless Services to interact with non-Kubernetes SD solutions.

- `NodePort`: When we use a `NodePort` Service, we associate a set of hosts' ports with the Service `clusterIP`'s defined IP address. This makes the Service accessible from outside a cluster. We can connect from a client computer to any of the cluster hosts in the defined port, and Kubernetes will route requests to the Service, associated with the `ClusterIP` address via internal DNS, no matter which node received the request. Thus, the Pods associated with the Service receive network traffic from the client.

- `LoadBalancer`: The `LoadBalancer` Service type is used to publish a defined Service in an external load balancer. It uses the external load balancer's API to define the required rules to reach the cluster, and indeed, this model uses a `NodePort` Service type to reach the Service cluster-wide. This Service type is mainly used to publish Services in cloud providers' Kubernetes clusters, although some vendors also provide this feature on-premises.

We have seen how Services can be published outside the Kubernetes cluster to be consumed by other external applications or even our users, but we can also do the opposite. We can include external Services, available in our real network, inside our Kubernetes clusters by using the `External` Service type.

The following schema represents a `NodePort` Service in which we publish port `7000`, attached to port `5000`, and exposed in the containers in this example:

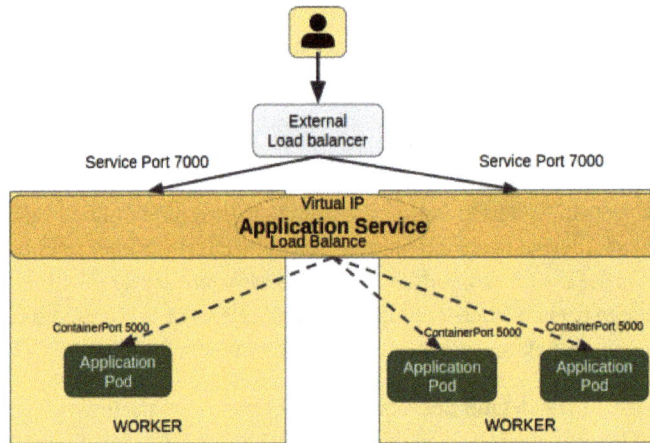

Figure 8.11 – NodePort Service schema

In this example, the external requests from users are load-balanced to port `7000`, listening on all cluster hosts. All traffic from the users will be internally load-balanced to port `5000`, making it available on all Services' assigned Pods.

The following example shows the manifest of a Kubernetes Service, obtained by using the **imperative method** to retrieve the output only:

```
PS >   kubectl  expose  pod  webserver  -o yaml
--port 80  --dry-run=client
apiVersion: v1
kind: Service
metadata:
  creationTimestamp: null
  labels:
    run: webserver
```

```
      name: webserver
  spec:
    ports:
    - port: 80
      protocol: TCP
      targetPort: 80
    selector:
      run: webserver
  status:
    loadBalancer: {}
```

In this example, the Service resource isn't created because we added the `-o yaml` argument to show the output in the YAML format and `-dry-run=client`. This option shows the output of the creation command executed against kube-apiserver.

Let's move on now to learn how to deploy workloads in a cluster because Pods don't provide resilience; they run as unmanaged standalone workloads without a controller.

Deploying orchestrated resources

Now that we know how to deploy Pods using imperative and declarative modes, we will define new resources that can manage the Pods' life cycle. Pods executed directly with the `kubectl` command line are not recreated if their containers die. To control the workloads within a Kubernetes cluster, we will need to deploy additional resources, managed by Kubernetes controllers. These controllers are control loops that monitor the state of different Kubernetes resources and make or request changes when needed. Each controller tracks some resources and tries to maintain their defined state. Kubernetes' kube-controller-manager manages these controllers that maintain the overall desired state of different cluster resources. Each controller can be accessed via an API, and we will use `kubectl` to interact with them.

In this section, we will learn the basics of Kubernetes controllers and dive deep into how to use them in *Chapter 9, Implementing Architecture Patterns*.

ReplicaSets

The most simple resource that allows us to maintain a defined state for our application is a ReplicaSet. It will keep a set of replica Pods running. To create a ReplicaSet, we will use a Pod manifest as a template to create multiple replicas. Let's see a quick example:

```
apiVersion: apps/v1
kind: ReplicaSet
metadata:
  name: replicated-webserver
spec:
```

```
replicas: 3
selector:
  matchLabels:
    application: webserver
template:
  metadata:
      application: webserver
  spec:
    containers:
    - name: webserver-container
      image: docker.io/nginx:alpine
```

This ReplicaSet will run three Pods with the same `docker.io/nginx:alpine` image; the `template` section defines the specifications for these three Pod resources, with one container each. The ReplicaSet identifies Pods to manage by using the defined `application` label and its `webserver` value, defined in the Pod's `template` section of the manifest.

When we deploy this ReplicaSet, the cluster creates these three Pods, and whenever any of them dies, the controller manages this change and triggers the creation of a new one.

We will continue to review more resources, but keep in mind that the basic idea of a `template` section embedded inside a more general definition applies to all of them.

Deployments

We may think of a Deployment as an evolution of a ReplicaSet. It allows us to update them because a deployment manages a set of replicas but only runs one. Every time we create a new deployment, we create an associated ReplicaSet. And when we update this deployment, a new ReplicaSet is created with a new definition, reflecting the changes from the previous resource.

DaemonSets

With a DaemonSet, we can ensure that all cluster nodes get one replica of our workload, but we cannot define the number of replicas in a DaemonSet.

StatefulSets

A StatefulSet allows more advanced features in our workloads. It allows us to manage the order and uniqueness of Pods, ensuring that each replica gets its own unique set of resources, such as volumes. Although Pods are created from the same template section, a StatefulSet maintains a different identity for each Pod.

Jobs

A Job creates one or more Pods at a time, but it will continue creating them until a defined number of them terminate successfully. When a Pod exits, the controller verifies whether the number of required completions was reached, and if not, it creates a new one.

CronJobs

We can schedule Pods by using CronJobs, as they schedule jobs. When the execution time comes, a Job is created and triggers the creation of defined Pods. As we will learn in *Chapter 9*, *Implementing Architecture Patterns*, CronJobs manifests include two `template` sections – one to create jobs and another one to define how Pods will be created.

ReplicationControllers

We can consider ReplicationControllers a previous version of the current ReplicaSet resource types. They work similarly to how we keep a number of Pod replicas alive, but they differ in how they group the monitored Pods because ReplicationControllers do not support set-based selectors. This selector method allows ReplicaSets to acquire the state management of Pods created outside of their own manifest; hence, if a Pod already running matches the ReplicaSet label's selection, it will be automatically included in the pool of replicated Pods.

Now that we have an overview of different resources that allow us to create orchestrated resources cluster-wide, we can learn some of the Kubernetes features that can improve our applications' security.

Improving your applications' security with Kubernetes

Applications running in containers offer many new different features. We can run multiple applications' releases at a time in a host; they start and stop in seconds. We can scale components easily, and different applications can coexist without even interaction between them. An application's resilience is also inherited from the container runtime features (exited containers autostarting).

However, we can also improve our applications by running them in Kubernetes. Each Kubernetes cluster is composed of multiple container runtimes running together and in coordination. Container runtimes isolate the hosts' resources thanks to kernel namespaces and **control groups** (**cgroups**), but Kubernetes adds some interesting features:

- **Namespaces**: Namespaces are Kubernetes resources that group other resources and are designed to distribute Kubernetes resources between multiple users, grouped in teams or projects.

- **Authentication strategies**: Requests from Kubernetes clients may use different authentication mechanisms, such as client certificates, bearer tokens, or an authenticating proxy to authenticate them.

- **Authorization requests**: Users request the Kubernetes API after passing authentication, authorization, and different admission controllers. The authorization phase involves granting permission to access Kubernetes' resources and features. Requests' attributes are evaluated against policies, and they are allowed or denied. The user, group, API, request path, namespace, verb, and so on that are provided in the requests are used for these validations.

- **Admission controllers**: Once the authentication and authorization have been passed, other mechanisms are in place. Validation and mutation admission controllers decide whether or not requests are accepted or even modify them to accomplish certain defined rules. For example, we can prevent the execution of any Pod or container executed as `root`, or simply change the final resulting user to a non-privileged user in order to maintain cluster security.

- **Role-based access control (RBAC)**: Several authorization models are available in Kubernetes (a node, **attribute-based access control** (**ABAC**), RBAC, and Webhooks). RBAC is used to control a resource's permissions cluster-wide by assigning roles to Kubernetes users. We can have internal users (with cluster scope), such as **ServiceAccounts**, and external users, such as those who execute tasks and workloads in the cluster using their clients. These roles and their users' binding can be defined at the namespace level and cluster-wide. Also, we can use `kubectl` to check whether some verbs are available for us or even for another user, by using impersonation:

```
$ kubectl auth can-i create pods –namespace dev
yes
$ kubectl auth can-i create svc -n prod -as dev-user
no
```

- **Secrets and ConfigMaps**: We already learned how to deploy certain configurations in orchestrated environments in *Chapter 7, Orchestrating with Swarm*. In Kubernetes, we also have Secrets and ConfigMaps resources, but they can be retrieved by users if they are allowed (RBAC). It is important to understand that Secrets are packaged in the Base64 format; hence, sensitive data can be accessed if we don't prepare appropriate roles. The kubelet Kubernetes component will mount Secrets and ConfigMaps automatically for you, and we can use them as files or environment variables in our application deployments. Kubernetes can encrypt Secrets at rest to ensure that operating systems administrators can't retrieve them from the etcd database files, but this capability is disabled by default.

- **Security contexts**: Security contexts can be defined either at the Pod or container level and allow us to specify the security features to be applied to our workloads. They are key in maintaining security in Kubernetes. We can specify whether our application requires some special capabilities, whether the containers run in read-only mode (the root filesystem keeps immutable), or the ID used for the main process execution, among many other features. We will dive deep into how they can help us protect our applications in *Chapter 9, Implementing Architecture Patterns*. It is important to know that we can use admission controllers to enforce defined `securityContext` profiles, and this is essential because we can ensure that a Pod doesn't run as root, privileged, or use non-read-only containers on any defined namespace.

- **Network policies**: Kubernetes deploys a flat network by default; hence, all containers can reach each other. To avoid such a situation, Kubernetes also provides NetworkPolicies and GlobalNetworkPolicies (applied at the cluster level). Not all CNIs are able to implement this feature. Kubernetes only provides **custom resource** (**CR**) types, which will be used to implement the **network policies**. Verify that your network provider can implement them to be able to use this feature (lots of popular CNI plugins such as Calico, Canal, and Cilium are completely capable). It is a good recommendation to implement some default global policies to drop all external accesses and allow the required communications for each application at the namespace level. Network policies define both Ingress and Egress rules. These rules work at the connectivity level; hence, we don't have raw packet logging (although some CNI plugins provide some logging features). We will learn how to implement rules following best practices in *Chapter 11*, *Publishing Applications*.

Now that we have an overview of the most important features available in Kubernetes that help us to protect both our applications and the entire cluster, we can continue by creating some simple labs that will cover the basic usage of a Kubernetes environment.

Labs

Now, we will have a short lab section that will help us to learn and understand the basics of deploying a local Kubernetes environment with Minikube, testing some of its resource types to validate the cluster.

The code for the labs is available in this book's GitHub repository at `https://github.com/PacktPublishing/Containers-for-Developers-Handbook.git`. Ensure you have the latest revision available by simply executing `git clone https://github.com/PacktPublishing/Containers-for-Developers-Handbook.git` to download all its content or `git pull` if you already downloaded the repository before. All commands and content used in these labs will be located inside the `Containers-for-Developers-Handbook/Chapter8` directory.

We will start by deploying a Minikube cluster with two nodes (one master and one worker). We will deploy them with 3 GB of RAM each, which will be more than enough to test application behavior when some of the cluster node dies, but you will probably not need two nodes for your daily usage.

Deploying a Minikube cluster with two nodes

In this lab, we will deploy a fully functional Kubernetes cluster locally, for testing purposes. We will continue working on a Windows 10 laptop with 16 GB of RAM, which is enough for the labs in this book. Follow these steps:

1. Install Minikube. First, download it from `https://minikube.sigs.k8s.io/docs/start/`, choose the appropriate installation method, and follow the simple installation steps for your operating system. We will use Hyper-V; hence, it must be enabled and running on your desktop computer or laptop.

2. Once Minikube is installed, we will open an administrator PowerShell terminal. Minikube deployments using Hyper-V require execution with administrator privileges. This is due to the Hyper-V layer; hence, admin privileges won't be required if you use VirtualBox as your hypervisor or Linux as your operating system (other hypervisors can be used, such as KVM, which works very nicely with Minikube). Admin rights are also required to remove the Minikube cluster. Once the PowerShell terminal is ready, we execute the following command:

```
minikube start -nodes 2 -memory 3G -cpus 2 `
--kubernetes-version=stable -driver=hyperv `
--cni=calico `
--addons=ingress,metrics-server,csi-hostpath-driver
```

Notice the ` character for breaking lines in Powershell. The preceding code snippet will create a cluster with two nodes (each with 3 GB of memory) with the current stable Kubernetes release (1.26.3 at the time of writing), using Calico as the CNI and the hosts' storage for volumes. It also delivers automatically an Ingress controller for us.

Let's create a two-node cluster using `minikube start`:

```
PS C:\Windows\system32> cd c:\
PS C:\>  minikube start -nodes 2 -memory 3G `
--cpus 2 -kubernetes-version=stable `
--driver=hyperv -cni=calico `
--addons=ingress,metrics-server,csi-hostpath-driver
* minikube v1.30.1 on Microsoft Windows 10 Pro 10.0.19045.2965
Build 19045.2965
* Using the hyperv driver based on user configuration
* Starting control plane node minikube in cluster minikube
* Creating hyperv VM (CPUs=2, Memory=3072MB, Disk=20000MB) …
* Preparing Kubernetes v1.26.3 on Docker 20.10.23 ...
  - Generating certificates and keys ...
...
  - Using image registry.k8s.io/metrics-server/metrics-
server:v0.6.3
* Verifying ingress addon...
...
* Starting worker node minikube-m02 in cluster minikube
...
* Done! kubectl is now configured to u"e "minik"be" cluster a"d
"defa"lt" namespace by default
PS C:\>
```

3. We can now verify the cluster status by executing `kubectl get nodes`:

```
PS C:\> kubectl get nodes
NAME            STATUS    ROLES           AGE    VERSION
minikube        Ready     control-plane   23m    v1.26.3
minikube-m02    Ready     <none>          18m    v1.26.3
```

As you can see, the `minikube-m02` node does not show any role. This is due to the fact that everything in Kubernetes is managed by labels. Remember that we saw how selectors are used to identify which Pods belong to a specific ReplicaSet.

4. We can review the node labels and create a new one for the worker node. This will show us how we can modify the resource's behavior by using labels:

```
PS C:\> kubectl get nod- --show-labels
NAME            STATUS    ROLES           AGE    VERSION   LABELS
minikube        Ready     control-plane   27m    v1.26.3   beta.
kubernetes.io/arch=amd64,beta.kubernetes.io/
os=linux,kubernetes.io/arch=amd64,kubernetes.io/
hostname=minikube,kubernetes.io/os=linux,minikube.k8s.io/
commit=08896fd1dc362c097c925146c4a0d0dac715ace0,minikube.
k8s.io/name=minikube,minikube.k8s.io/primary=true,minikube.
k8s.io/updated_at=2023_05_31T10_59_55_0700,minikube.k8s.io/
version=v1.30.1,node-role.kubernetes.io/control-plane=,node.
kubernetes.io/exclude-from-external-load-balancers=,topology.
hostpath.csi/node=minikube

minikube-m02    Ready     <none>          22m    v1.26.3   beta.
kubernetes.io/arch=amd64,beta.kubernetes.io/os=linux,kubernetes.
io/arch=amd64,kubernetes.io/hostname=minikube-m02,kubernetes.io/
os=linux,topology.hostpath.csi/node=minikube-m02
```

Lots of labels are assigned to both nodes, but here, we missed one defining the role of the worker node, `node-role.kubernetes.io/worker`. This label is not required, and that's why it is not included by default, but it is good to include it to identify nodes in the node cluster review (`kubectl get nodes`).

5. We now add a new label to the worker node by using a `kubectl` label, `<RESOURCE>` `<LABEL_TO_ADD>`:

```
PS C:\> kubectl label node minikube-m02        `
node-role.kubernetes.io/worker=
node/minikube-m02 labeled
PS C:\> kubectl get nodes
NAME            STATUS    ROLES           AGE    VERSION
minikube        Ready     control-plane   33m    v1.26.3
minikube-m02    Ready     worker          28m    v1.26.3
```

We can use the `kubectl` label to add any label to any resource. In the specific case of `node-role.kubernetes.io`, it is used by Kubernetes to show the ROLES column, but we can use any other label to identify a set of nodes. This will help you in a production cluster to run your applications in nodes that best fit your purposes – for example, by selecting nodes with fast solid-state disks, or those with special hardware devices. You may need to ask your Kubernetes administrators about these nodes' special characteristics.

We will now show you how you can use the deployed Kubernetes cluster.

Interacting with the Minikube deployed cluster

In this lab, we will interact with the current cluster, reviewing and creating some new resources:

1. We will start by listing all Pods deployed in the cluster using `kubectl get pods --A`. This will list all Pods in the cluster. We are able to list them after the Minikube installation because we connect as administrators. The following screenshot shows the output of `kubectl get pods -A`, followed by a list of the current namespaces, using `kubectl get namespaces`:

```
Administrator: Windows PowerShell                                              —   □   ×
PS C:\> kubectl get pods -A
NAMESPACE        NAME                                           READY   STATUS      RESTARTS   AGE
ingress-nginx    ingress-nginx-admission-create-k1vd9           0/1     Completed   0          45m
ingress-nginx    ingress-nginx-admission-patch-nvvmw            0/1     Completed   3          45m
ingress-nginx    ingress-nginx-controller-6cc5ccb977-vtdvh      1/1     Running     0          45m
kube-system      calico-kube-controllers-7bdbfc669-htwvp        1/1     Running     0          45m
kube-system      calico-node-7w6jk                              1/1     Running     0          45m
kube-system      calico-node-z59z5                              1/1     Running     0          41m
kube-system      coredns-787d4945fb-hc2db                       1/1     Running     0          45m
kube-system      csi-hostpath-attacher-0                        1/1     Running     0          45m
kube-system      csi-hostpath-resizer-0                         1/1     Running     0          45m
kube-system      csi-hostpathplugin-6sgqk                       6/6     Running     0          40m
kube-system      csi-hostpathplugin-kfnhc                       6/6     Running     0          45m
kube-system      etcd-minikube                                  1/1     Running     0          45m
kube-system      kube-apiserver-minikube                        1/1     Running     0          45m
kube-system      kube-controller-manager-minikube               1/1     Running     0          45m
kube-system      kube-proxy-cgk68                               1/1     Running     0          41m
kube-system      kube-proxy-j64j5                               1/1     Running     0          45m
kube-system      kube-scheduler-minikube                        1/1     Running     0          45m
kube-system      metrics-server-6588d95b98-16p9g                1/1     Running     0          45m
kube-system      storage-provisioner                            1/1     Running     0          45m
PS C:\> kubectl get namespace
NAME              STATUS   AGE
default           Active   46m
ingress-nginx     Active   46m
kube-node-lease   Active   46m
kube-public       Active   46m
kube-system       Active   46m
PS C:\>
```

Figure 8.12 – The output of the kubectl get pods –A and kubectl get namespace commands

2. Let's create a new namespace, `chapter8`, by using `kubectl create ns chapter8`:

```
PS C:\> kubectl create ns chapter8
namespace/chapter8 created
```

3. We can list the Pods in the `chapter8` namespace by adding `--namespace` or just `--n` to `kubectl get pods`:

```
PS C:\> kubectl get pods -n chapter8
No resources found in chapter8 namespace.
```

4. We can try now out the `ingress-nginx` namespace. We will list all the resources deployed in this namespace using `kubectl get all`, as we can see in the following screenshot:

```
Administrator: Windows PowerShell                                                                    —    □    ×
PS C:\> kubectl get all -n ingress-nginx
NAME                                            READY   STATUS      RESTARTS   AGE
pod/ingress-nginx-admission-create-klvd9        0/1     Completed   0          83m
pod/ingress-nginx-admission-patch-nvvmw         0/1     Completed   3          83m
pod/ingress-nginx-controller-6cc5ccb977-vtdvh   1/1     Running     0          83m

NAME                                       TYPE        CLUSTER-IP      EXTERNAL-IP   PORT(S)                      AGE
service/ingress-nginx-controller           NodePort    10.105.176.61   <none>        80:32584/TCP,443:30998/TCP   83m
service/ingress-nginx-controller-admission ClusterIP   10.103.53.10    <none>        443/TCP                      83m

NAME                                       READY   UP-TO-DATE   AVAILABLE   AGE
deployment.apps/ingress-nginx-controller   1/1     1            1           83m

NAME                                                  DESIRED   CURRENT   READY   AGE
replicaset.apps/ingress-nginx-controller-6cc5ccb977   1         1         1       83m

NAME                                         COMPLETIONS   DURATION   AGE
job.batch/ingress-nginx-admission-create     1/1           93s        83m
job.batch/ingress-nginx-admission-patch      1/1           103s       83m
PS C:\>
```

Figure 8.13 – The output of kubectl get all –n ingress-nginx

Now, we know how we can filter resources associated with a specific namespace.

5. Let's now create a simple Pod in the `chapter8` namespace by using the imperative format. We will execute `kubectl run webserver --image=nginx:alpine` to run a Pod with one container using the `docker.io/nginx:alpine` image:

```
PS C:\> kubectl run webserver --image=nginx:alpine `
-n chapter8
pod/webserver created
PS C:\> kubectl get pods -n chapter8
NAME        READY   STATUS    RESTARTS   AGE
webserver   1/1     Running   0          11s
```

Note that we used an external registry to store the images. Kubernetes does not use local stores to pull images. As we expect in any other container orchestrator environment, a registry is needed, and the hosts' kubelet component will pull images from this. We will never synchronize images manually between nodes in a cluster; we will use container image registries to store and pull the required container images.

6. Let's review the resource manifest now. It is important to understand that the `kubectl` command talks with the Kubernetes API using the credentials from our local `kubeconfig` file (this file is located in your home directory in the `.kube` directory; you can use `$env:USERPROFILE\.kube` in Microsoft Windows), and kube-apiserver gets this information from etcd before it is presented in our terminal. The following screenshot shows part of the output we get by using `kubectl get pod <PODNAME> -o yaml`. The `-o yaml` modifier shows the output

from a current resource in the YAML format. This really helps us to understand how objects are created and managed by Kubernetes:

```
Administrator: Windows PowerShell                                              −   □   ×
PS C:\> kubectl get pods -n chapter8 -o yaml webserver
apiVersion: v1
kind: Pod
metadata:
  annotations:
    cni.projectcalico.org/containerID: 3221683d0d94bb235bf5518c134c26a60b2e5e6b474286b7389900319e7a15cd
    cni.projectcalico.org/podIP: 10.244.205.194/32
    cni.projectcalico.org/podIPs: 10.244.205.194/32
  creationTimestamp: "2023-05-31T10:33:07Z"
  labels:
    run: webserver
  name: webserver
  namespace: chapter8
  resourceVersion: "7837"
  uid: d62f036c-20e6-4295-b0e5-43e72d2b5f92
spec:
  containers:
  - image: nginx:alpine
    imagePullPolicy: IfNotPresent
    name: webserver
    resources: {}
    terminationMessagePath: /dev/termination-log
    terminationMessagePolicy: File
    volumeMounts:
    - mountPath: /var/run/secrets/kubernetes.io/serviceaccount
      name: kube-api-access-mrl4s
      readOnly: true
  dnsPolicy: ClusterFirst
  enableServiceLinks: true
  nodeName: minikube-m02
  preemptionPolicy: PreemptLowerPriority
  priority: 0
  restartPolicy: Always
  schedulerName: default-scheduler
```

Figure 8.14 – The output of kubectl get pods --namespace chapter8 -o yaml webserver

7. Let's see which node executes our Pod by using either `kubectl get pods -o wide`, which shows extended information, or by filtering the `hostIP` key from the YAML output:

```
PS C:\> kubectl get pods -n chapter8 -o wide
NAME            READY    STATUS       RESTARTS    AGE    IP              NODE
NOMINATED NODE    READINESS GATES
webserver    1/1        Running    0                  27m    10.244.205.194
minikube-m02
<none>                    <none>
```

This can also be done by using the JSON path template (`https://kubernetes.io/docs/reference/kubectl/jsonpath/`):

```
PS C:\> kubectl get pods -n chapter8 `
-o jsonpath='{ .status.hostIP }' webserver
172.19.146.184
```

> **Important note**
>
> You can see that the node name is also available in the spec section (spec.nodeName), but this section is where Pod specifications are presented. We will learn in the next chapter how we can change the workload behavior by changing the specifications from the online manifests, directly in Kubernetes. The status section is read-only because it shows the actual state of the resource, while some of the sections in either the metadata or spec sections can be modified – for example, by adding new labels or annotations.

Before ending the labs from this chapter, we will expose the deployed Pod by adding a NodePort Service, which will guide our requests to the running web server Pod.

Exposing a Pod with a NodePort Service

In this quick lab, we will use the imperative model to deploy a NodePort Service to expose the already deployed web server Pod:

1. Because we haven't defined the container port in the webserver Pod, Kubernetes will not know which port must be associated with the Service; hence, we need to pass the --target-port 80 argument to specify that the Service should link the NGINX container port that is listening. We will use port 8080 for the Service, and we will let Kubernetes choose one NodePort port for us:

   ```
   PS C:\> kubectl expose pod webserver -n chapter8 `
   --target-port 80 --port 8080 --type=NodePort
   service/webserver exposed
   ```

2. We can now review the resources in the chapter8 namespace:

   ```
   PS C:\> kubectl get all -n chapter8
   NAME                 READY   STATUS    RESTARTS   AGE
   pod/webserver        1/1     Running   0          50m

   NAME                 TYPE        CLUSTER-IP     EXTERNAL-IP
   PORT(S)              AGE
   service/webserver    NodePort    10.103.82.252  <none>
   8080:32317/TCP       72s
   ```

   Note that the hosts' port 32317 is associated with the Service's port, 8080, which is associated with the webserver Pod's port, 80 (the NGINX container listens on that port).

3. We can now access the published NodePort port on any host, even if it does not run any Service-related Pod. We can use the IP address of any of the Minikube cluster nodes or use minikube service -n chapter8 webserver to automatically open our default web browser in the associated URL.

The following screenshot shows the output in both cases. First, we obtained the host's IP addresses by using `kubectl get nodes -o wide`. We used PowerShell's `Invoke-WebRequest` command to access a combination of IP addresses of the nodes and the `NodePort`-published port. Then, we used Minikube's built-in DNS to resolve the Service's URL by using the `minikube` Service:

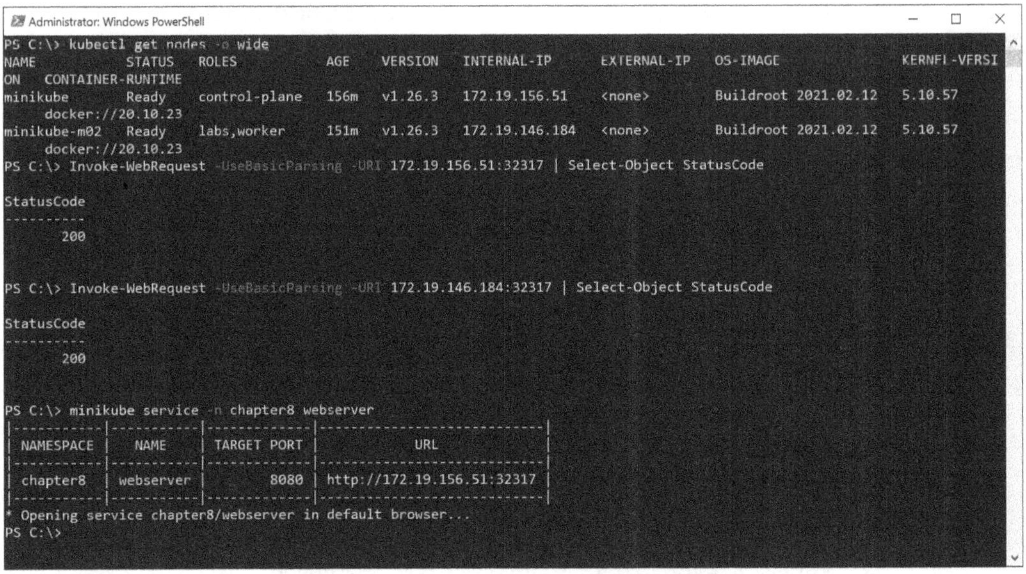

Figure 8.15 – The output of kubectl get nodes, different tests using the
cluster nodes, and the minikube Service URL resolution

As you can see, we used the IP addresses of both the master and worker nodes for the tests, and they worked, even though the Pod only ran on the worker node. This output also shows how easy it is to test Services by using Minikube's integrated Services resolution. It automatically opened our default web browser, and we can access our Service directly, as we can see in the following screenshot:

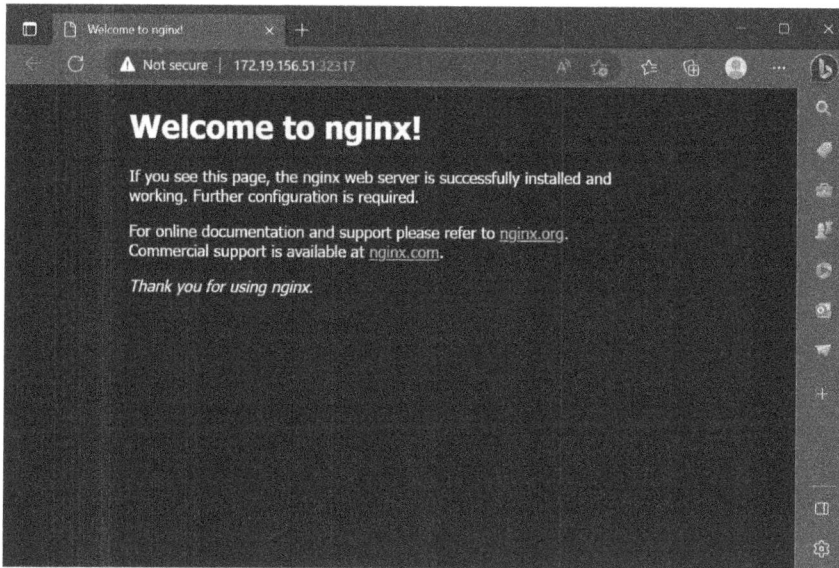

Figure 8.16 – The default web browser accessing the webserver Service's NodePort port

4. We can now remove all the resources created in this chapter. It is important to first remove the Pod, the Service, and then the namespace. Removing the namespace first triggers the removal of all associated resources in cascade, and there may be some issues if Kubernetes isn't able to remove some resources. It will never happen in this simple lab, but it is a good practice to remove resources inside a namespace before deleting the namespace itself:

```
PS C:\> kubectl delete service/webserver pod/webserver -n
chapter8
service "webserver" deleted
pod "webserver" deleted
PS C:\> kubectl delete ns chapter8
namespace "chapter8" deleted
```

You are now ready to learn more advanced Kubernetes topics in the next chapter.

Summary

In this chapter, we explored Kubernetes, the most popular and extended container orchestrator. We had an overall architecture review, describing each component and how we can implement an environment with HA, and we learned the basics of some of the most important Kubernetes resources. To be able to prepare our applications to run in Kubernetes clusters, we learned some applications that will help us to implement fully functional Kubernetes environments on our desktop computers or laptops.

In the next chapter, we will deep dive into the resource types we will use to deploy our applications, reviewing interesting use cases and examples and learning different architecture patterns to apply to our applications' components.

Part 3: Application Deployment

This part will describe how applications run in production, and we will use different models and Kubernetes features to help us deliver reliable applications securely.

This part has the following chapters:

- *Chapter 9, Implementing Architecture Patterns*
- *Chapter 10, Leveraging Application Data Management in Kubernetes*
- *Chapter 11, Publishing Applications*
- *Chapter 12, Gaining Application Insights*

9

Implementing Architecture Patterns

Kubernetes is the most popular container orchestrator in production. This platform provides different resources that allow us to deploy our applications with high resilience and with their components distributed cluster-wide while the platform itself runs with high availability. In this chapter, we will learn how these resources can provide different application architecture patterns, along with use cases and best practices to implement them. We will also review different options for managing application data and learn how to manage the health of our applications to make them respond to possible health and performance issues in the most effective way. At the end of this chapter, we will review how Kubernetes provides security patterns that improve application security. This chapter will give you a good overview of which Kubernetes resources will fit your applications' requirements most accurately. The following topics will be covered in this chapter:

- Applying Kubernetes resources to common application patterns
- Understanding advanced Pod application patterns
- Verifying application health
- Resource management and scalability
- Improving application security with Pods

Technical requirements

You can find the labs for this chapter at `https://github.com/PacktPublishing/ Containers-for-Developers-Handbook/tree/main/Chapter9`, where you will find some extended explanations that have been omitted in this chapter's content to make it easier to follow. The *Code In Action* video for this chapter can be found at `https://packt.link/JdOIY`.

We will start this chapter by reviewing some of the resource types that were presented in *Chapter 8, Deploying Applications with the Kubernetes Orchestrator*, and present some common use cases.

Applying Kubernetes resources to common application patterns

The Kubernetes container orchestrator is based on resources that are managed by different controllers. By default, our applications can use one of the following to run our processes as containers:

- Pods
- ReplicaSets
- ReplicaControllers
- Deployments
- StatefulSets
- DaemonSets
- Jobs
- CronJobs

This list shows the standard or default resources that are allowed on every Kubernetes installation, but we can create custom resources to implement any non-standard or more specific application behavior. In this section, we are going to learn about these standard resources so that we can decide which one will fit our application's needs.

Pods

Pods are the minimal unit for deploying workloads on a Kubernetes cluster. A Pod can contain multiple containers, and we have different mechanisms for doing this.

The following applies to all the containers running inside a Pod by default:

- They all share the network namespace, so they all refer to the same localhost and run with the same IP address.
- They are all scheduled together in the same host.
- They all share the namespaces that can be defined at the container level. We will define resource limits (using cgroups) for each container, although we can also define resource limits at the Pod level. Pod resource limits will be applied for all the containers.
- Volumes attached to a Pod are available to all containers running inside.

So, we can see that Pods are groups of containers running together that share kernel namespaces and compute resources.

You will find a lot of Pods running a single container, and this is completely fine. The distribution of containers inside different Pods depends on your application's components distribution. You need to ask yourself whether or not some processes must run together. For example, you can put together two containers that need to communicate fast, or you may need to keep track of files created by one of them without sharing a remote data volume. But it is very important to understand that all the containers running in a Pod scale and replicate together. This means that multiple replicas of a Pod will execute the same number of replicas of their containers, and thus your application's behavior can be impacted because the same type of process will run multiple times, and it will also access your files at the same time. This, for example, will break the data in your database or may lead to data inconsistency. Therefore, you need to decide wisely whether or not your application should distribute your application's containers into different Pods or run them together.

Kubernetes will keep track of the status of each container running inside a Pod by executing the Pod's defined probes. Each container should have its own probes (different types of probes exist, as we will learn in the *Verifying application health* section). But at this point, we have to understand that the health of all the containers inside a Pod controls the overall behavior of the Pod. If one of the containers dies, the entire Pod is set as unhealthy, triggering the defined Kubernetes event and resource's behavior. Thus, we can execute multiple Service containers in parallel or prepare our application by executing some pre-processes that can, for example, populate some minimal filesystem resources, binaries, permissions, and so on. These types of containers, which run before the actual application processes, are called **initial containers**. The **main containers** (those running in parallel as part of the Pod) will start after the complete execution of all the initial containers and only if all of them end correctly. The kubelet will run **initial containers** (or `initContainers`, which is the key that's used to define them) sequentially before any other container; therefore, if any of these initial containers fail, the Pod will not run as expected.

> **Important note**
>
> The Kubernetes 1.25 release introduced **ephemeral containers**, which is a new and different concept. While Pods are intended to run applications' processes, ephemeral containers will be created to provide debugging facilities to the actual Pod containers. They run by using the `kubectl debug` action followed by the name of the Pod to which your terminal should attach (shared kernel namespaces). We will provide a quick example of this in the *Labs* section.

Let's review how we write the manifest required for creating a Pod:

```yaml
apiVersion: v1
kind: Pod
metadata:
    name: webserver
spec:
    terminationGracePeriodSeconds: 10
    containers:
    - name: web
      image: nginx:alpine
      ports:
      - containerPort: 80
```

Figure 9.1 – Pod manifest

All the manifests for Kubernetes have at least the `apiVersion`, `kind`, and `metadata` keys, and these are used to define which API version will be used to reach the associated API server path, which type of resource we are defining, and information that uniquely describes the resource within the Kubernetes cluster, respectively. We can access all the resource manifest information via the Kubernetes API by using the JSON or YAML keys hierarchy; for example, to retrieve a Pod's name, we can use `.metadata.name` to access its key. The properties of the resource should usually be written in the `spec` or `data` section. Kubernetes roles, role bindings (in cluster and namespaces scopes), Service accounts, and other resources do not include either `data` or `spec` keys for declaring their functionality. And we can even create custom resources, with custom definitions for declaring their properties. In the default workload resources, we will always use the `spec` section to define the behavior of our resource.

Notice that in the previous code snippet, the `containers` key is an array. This allows us to define multiple containers, as we already mentioned, and the same happens with initial containers; we will define a list of containers in both cases and we will need at least the image that the container runtime must use and the name for the container.

> **Important note**
>
> We can use `kubectl explain pod.spec.containers --recursive` to retrieve all the existent keys under the `spec` section for a defined resource. The `explain` action allows you to retrieve all the keys for each resource directly from the Kubernetes cluster; this is important as it doesn't depend on your `kubectl` binary version. The output of this action also shows which keys can be changed at runtime, once the resource has been created in the cluster.

It is important to mention here that Pods by themselves don't have cluster-wide auto-healing. This means that when you run a Pod and it is considered unhealthy for whatever reason (any of the containers is considered unhealthy) in a host within the cluster, it will not execute on another host. Pods include the `restartPolicy` property to manage the behavior of Pods when they die. We can set this property to `Always` (always restart the container's Pods), `OnFailure` (only restart containers when they fail), or `Never`. A new Pod will never be recreated on another cluster host. We will need more advanced resources for managing the containers' life cycle cluster-wide; these will be discussed in the subsequent sections.

Pods are used to run a test for an application or one of its components, but we never use them to run actual Services because Kubernetes just keeps them running; it doesn't manage their updates or reconfiguration. Let's review how ReplicaSets solve these situations when we need to keep our application's containers up and running.

ReplicaSets

A ReplicaSet is a set of Pods that should be running at the same time for an application's components (or for the application itself if it just has one component). To define a ReplicaSet resource, we need to write the following:

- A `selector` section in which we define which Pods are part of the resource

- The number of replicas required to keep the resource healthy

- A Pod template in which we define how new Pods should be created when one of the Pods in the set dies

Let's review the syntax of these resources:

```
apiVersion: apps/v1

kind: ReplicaSet

metadata:

  name: webserver

  labels:

    app: guestbook

    component: webserver

spec:

  replicas: 3

  selector:

    matchLabels:

      component: webserver
```

```
template:

  metadata:

    labels:

      component: webserver

  spec:

    containers:

    - name: webserver

      image: docker.io/nginx:alpine
```

Figure 9.2 – ReplicaSet manifest

As you can see, the `template` section describes a Pod definition inside the ReplicaSet's `spec` section. This `spec` section also includes the `selector` section, which defines what Pods will be included. We can use `matchLabels` to include exact label-key pairs from Pods, and `matchExpressions` to include advanced rules such as the existence of a defined label or its value included in a list of strings.

> **Important note**
>
> Selectors from ReplicaSet resources apply to running Pods too. This means that you have to take care of the labels that uniquely identify your application's components. ReplicaSet resources are namespaced, so we can use `kubectl get pods --show-labels` before actually creating a ReplicaSet to ensure that the right Pods will be included in the set.

In the Pod template, we will define the volumes to be attached to the different containers created by the ReplicaSet, but it is important to understand that these volumes are common to all the replicas. Therefore, all container replicas will attach the same volumes (in fact, the hosts where they run mount the volumes and the kubelet makes them available to the Pods' containers), which may generate issues if your application does not allow such a situation. For example, if you are deploying a database, running more than one replica that's attaching the same volume will probably break your data files. We should ensure that our application can run more than one replicated process at a time, and if not, ensure we apply the appropriate `ReadWriteOnce` mode flag in the `accessMode` key. We will deep dive into this key, its importance, and its meaning for our workloads in *Chapter 10, Leveraging Application Data Management in Kubernetes*.

The most important key in ReplicaSets is the `replicas` key, which defines the number of active healthy Pods that should be running. This allows us to scale up or down the number of instances for our application's processes. The names of the Pods associated with a ReplicaSet will follow `<REPLICASET_NAME>-<POD_RANDOM_UNIQUE_GENERATED_ID>`. This helps us understand which ReplicaSet generated them. We also can review the ReplicaSet creator by using `kubectl get pod -o yaml`. The `metadata.OwnerReferences` key shows the ReplicaSet that finally created each Pod resource.

We can modify the number of replicas of a running ReplicaSet resource using any of the following methods:

- Editing the running ReplicaSet resource directly in Kubernetes using `kubectl edit <REPLICASET_NAME>`
- Patching the current ReplicaSet resource using `kubectl patch`
- Using the `scale` action with `kubectl`, setting the number of replicas: `kubectl scale rs --replicas <NUMBER_OF_REPLICAS> <REPLICASET_NAME>`

Although changing the number of replicas works automatically, other changes don't work so well. In the Pod template, if we change the image to use for creating containers, the resource will show this change, but the current associated Pods will not change. This is because ReplicaSets do not manage their changes; we need to work with Deployment resources, which are more advanced. To make any change available in a ReplicaSet, we need to recreate the Pods manually by just removing the current Pods (using `kubectl delete pod <REPLICASET_POD_NAMES>`) or scaling the replicas down to zero and scaling up after all are deleted. Any of these methods will create fresh new replicas, using the new ReplicaSet definition.

> **Important note**
>
> You can use `kubectl delete pod --selector <LABEL_SELECTOR>`, with the current ReplicaSet selectors that were used to create them, to delete all associated Pod resources.

ReplicaSets, by default, don't publish any Service; we need to create a Service resource to consume the deployed containers. When we create a Service associated with a ReplicaSet (using the Service's label selector with the appropriate ReplicaSet's labels), all the ReplicaSet instances will be accessible by using the Service's `ClusterIP` address (default Service mode). All replicas get the same number of requests because the internal load balancing provides round-robin access.

We will probably not use ReplicaSets as standalone resources in production as we have seen that any change in their definition requires additional interaction from our side, and that's not ideal in dynamic environments such as Kubernetes.

Before we look at Deployments, which are advanced resources for deploying ReplicaSets, we will quickly review ReplicationControllers, which are quite similar to ReplicaSets.

ReplicationControllers

The ReplicationController was the original method for Pod replication in Kubernetes, but now, it has been almost completely replaced by ReplicaSet resources. We can consider a ReplicationController as a less configurable ReplicaSet. Nowadays, we don't directly create ReplicationControllers as we usually create Deployments for deploying application's components running on Kubernetes. We learned that ReplicaSets have two options for selecting associated labels. The `labelSelector` key can be either a simple label search (`matchLabels`) or a more advanced rule that uses `matchExpressions`. ReplicationController manifests can only look for specific labels in Pods, which makes them simpler to use. The Pod template section looks similar in both ReplicaSets and ReplicaControllers. However, there is also a fundamental difference between ReplicationControllers and ReplicaSets. We can execute application upgrades by using rolling-update actions. These are not available for ReplicaSets but upgrades are provided in such resources thanks to the use of Deployments.

Deployments

We can say that a Deployment is an advanced ReplicaSet. It adds the life cycle management part we missed by allowing us to upgrade the specifications of the Pods by creating a new ReplicaSet resource. This is the most used workload management resource in production. A Deployment resource creates and manages different ReplicaSets resources. When a Deployment resource is created, an associated

ReplicaSet is also dynamically created, following the <DEPLOYMENT_NAME>-<RS_RANDOM_UNIQUE_GENERATED_ID> nomenclature. This dynamically created ReplicaSet will create associated Pods that follow the described nomenclature, so we will see Pod names such as <DEPLOYMENT_NAME>-<RS_RANDOM_UNIQUE_GENERATED_ID>-<POD_RANDOM_UNIQUE_GENERATED_ID> in the defined namespace. This will help us follow which Deployment generates which Pod resources. Deployment resources manage the complete ReplicaSet life cycle. To do this, whenever we change any Deployment template specification key, a new ReplicaSet resource is created and this triggers the creation of new associated Pods. The Deployment resource keeps track of all associated ReplicaSets, which makes it easy to roll back to a previous release, without the latest resource modifications. This is very useful for releasing new application updates. Whenever an issue occurs with the updated resource, we can go back to any previous version in a few seconds thanks to Deployment resources – in fact, we can go back to any previous existing ReplicaSet resource. We will deep dive into rolling updates in *Chapter 13*, *Managing the Application Life Cycle*.

The following code snippet shows the syntax for these resources:

```
apiVersion: apps/v1
kind: Deployment
metadata:
  name: webserver
  labels:
    component: webserver
spec:
  strategy: RollingUpdate
  replicas: 3
  selector:
    matchLabels:
      component: webserver

template:
  metadata:
    labels:
      component: webserver
  spec:
    containers:
    - name: webserver
      image: docker.io/nginx:alpine
```

Figure 9.3 – Deployment manifest

The strategy key allows us to decide whether our new containers try to start before the old ones die (the RollingUpdate value, which is used by default) or completely recreate the associated ReplicaSet (the Recreate value), which is needed when only one container can access attached volumes in write mode at the time.

We will use Deployments to deploy stateless or stateful application workloads in which we don't require any special storage attachment and all replicas can be treated in the same way (all replicas are the same). Deployments work very well for deploying web Services with static content and dynamic ones when session persistence is managed in a different application component. We can't use Deployment resources to deploy our application containers when each replica has to attach its own specific data volume or when we need to execute processes in order.

We will now learn how StatefulSet resources help us solve these specific situations.

StatefulSets

StatefulSet resources are designed to manage stateful application components – those where persistent data must be unique between replicas. These resources also allow us to provide an order to different replicas when processes are executed. Each replica will receive a unique ordered identifier (an ordinal number starting from 0) and it will be used to scale the number of replicas up or down.

The following code snippet shows an example of a StatefulSet resource:

```yaml
apiVersion: apps/v1
kind: StatefulSet
metadata:
  name: database
spec:
  selector:
    matchLabels:
      app: mongodb
  serviceName: "database"
  replicas: 3
  template:
    metadata:
      labels:
        app: mongodb
    spec:
      containers:
      - name: db
        image: mongodb:alpine
        volumeMounts:
        - name: data
          mountPath: /data
  volumeClaimTemplates:
  - metadata:
      name: data
    spec:
      accessModes: [ "ReadWriteOnce" ]
      resources:
        requests:
          storage: 1Gi
```

Figure 9.4 – StatefulSet manifest

The preceding code snippet shows `template` sections for both the Pod resources and the volume resources.

The names for each Pod will follow `<STATEFULSET_NAME>-<REPLICA_NUMBER>`. For example, if we create a `database` StatefulSet resource with three replicas, the associated Pods will be `database-0`, `database-1`, and `database-2`. This name structure is also applied to the volumes defined in the StatefulSet's `volumeClaimTemplates` template section.

Notice that we also included the `serviceName` key in the previous code snippet. A headless Service (without `ClusterIP`) should be created to reference the ReplicaSet's Pods in the Kubernetes internal DNS, but this key tells Kubernetes to create the required DNS entries. For the example presented, the first replica will be announced to the cluster DNS as `database-0.database.NAMESPACE.svc.<CLUSTER_NAME>`, and all other replicas will follow the same name schema. These names can be integrated into our application to create an application cluster or even configure advanced load-balancing mechanisms other than the default (used for ReplicaSets and Deployments).

When we use StatefulSet resources, Pods will be created in order, which may introduce extra complexity when we need to remove some replicas. We will need to guarantee the correct execution of processes that may resolve dependencies between replicas; therefore, if we need to remove a StatefulSet replica, it will be safer to scale down the number of replicas instead of directly removing it. Remember, we have to prepare our application to manage unique replicas completely, and this may need some application process to remove an application's cluster component, for example. This situation is typical when you run distributed databases with multiple instances and decommissioning one instance requires database changes, but this also applies to any ReplicaSet manifest updates. You have to ensure that the changes are applied in the right order and, usually, it is preferred to scale down to zero and then scale up to the required number of replicas.

In the StatefulSet example presented in the preceding code snippet, we specified a `volumeClaimTemplate` section, which defines the properties that are required for a dynamically provisioned volume. We will learn how dynamic storage provisioning works in *Chapter 10, Leveraging Application Data Management in Kubernetes*, but it is important to understand that this `template` section will inform the Kubernetes API that every replica requires its own ordered volume. This requirement for dynamic provisioning will usually be associated with the use of `StorageClass` resources.

Once these volumes (associated with each replica) are provisioned and used, deleting a replica (either directly by using `kubectl delete pod` or by scaling down the number of replicas) will never remove the associated volume. You can be sure that a database deployed via a ReplicaSet will never lose its data.

> **Important note**
> The ReplicaSet's associated volumes will not be automatically removed, which makes these resources interesting for any workload if you need to ensure that data will not be deleted if you remove the resource.

We can use StatefulSet to ensure that a replicated Service is managed uniquely. Software such as Hashicorp's Consul runs clusterized on several predefined Nodes; we can deploy it on top of Kubernetes using containers, but Pods will need to be deployed in order and with their specific storage as if they were completely different hosts. A similar approach has to be applied in database Services because the replication of their processes may lead to data corruption. In these cases, we can use StatefulSet replicated resources, but the application should manage the integration between the different deployed replicas and the scaling up and down procedure. Kubernetes just provides the underlying architecture that guarantees the data's uniqueness and the replica execution order.

DaemonSets

A DaemonSet resource will execute exactly one associated Pod in each Kubernetes cluster Node. This ensures that any newly joined Node will get its own replica automatically.

The following code snippet shows a DaemonSet manifest example:

```
apiVersion: apps/v1                          containers:
kind: DaemonSet                              - name: fluentd-elasticsearch
metadata:                                      image: fluentd:v2.5.2
  name: fluentd-elasticsearch                  resources:
  namespace: kube-system                         limits:
  labels:                                          memory: 200Mi
    k8s-app: fluentd-logging                     requests:
spec:                                              cpu: 100m
  selector:                                        memory: 200Mi
    matchLabels:                               volumeMounts:
      name: fluentd-elasticsearch            - name: varlog
  template:                                      mountPath: /var/log
    metadata:                                volumes:
      labels:                                - name: varlog
        name: fluentd-elasticsearch            hostPath:
      spec:                                        path: /var/log
        tolerations:
        - key: node-role.kubernetes.io/control-plane
          operator: Exists
          effect: NoSchedule
```

Figure 9.5 – DaemonSet manifest

As you may have noticed, we use label selectors to match and associate Pods. In the preceding example, we also introduced the `tolerations` key. Let's quickly introduce how **taints** and **tolerations** work with any kind of workload manifest. They are used to avoid the execution of Pods on inappropriate Nodes. While taints are always associated with Nodes, tolerations are defined for Pods. A Pod must include all the taints associated with a Node as tolerations to be able to run on that particular Node. Node taints can produce three different effects: `NoSchedule` (only Pods with appropriate tolerations for the Node are allowed), `PreferNoSchedule` (Pods will not run on the Node unless no other one is available), or `NoExecute` (Pods will be evicted from the Node if they don't have the appropriate tolerations). Taints and tolerations must match, and this allows us to dedicate Nodes for certain tasks and avoid the execution of any other workloads on them. The kubelet will use dynamic taints to evict Pods when issues are found on a cluster Node – for example, when too much memory is in use or the disk is getting full. In our example, we add a toleration to execute the DaemonSet Pods on Nodes with the `node-role.kubernetes.io/control-plane=NoSchedule` taint.

DaemonSets are often used to deploy applications that should run on all Nodes, such as those running as software agents for monitoring or logging purposes.

> **Important note**
>
> Although it isn't too common, it is possible to use static Pods to run Node-specific processes. This is the mechanism that's used by Kubernetes kubeadm-based Deployments. Static Pods are Pods associated with a Node, executed directly by the kubelet, and thus, they are not managed by Kubernetes. You can identify these Pods by their name because they include the host's name. Manifests for executing static Pods are located in the `/etc/kubernetes/manifests` directory in kubeadm clusters.

At this point, we have to mention that none of the workload management resources presented so far provide a mechanism to run a task that shouldn't be maintained during its execution time. We will now review Job resources, which are specifically created for this purpose.

Jobs

A Job resource is in charge of executing a Pod until we get a successful termination. The Job resource also tracks the execution of a set of Pods using template selectors. We configure a required number of successful executions and the Job resource is considered *Completed* when all the required Pod executions are successfully finished.

In a Job resource, we can configure parallelism for executing more than one Pod at a time and being able to reach the required number of successful executions faster. Pods related to a Job will remain in our Kubernetes cluster until we delete the associated Job or remove them manually.

A Job can be suspended, which will delete currently active Pods (in execution) until we resume it again.

We can use Jobs to execute one-time tasks, but they are usually associated with periodic executions thanks to `CronJob` resources. Another common use case is the execution of certain one-time tasks from applications directly in the Kubernetes cluster. In these cases, your application needs to be able to reach the Kubernetes API internally (the `kubernetes` Service in the `default` namespace) and the appropriate permissions for creating Jobs. This is usually achieved by associating a namespaced `Role`, which allows such actions, with the `ServiceAccount` resource that executes your application's Pod. This association is established using a namespaced `RoleBinding`.

The following code snippet shows a `Job` manifest example:

```
apiVersion: batch/v1           template:
kind: Job                        spec:
metadata:                          containers:
  name: pi                         - name: pi
spec:                                image: perl:5.34.0
  backoffLimit: 4                    command:
  completions: 3                       - perl
                                       - -Mbignum=bpi
                                       - -wle
                                       - "print bpi(2000)"

                                     restartPolicy: Never
```

Figure 9.6 – Job manifest

Here, we defined the number of successful completitions and the number of failures that will set the Job as failed by setting the `completions` and `backoffLimit` keys. At least three Pods must exit successfully before the limit of four failures is reached. Multiple Pods can be executed in parallel to speed up the completion by setting the `parallelism` key, which defaults to 1.

The *TTL-after-finished* controller provides a **time-to-live** (**TTL**) mechanism to limit the lifetime of completed Job resources. This will clean up finished jobs to remove old executions and maintain a clear view of current executions. This is very important when we continuously execute a lot of tasks. The controller will review completed Jobs and remove them when the time since their completion is greater than the value of the `ttlSecondsAfterFinished` key. Since this key is based on a date-time reference, it is key to maintain our clusters' time according to our time zone.

Jobs are commonly used within CronJobs to define tasks that should be executed at certain periods – for example, for executing backups. Let's learn how to implement CronJobs so that we can schedule Jobs periodically.

CronJobs

CronJob resources are used to schedule Jobs at specific times. The following code snippet shows a `CronJob` manifest example:

```
apiVersion: batch/v1              template:

kind: CronJob                       spec:

metadata:                             containers:

  name: daily-pi                      - name: pi

spec:                                   image: perl:5.34.0

  schedule: "0 0 * * *"                 command:

  failedJobsHistoryLimit: 5               - perl
                                          - -Mbignum=bpi
  successfulJobsHistoryLimit: 3           - -wle
                                          - "print bpi(2000)"

                                      restartPolicy: OnFailure
```

Figure 9.7 – CronJob manifest

To be able to review logs from executed Pods (associated with the Jobs created), we can set failedJobsHistoryLimit and successfulJobsHistoryLimit to the desired number of Jobs to keep to be able to review the Pods' logs. Notice that we planned the example Job daily, at 00:00, using the common *Unix Crontab* format, as shown in the following schema:

```
┌──────────── minute (0 - 59)
│ ┌────────── hour (0 - 23)
│ │ ┌──────── day of the month (1 - 31)
│ │ │ ┌────── month (1 - 12)
│ │ │ │ ┌──── day of the week (0 - 6) (Sunday to Saturday;
│ │ │ │ │                       7 is also Sunday on some systems)
│ │ │ │ │                       OR sun, mon, tue, wed, thu, fri, sat
│ │ │ │ │
│ │ │ │ │
* * * * *
```

Figure 9.8 – Unix Crontab format

The schedule key defines when the Job will be created and associated Pods will run. Remember to always quote your value to avoid problems.

> **Important note**
> CronJob resources use the Unix Crontab format, hence values such as @hourly, @daily, @monthly, or @yearly can be used.

CronJobs can be suspended, which will affect any new Job creation if we change the value of the `suspend` key to `true`. To enable the CronJob again, we need to change this key to `false`, which will continue with the normal scheduling for creating new Jobs.

A common use case for CronJobs is the execution of backup tasks for applications deployed on Kubernetes. With this solution, we avoid opening internal applications externally if user access isn't required.

Now that we understand the different resources we can use to deploy our workloads, let's quickly review how they will help us provide resilience and high availability to our applications.

Ensuring resilience and high availability with Kubernetes resources

Pod resources provide resilience out of the box as we can configure them to always restart if their processes fail. We can use the `spec.restartPolicy` key to define when they should restart. It is important to understand that this option is limited to the host's scope, so a Pod will just try to restart on the host on which it was previously running. Pod resources do not provide high availability or resilience cluster-wide.

Deployments, and therefore ReplicaSets, and StatefulSets are prepared for applying resilience cluster-wide because resilience doesn't depend on hosts. A Pod will still try to restart on the Node where it was previously running, but if it is not possible to run it, it will be scheduled to a new available one. This will allow Kubernetes administrators to perform maintenance tasks on Nodes moving workloads from one host to another, but this may impact your applications if they are not ready for such movements. In other words, if you only have one replica of your processes, they will go down for seconds (or minutes, depending on the size of your image and the time required by your processes to start), and this will impact your application. The solution is simple: deploy more than one replica of your application's Pods. However, it is important to understand that your application needs to be prepared for multiple replicated processes working in parallel.

StatefulSets' replicas will never use the same volume, but this isn't true for Deployments. All the replicas will share the volumes, and you must be aware of that. Sharing static content will work like a charm, but if multiple processes are trying to write the same file at the same time, you may encounter problems if your code doesn't manage concurrency.

DaemonSets work differently and we don't have to manage any replication; just one Pod will run on each Node, but they will share volumes too. Because of the nature of such resources, it is not common to include shared volumes in these cases.

But even if our application runs in a replicated manner, we can't ensure that all the replicas die at the same time without configuring a **Pod disruption policy**. We can configure a minimum number of Pods to be available at the same time, ensuring not only resilience but also high availability. Our application will have some impact, but it will continue serving requests (high availability).

To configure a disruption policy, we must use `PodDisruptionBudget` resources to provide the logic we need for our application. We will be able to set up the number of Pods that are required for our application workload under all circumstances by configuring the `minAvailable` or `maxUnavailable` keys. We can use integers (the number of Pods) or a percentage of the configured replicas. `PodDisruptionBudget` resources use selectors to choose between the Pods in the namespace (which we already use to create Deployments, ReplicaSets, and more). The following code snippet shows an example:

```
apiVersion: policy/v1
kind: PodDisruptionBudget
metadata:
  name: webserver-pdb
spec:
  minAvailable: 2
  selector:
    matchLabels:
      app: webserver
```

In this example, a minimum of two Pods with the `app=webserver` label are being monitored. We will define the number of replicas in our Deployment, but the `PodDisruptionBudget` resource will not allow us to scale down below two replicas. Therefore, two replicas will be running even if we decide to execute `kubectl drain node1` (assuming, in this example, that the `webserver` Deployment matches the `app=webserver` Pod's labels and `node1` and `node2` have one replica each). `PodDisruptionBudget` resources are namespaced, so we can show all these resources in the namespace by executing `kubectl get poddisruptionbudgets`.

In the following section, we will review some interesting ideas for solving common application architecture patterns using Pod features.

Understanding advanced Pod application patterns

In this section, we are going to discuss some interesting patterns using simple Pods. All the patterns we are going to review are based on the special mechanisms offered by Kubernetes for sharing kernel namespaces in a Pod, which allow containers running inside to mount the same volumes and interconnect via localhost.

Init containers

More than one Pod can run inside a container. Pods allow us to isolate different application processes that we want to maintain separately in different containers. This helps us, for example, to maintain different images that can be represented by separated code repositories and build workflows.

Init containers run before the main application container (or containers, if we run more in parallel). These init containers can be used to set permissions on shared filesystems presented as volumes, create database schemas, or any other procedure that helps initialize our application. We can even use them to check dependencies before a process starts or even provision required files by retrieving them from an external source.

We can define many init containers, and they will be executed in order, one by one, and all of them must end successfully before the actual application containers start. If any of the init containers fails, the Pod fails, although these containers don't have associated probes for verifying their state. The processes executed by them must include verification if something goes wrong.

It is important to understand that the total CPU and memory resources consumed by a Pod are calculated from the initialization of the Pod, hence init containers are checked. You keep the resource usage between the defined limits for your Pod (which includes the usage of all the containers running in parallel).

Sidecar containers

Sidecar containers run in parallel with the main containers without modifying their main behavior. These containers may change some of the features of the main containers without impacting their processes. We can use them, for example, to include a new container for monitoring purposes, present a component using a web server, sync content between application components, or even include some libraries in them to debug or instrument your processes. The new container can be injected into a running Deployment, triggering its Pods to recreate with this new item included, adding the desired feature. We can do this by patching (using `kubectl patch`) to modify the running Deployment resource manifest.

Some modern monitoring applications, designed to integrate into Kubernetes, also use sidecar containers to deploy an application-specific monitoring component, which retrieves application metrics and exposes them as a new Service.

The next few patterns we are going to review are based on this sidecar container concept.

Ambassador containers

The **Ambassador** applications pattern is designed to offload common client connectivity tasks, helping legacy applications implement more advanced features without changing any of their old code. With this design, we can improve the application's routing, communications security, and resilience by adding additional load balancing, API gateways, and SSL encryption.

We can deploy this pattern within Pods by adding special containers, designed for delivering light reverse-proxy features. In this way, Ambassador containers are used for deploying service mesh solutions, intercepting application process communications, and securing the interconnection with other application components by enforcing encrypted communications and managing application routes, among other features.

Adaptor containers

The **Adaptor** container pattern is used, for example, when we want to include monitoring or retrieve logs from legacy applications without changing their code. To avoid this circumstance, we can include a second container in our application's Pod to get the metrics or the logs from our application without modifying any of its original code. This also allows us to homogenize the content of a log or send it to a remote server. Well-prepared containers will redirect processes' standard and error output to the foreground, and this allows us to review their log, but sometimes, the application can't redirect the log or more than one log is created. We can unify them in one log or redirect their content by adding a second process (the Adaptor container), which formats (adding some custom columns, date format, and so on) and redirects the result to the standard output or a remote logging component. This method does not require special access to the host's resources and it may be transparent for the application.

Prometheus is a very popular open source monitoring solution and is extended in Kubernetes environments. Its main component will poll agent-like components and retrieve metrics from them, and it's very common to use this Adaptor container pattern to present the application's metrics without modifying its standard behavior. These metrics will be exposed in the application Pod in a different port, and the Prometheus server will connect to it to obtain its metrics.

Let's learn how containers' health is verified by Kubernetes to decide the Pod's status.

Verifying application health

In this section, we are going to review how application Pods are considered healthy. Pods always start in the `Pending` state and continue to the `Running` state once the main container is considered healthy. If the Pod executes a Service process, it will stay in this `Running` state. If the Pod is associated with a Job resource, it may end successfully (the `Succeeded` state) or fail (the `Failed` state).

If we remove a Pod resource, it will go to `Terminating` until it is completely removed from Kubernetes.

> **Important note**
>
> If Kubernetes cannot retrieve the Pod's status, its state will be `Unknown`. This is usually due to communication issues between the hosts' kubelet and the API server.

Kubernetes reviews the state of the containers to set the Pod's state, and containers can be either `Waiting`, `Running`, or `Terminated`. We can use `kubectl describe pod <POD_NAME>` to review the details of these phases. Let's quickly review these states:

- `Waiting` represents the state before `Running`, where all the pre-container-execution processes appear. In this phase, the container image is pulled from the registry and different volume mounts are prepared. If the Pod can't run, we can have a `Pending` state, which will indicate a problem with deploying the workload.

- `Running` indicates that the containers are running correctly, without any issues.

- The `Terminated` state is considered when the containers are stopped.

If the Pod was configured with a `restartPolicy` property of the `Always` or `OnFailure` type, all the containers will be restarted on the node where they stopped. That's why a Pod resource does not provide either high availability or resilience if the node goes down.

Let's review how the Pod's status is evaluated in these phases thanks to the execution of **probes**.

Understanding the execution of probes

The kubelet will execute probes periodically by executing some code inside the containers or by directly executing network requests. Different probe types are available depending on the type of check we need for our application's components:

- `exec`: This executes a command inside the container and the kubelet verifies whether this command exits correctly.

- `httpGet`: This method is probably the most common as modern applications expose Services via the REST API. This check's response must return 2XX or 3XX (redirects) codes.

- `tcpSocket`: This probe is used to check whether the application's port is available.

- `grpc`: If our application is consumed via modern **Google Remote Procedure Calls (gRPCs)**, we can use this method to verify the container's state.

Probes must return a valid value to consider the container healthy. Different probes can be executed one after another through the different phases of their lives. Let's consider the different options available to verify whether the container's processes are starting or serving the application itself.

Startup probes

Startup probes are the first in the execution list if they are defined. They are executed when the container is started, and all other probes will wait until this probe ends successfully before they execute. If the probe fails, the kubelet will restart the Pod if it has defined `Always` or `OnFailure` in its `restartPolicy` key.

We will set up these probes if our processes take a lot of time before they are ready – for example, when we start a database server and it must manage previous transactions in its data before it is ready, or when our processes already integrate some sequenced checks before the final execution of the main process.

Liveness probes

Liveness probes check whether the container is running. If they fail, the kubelet will follow the `restartPolicy` value. They are used when it's hard to manage the failure of your main process within the process itself. It may be easier to integrate an external check via the `livenessProbe` key, which verifies whether or not the main process is healthy.

Readiness probes

Readiness probes are the final probes in the sequence because they indicate whether or not the container is ready to accept requests. The Services that match this Pod in their Pod `selector` section will not mark this Pod as ready for requests until this probe ends successfully. The same happens when the probe fails; it will be removed from the list of available endpoints for the Service resource.

Readiness probes are key to managing traffic to the Pods because we can ensure that the application component will correctly manage requests. This probe should always be set up to improve our microservices' interactions.

There are common keys that can be used at the `spec.containers` level that will help us customize the behavior of the different probe types presented. For example, we can configure the number of failed checks required to consider the probe as failed (`failureThreshold`) or the period between the execution of a probe type (`periodSeconds`). We can also configure some delay before any of these probes start by setting the `initialDelaySeconds` key, although it is recommended to understand how the application works and adjust the probes to fit our initial sequence. In the *Labs* section of this chapter, we will review some of the probes we've just discussed.

Now that we know how Kubernetes (the kubelet component) verifies the health of the Pods starting or running in the cluster, we must understand the *stop* sequence when they are considered `Completed` or `Failed`.

Termination of Pods

We can use the `terminationGracePeriodSeconds` key to set how much time the kubelet will wait if the Pod's processes take a long time to end. When a Pod is deleted, the kubelet sends it a `SIGTERM` signal, but if it takes too long, the kubelet will send a `SIGKILL` signal to all container processes that are still alive when the `terminationGracePeriodSeconds` configured time is reached. This time threshold can also be configured at the probe level.

To remove a Pod immediately, we can force and change this Pod-level defined grace period by using `kubelet delete pod <POD_NAME> --force` along with `--grace-period=0`. Forcing the deletion of a Pod may result in unexpected consequences for your applications if you don't understand how it works. The kubectl client sends the `SIGKILL` signal and doesn't wait for confirmation, informing the API server that the Pod is already terminated. When the Pods are part of a StatefulSet, this may be dangerous as the Kubernetes cluster will try to execute a new Pod without confirming whether it has already been terminated. To avoid these situations, it is better to scale down to the replicas and scale up to do a full restart.

Our applications may need to execute some specific processes to manage the interactions between different components when we update some of them, or even if they fail with an error. We can include some triggers when our containers start or stop – for example, to reconfigure a new master process in a clusterized application.

Container life cycle hooks

Containers within a Pod can include a **life cycle hook** in their specifications. Two types are available:

- **PostStart** hooks can be used to execute a process *after* a container is created.

- **PreStop** hooks are executed *before* the container is terminated. The grace period starts when the kubelet receives a stop action, so this hook may be affected if the defined process takes too long.

Pods can be scaled up or down manually whenever our application needs it and it's supported, but we can go further and manage replicas automatically. The following section will show us how to make it possible.

Resource management and scalability

By default, Pods run without compute resource limits. This is fine for learning how your application behaves, and it can help you define its requirements and limits.

Kubernetes cluster administrators can also define quotas that can be configured at different levels. It is usual to define them at the namespace level and your applications will be confined with limits for CPU and memory. But these quotas can also identify some special resources, such as GPUs, storage, or even the number of resources that can be deployed in a namespace. In this section, we will learn how to limit resources in our Pods and containers, but you should always ask your Kubernetes administrators if any quota is applied at the namespace level to prepare your deployments for such compliance. More information about resource quota configurations can be found in the Kubernetes official documentation: `https://kubernetes.io/docs/concepts/policy/resource-quotas`.

We will use the `spec.resources` section to define the limits and requests associated with a Pod. Let's look at how they work:

- **Resource requests** are used to ensure workloads will run on the cluster. By setting the memory and CPU requests (`spec.resources.requests.memory` and `spec.resources.requests.cpu`, respectively), we can define the minimum resources required in any cluster host to run our Pod.

- **Resource limits**, on the other hand, are used to define the maximum resources allowed for a Pod. We will use `spec.resources.limits.memory` and `spec.resources.limits.cpu` to configure the maximum memory and number of CPUs allocable, respectively.

Resources can be defined either at the Pod or container level and they must be compliant with each other. The sum of all the container resource limits must not exceed the Pod values. If we omit the Pod resources, the sum of the defined container resources will be used. If any of the containers do not contain a resource definition, the Pod limits and requests will be used. The container's equivalent key is `spec.containers[].resources`.

Memory limits and requests will be configured in bytes and we can use suffixes such as `ki`, `Mi`, `Gi`, and `Ti` for multiples of 1,000, or `k`, `M`, and `T` for multiples of 1,024. For example, to specify a limit of 100 MB of memory, we will use `100M`. When the limited memory allowed is reached, `OOMKiller` will be triggered in the execution host and the Pod or container will be terminated.

For the CPU, we will define the number of CPUs (it doesn't matter whether they are physical or virtual) to be allowed or requested, if we are defining a request limit. When the CPU limit is reached, the container or Pod will not get more CPU resources, which will probably make your Pod to be considered unhealthy because checks will fail. CPU resources must be configured in either integers or fractionals, and we can add `m` as a suffix to represent millicores; hence, 0.5 CPUs can also be written as `500m`, and 0.001 CPUs will be represented as `1m`.

> **Important note**
>
> When we are using Linux nodes, we can request and limit huge page resources, which allows us to define the page size for memory blocks allocated by the kernel. Specific key names must be used; for example, `spec.resources.limits.hugepages-2Mi` allows us to define the limit of memory blocks allocated for 2 MiB huge pages.

Your administrators can prepare for some `LimitRange` resources, which will define constraints for the limits and requests associated with your Pod resources.

Now that we know how we can limit and ask for resources, we can vertically scale a workload by increasing its limits. Horizontal scaling, on the other hand, will require the replication of Pods. We can now continue and learn how to dynamically and horizontally scale Pods related to a running workload.

> **Important note**
>
> **Vertical Pod autoscaling** is also available as a project inside Kubernetes. It is less popular because vertical scaling impacts your current Deployments or StatefulSets as it requires scaling the number of resources on your running replicas up or down. This makes them hard to apply and it is better to fine-grain resources in your applications and use horizontal Pod autoscaling, which does not modify current replica specifications.

Horizontal Pod autoscaling

HorizontalPodAutoscaler works as a controller. It scales Pods up or down when the load associated with the workload is increased or decreased. Autoscaling is only available for Deployments (by scaling and modifying their ReplicaSets) and StatefulSets. To measure the consumption of resources associated with a specific workload, we have to include a tool such as **Kubernetes Metrics Server** in our cluster. This server will be used to manage the standard metrics. This can be easily deployed using its manifests at `https://github.com/kubernetes-sigs/metrics-server`. It can also be executed as a pluggable add-on if you are using Minikube on your laptop or desktop computer.

We will define a `HorizontalPodAutoscaler` (hpa) resource; the controller will retrieve and analyze the metrics for a workload resource specified in the hpa definition.

Different types of metrics can be used for the hpa resource, although the most common is the Pod's CPU consumption.

Metrics related to Pods can be defined and thus the controller checks their metrics and analyzes them using an algorithm that combines these metrics with cluster available resources and Pod states (`https://kubernetes.io/docs/tasks/run-application/horizontal-pod-autoscale/#algorithm-details`) and then decides whether or not the associated resource (Deployment or StatefulSet) should be scaled.

To define an hpa resource, we will set up a metric to analyze and a range of replicas to use (max and min replicas). When this value is reached, the controller reviews the current replicas, and if there's still room for a new one, it will be created. hpa resources can be defined in either imperative or declarative format. For example, to manage a minimum of two Pods and a maximum of 10 when more than 50% of the CPU consumption is reached for the current Pods, we can use the following syntax:

```
kubectl autoscale <RESOURCE_TYPE> <RESOURCE_NAME> --cpu-percent=50
--min=2 --max=10
```

When a resource's CPU consumption is more than 50%, then a replica is created, while one replica is decreased when this metric is below that value; however, we will never execute more than 10 replicas or less than two.

> **Important note**
>
> We can review the manifest that's used for creating any resource by adding `-o yaml`. The manifest will be presented and we will be able to verify its values. As an example, we can use `kubectl autoscale deploy webserver --cpu-percent=50 --min=2 --max=10 -o yaml`.
>
> If we want to review values before creating the resource, we can add the `--dry-run=client` argument to only show the manifest, without actually creating the resource.

As hpa resources are namespaced, we can get all the already deployed hpa resources by executing `kubectl get hpa -A`.

With that, we have seen how Kubernetes provides resilience, high availability, and autoscaling facilities out of the box by using specific resources. In the next section, we will learn how it also provides some interesting security features that will help us improve our application security.

Improving application security with Pods

In a Kubernetes cluster, we can categorize the applications' workloads distributed cluster-wide as either privileged or unprivileged. Privileged workloads should always be avoided for normal applications unless they are strictly necessary. In this section, we will help you define the security of your applications by declaring your requirements in your workload manifests.

Security contexts

In a security context, we define the privileges and security configuration required for a Pod or the containers included in it. Security contexts allow us to configure the following security features:

- `runAsUser`/`runAsGroup`: These options manage the `userID` and `groupID` properties that run the main process with containers. We can add more groups by using the `supplementalGroups` key.

- `runAsNonRoot`: This key can control whether we allow the process to run as `root`.

- `fsGroup`/`fsGroupChangePolicy`: These options manage the permissions of the volumes included within a Pod. The `fsGroup` key will set the owner of the filesystems mounted as volumes and the owner of any new file. We can use `fsGroupChangePolicy` to only apply the ownership change if the permissions don't match the configured `fsGroup`.

- `seLinuxOptions`/`seccompProfile`: These options allow us to overwrite default SELinux and `seccomp` settings by configuring special SELinux labels and a special `seccomp` profile.

- `capabilities`: Kernel capabilities can be added or removed (`drop`) to only allow specific kernel interactions (containers share the host's kernel). You should avoid unnecessary capabilities in your applications.

- `privileged`/`AllowPrivilegeEscalation`: We can allow processes inside a container to be executed as `privileged` (with all the capabilities) by setting the `privileged` key to `true` or to be able to gain privileges, even if this key was set to `false`, by setting `AllowPrivilegeEscalation` to `true`. In this case, container processes do not have all capabilities but they will allow internal processes to run as if they had the `CAP_SYS_ADMIN` capability.

- `readOnlyRootFilesystem`: It is always a very good idea to run your containers with their root filesystem in read-only mode. This won't allow processes to make any changes in the container. If you understand the requirements of your application, you will be able to identify any directory that may be changed and add an appropriate volume to run your processes correctly. It is quite usual, for example, to add `/tmp` as a separate temporal filesystem (`emptyDir`).

Some of these keys are available at the container or Pod level or both. Use `kubectl explain pod.spec.securityContext` or `kubectl explain pod.spec.containers.securityContext` to retrieve a detailed list of the options available in each scope. You have to be aware of the scope that's used because Pod specifications apply to all containers unless the same key exists under the container scope – in which case, its value will be used.

Let's review the best settings we can prepare to improve our application security.

Security best practices

The following list shows some of the most used settings for improving security. You, as a developer, can improve your application security if you ensure the following security measures can be enabled for your Pods:

- `runAsNonRoot` must always be set to `true` to avoid the use of `root` on your containers. Ensure you also configure `runAsUser` and `runAsGroup` to IDs greater than `1000`. Your Kubernetes administrators can suggest some IDs for your application. This will help control application IDs cluster-wide.

- Always drop all capabilities and enable only those required by your application.

- Never use privileged containers for your applications unless it is strictly necessary. Usually, only monitoring- or kernel-related applications require special privileges.

- Identify the filesystem's requirement for your application and always set `readOnlyRootFilesystem` to `true`. This simple setting improves security, disabling any unexpected changes. Required filesystems can be mounted as volumes (many options are available, as we will learn in *Chapter 10, Leveraging Application Data Management in Kubernetes*).

- Ask your Kubernetes administrators whether there are some SELinux settings you should consider to apply them on your Pods. This also applies to `seccomp` profiles. Your administrators may have configured a default profile. Ask your administrators about this situation to avoid any system call issues.

- Your administrators may have been using tools such as Kyverno or OPA Gatekeeper to improve cluster security. In these cases, they can enforce security context settings by using **admission controllers** in the Kubernetes cluster. The use of these features is outside the scope of this book but you may ask your administrators about the compliance rules required to execute applications in your Kubernetes platform.

In the next section, we will review how to implement some of the Kubernetes features we learned about in this chapter by preparing the multi-component application we used in previous chapters (*Chapter 5, Creating Multi-Container Applications*, and *Chapter 7, Orchestrating with Swarm*) to run on Kubernetes.

Labs

This section will show you how to deploy the `simplestlab` three-tier application in Kubernetes. Manifests for all its components have been prepared for you while following the techniques and Kubernetes resources explained in this chapter. You will be able to verify the usage of the different options and you will able to play with them to review the content and best practices described in this chapter.

The code for these labs is available in this book's GitHub repository at `https://github.com/PacktPublishing/Containers-for-Developers-Handbook.git`. Ensure you have the latest revision available by simply executing `git clone https://github.com/PacktPublishing/Containers-for-Developers-Handbook.git` to download all its content, or `git pull` if you've already downloaded the repository before. All the manifests and the steps required for running `simplestlab` are located inside the `Containers-for-Developers-Handbook/Chapter9` directory.

In the labs in GitHub, we will deploy the `simplestlab` application, which is used in previous chapters, on Kubernetes by defining appropriate resource manifests:

- The **database** component will be deployed using a StatefulSet resource
- The **application backend** component will be deployed using a Deployment resource
- The **load balancer** (or **presenter**) component will be deployed using a DaemonSet resource

In their manifests, we have included some of the mechanisms we learned about in this chapter for checking the component's health, replicating their processes, and improving their security by disallowing their execution as the root user, among other features. Let's start by reviewing and deploying the database component:

1. We will use a StatefulSet to ensure that replicating its processes (scaling up) will never represent a problem to our data. It is important to understand that a new replica starts empty, without data, and joins the pool of available endpoints for the Service, which will probably be a problem. This means that in these conditions, the Postgres database isn't scalable, so this component is deployed as a StatefulSet to preserve its data even in the case of a manual replication. This example only provides resilience, so do not scale this component. If you need to deploy a database with high availability, you will need a distributed database such as MongoDB. The full manifest for the database manifest can be found in `Chapter9/db.satatefulset.yaml`. Here is a small extract from this file:

```
apiVersion: apps/v1
kind: StatefulSet
metadata:
  name: db
  labels:
    component: db
    app: simplestlab
```

```
      spec:
        replicas: 1
        selector:
          matchLabels:
            component: db
            app: simplestlab
        template:
          metadata:
            labels:
              component: db
              app: simplestlab
          spec:
            securityContext:
              runAsNonRoot: true
              runAsUser: 10000
              runAsGroup: 10000
              fsGroup: 10000
...
        volumeClaimTemplates:
        - metadata:
            name: postgresdata
          spec:
            accessModes: [ "ReadWriteOnce" ]
            #storageClassName: "csi-hostpath-sc"
            resources:
              requests:
                storage: 1Gi
```

Here, we defined a template for the Pods to create and a separate template for the VolumeClaims (we will talk about them in *Chapter 10*). This ensures that each Pod will get its own volume. The volume that's created will be mounted in the database container as the /data filesystem and its size will be 1,000 MB (1 Gi). No other container is created. The POSTGRES_PASSWORD and PGDATA environment variables are set and passed to the container. They will be used to create the password for the Postgres user and the patch for the database data. The image that's used for the container is docker.io/frjaraur/simplestdb:1.0 and port 5432 will be used to expose its Service. Pods only expose their Services internally, in the Kubernetes network, so you will never be able to reach these Services from remote clients. We specified one replica and the controller will associate the pods with this StatefulSet by searching for Pods with component=db and app=simplestlab labels. We simplified the database's probes by just checking a TCP connection to port 5432. We defined a security context at the Pod's level, which will apply to all the containers by default:

```
        securityContext:
          runAsNonRoot: true
```

```
runAsUser: 10000
runAsGroup: 10000
fsGroup: 10000
fsGroupChangePolicy: OnRootMismatch
```

2. The database processes will run as `10000:10000 user:group`, hence they are secure (no root is required). We could have gone further if we set the container as read-only but in this case, we didn't as Docker's official Postgres image; however, it would have been better to use a full read-only filesystem.

    The Pod will get an IP address, though this may change if the Pod is recreated for any reason, which makes Pods' IP addresses impossible to use in such dynamic environments. We will use a Service to associate a *fixed* IP address with a Service and then with the endpoints of the Pods related to the Service.

3. The following is an extract from the Service manifest (you will find it as `Chapter9/db.service.yaml`):

```
apiVersion: v1
kind: Service
metadata:
  name: db
spec:
  clusterIP: None
  selector:
    component: db
    app: simplestlab
  ...
```

    This Service is associated with the Pods by using a selector (the `components=db` and `app=simplestlab` labels) and Kubernetes will route the traffic to the appropriate Pods. When a TCP packet reaches the Service's port, `5432`, it is load balanced to all the available Pod's endpoints (in this case, we will just have one replica) in port `5432`. In both cases, we used port `5432`, but you must understand that `targetPort` refers to the container port, while the port key refers to the Service's port, and they can be completely different. We are using a headless Service because it works very well with StatefulSets and their resolution in round-robin mode.

4. With the StatefulSet definition and the Service, we can deploy the database component:

```
PS Chapter9> kubectl create -f .\db.statefulset.yaml
statefulset.apps/db created
PS Chapter9> kubectl create -f .\db.service.yaml
service/db created
PS Chapter9> kubectl get pods
NAME     READY    STATUS     RESTARTS     AGE
db-0     1/1      Running    0            17s
```

We can now continue and review the app component.

5. The application (backend component) is deployed as a Deployment workload. Let's see an extract of its manifest:

```yaml
apiVersion: apps/v1
kind: Deployment
metadata:
  name: app
  labels:
    component: app
    app: simplestlab
spec:
  replicas: 3
  selector:
    matchLabels:
      component: app
      app: simplestlab
  template:
    metadata:
      labels:
        component: app
        app: simplestlab
    spec:
      securityContext:
        runAsNonRoot: true
        runAsUser: 10001
        runAsGroup: 10001
```

You can find the complete manifest in the Chapter9/app.deployment.yaml file.

6. For this component, we defined three replicas, so three Pods will be deployed cluster-wide. In this component, we are using the docker.io/frjaraur/simplestapp:1.0 image. We've configured two security contexts, one at the Pod's level:

```yaml
securityContext:
  runAsNonRoot: true
  runAsUser: 10001
  runAsGroup: 10001
```

The second is for enforcing the use of a read-only filesystem for the container:

```yaml
securityContext:
  readOnlyRootFilesystem: true
```

7. Here, we prepared `readinessProbe` using `httpGet` but we still keep `tcpSocket` for `livenessProbe`. We coded `/healthz` as the application's health endpoint for checking its healthiness.

8. In this component, we added a resource section for the app container:

```
resources:
    requests:
        cpu: 10m
        memory: 20M
    limits:
        cpu: 20m
        memory: 30Mi
```

In this case, we asked Kubernetes for at least 10 millicores of CPU and 20M of memory. The `limits` section describes the maximum CPU (20 millicores) and memory (30Mi). If the memory limit is reached, Kubelet will trigger the OOM-Killer procedure and it will kill the container. When the CPU limit is reached, the kernel does not provide more CPU cycles to the container, which may lead the probes to fail and hence the container will die. This component is stateless and it is running completely in read-only mode.

> **Important note**
>
> In the full YAML file manifest, you will see that we are using the environment variables for passing sensitive data. Always avoid passing sensitive data in environment variables as anyone with access to your manifest files will be able to read it. We will learn how to include sensitive data in *Chapter 10, Leveraging Application Data Management in Kubernetes*.

9. We will also add a Service for accessing the app `Deployment` workload:

```
apiVersion: v1
kind: Service
metadata:
  name: app
spec:
  selector:
    app: simplestlab
    component: app
  ports:
    - protocol: TCP
      port: 3000
      targetPort: 3000
```

10. We create both Kubernetes resources:

```
PS Chapter9> kubectl create -f .\app.deployment.yaml `
-f .\app.service.yaml
deployment.apps/app created
service/app created
PS Chapter9> kubectl get pods
NAME                     READY   STATUS    RESTARTS   AGE
db-0                     1/1     Running   0          96s
app-585f8bb87-r8dqh      1/1     Running   0          41s
app-585f8bb87-wsfm7      1/1     Running   0          41s
app-585f8bb87-t5gpx      1/1     Running   0          41s
```

We can now continue with the frontend component.

11. For the frontend component, we will deploy NGINX on each Kubernetes cluster node. In this case, we are using a DaemonSet to run the component cluster-wide; one replica will be deployed on each node. We will prepare this component to also run as a non-root user, so some special configurations are needed. Here, we prepared `configMap` with these special configurations for NGINX; you will find it as `Chapter9/ lb.configmap.yaml`. This configuration will allow us to run as user `101` (`nginx`). We created this configMap before the actual DaemonSet; although it is possible to do the opposite, it is important to understand the requirements and prepare them before the workload deployment. This configuration allows us to run NGINX as non-root on a port greater than `1024` (system ports). We will use port `8080` to publish the `loadbalancer` component.

Notice that we added a `proxy_pass` sentence to reroute the requests for / to `http://app:3000`, where app is the Service's name, resolved via internal DNS. We will use `/healthz` to check the container's healthiness.

12. Let's see an extract from the DaemonSet manifest:

```
apiVersion: apps/v1
kind: DaemonSet
metadata:
  name: lb
  labels:
    component: lb
    app: simplestlab
        image: docker.io/nginx:alpine
        ports:
        - containerPort: 8080
        securityContext:
          readOnlyRootFilesystem: true
        volumeMounts:
        - name: cache
```

```
        mountPath: /var/cache/nginx
      - name: tmp
        mountPath: /tmp/nginx
      - name: conf
        mountPath: /etc/nginx/nginx.conf
        subPath: nginx.conf
...
    volumes:
    - name: cache
      emptyDir: {}
    - name: tmp
      emptyDir: {}
    - name: conf
      configMap:
        name: lb-config
```

Notice that we added /var/cache/nginx and /tmp as emptyDir volumes, as mentioned previously. This component will be also stateless and run in read-only mode, but some temporal directories must be created as emptyDir volumes so that they can be written to without allowing the full container's filesystem.

13. The following security contexts are created:

- At the Pod level:

```
securityContext:
    runAsNonRoot: true
    runAsUser: 101
    runAsGroup: 101
```

- At the container level:

```
securityContext:
    readOnlyRootFilesystem: true
```

14. Finally, we have the Service definition, where we will use a NodePort type to quickly expose our application:

```
apiVersion: v1
kind: Service
metadata:
  name: lb
spec:
  type: NodePort
  selector:
    app: simplestlab
```

```
        component: lb
     ports:
        - protocol: TCP
          port: 80
          targetPort: 8080
          nodePort: 32000
```

15. Now, let's deploy all the `lb` component (frontend) manifests:

```
PS Chapter9> kubectl create -f .\lb.configmap.yaml
configmap/lb-config created
PS Chapter9> kubectl create -f .\lb.daemonset.yaml
daemonset.apps/lb created
PS Chapter9> kubectl create -f .\lb.service.yaml
service/lb created
```

Now, we can reach our application in any Kubernetes cluster host's port `32000`. Your browser should access the application and show something like this (if you're using Docker Desktop, you will need to use `http://localhost:32000`):

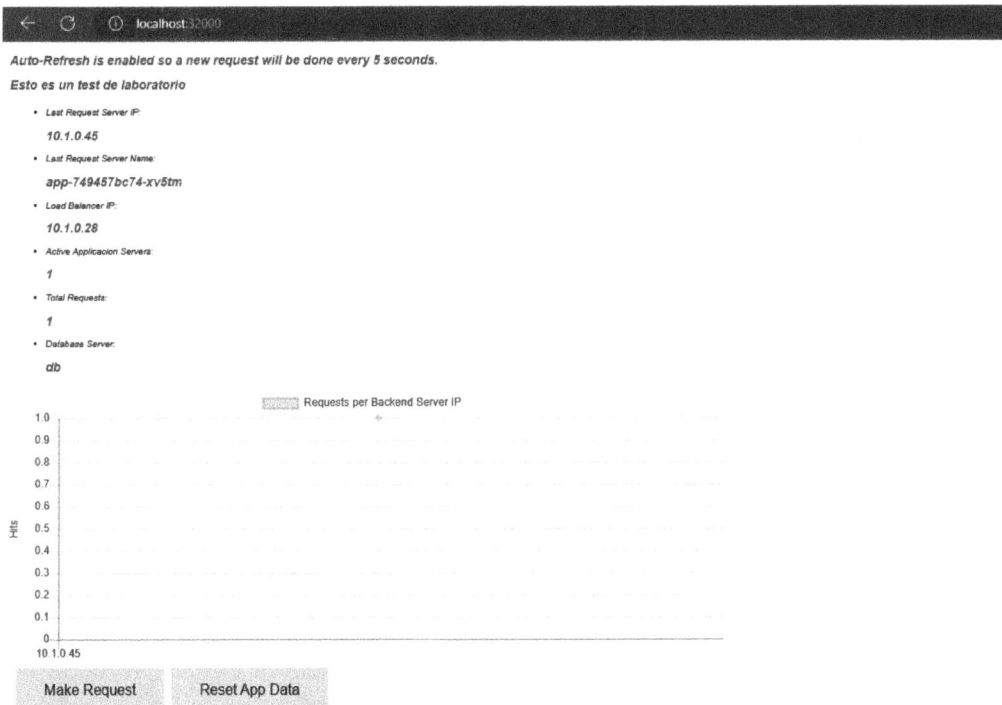

Figure 9.9 – simplestlab application web GUI

You can find additional steps for scaling up and down the application backend component in the `Chapter9` code repository. The labs included in this chapter will help you understand how to deploy an application using different Kubernetes resources.

Summary

In this chapter, we learned about the resources that can help us deploy application workloads in Kubernetes. We took a look at the different options for running replicated processes and verifying their health to provide resilience, high availability, and auto-scalability. We also learned about some of the Pod features that can help us implement advanced patterns and improve the overall application security. We are now ready to deploy our application using the best patterns and apply and customize the resources provided by Kubernetes, and we know how to implement appropriate health checks while limiting resource consumption in our platform.

In the next chapter, we will deep dive into the options we have for managing data within Kubernetes and presenting it to our applications.

10

Leveraging Application Data Management in Kubernetes

Deploying applications in Kubernetes helps in managing resilience, **high availability** (**HA**), and scalability by using replicated instances. But none of these features can be used without knowing how your application actually works and how to manage its data. In this chapter, we will review how to create and manage **Secrets**, **ConfigMaps**, and different **volume** options. While Secret and ConfigMap resources will be used to integrate different authentication options inside containers, volumes are used to manage an application's data, as we briefly introduced in *Chapter 8*, *Deploying Applications with the Kubernetes Orchestrator*. Applications can be either stateful, stateless or – as is usually the case – a combination of both. We will learn in this chapter about different options for managing data and separating it from the application's life cycle.

The following main concepts are reviewed in this chapter:

- Understanding the data within your application

- Applying configurations using ConfigMaps

- Managing sensitive data using Secret resources

- Managing stateless and stateful data

- Enhancing storage management in Kubernetes with **PersistentVolume** (**PV**) resources

Technical requirements

You can find the labs for this chapter at `https://github.com/PacktPublishing/ Containers-for-Developers-Handbook/tree/main/Chapter10`, where you will find some extended explanations, omitted in the chapter's content to make it easier to follow. The *Code In Action* video for this chapter can be found at `https://packt.link/JdOIY`.

As the data used by your application is very important, we will first review the different options we have and the resources we can use within Kubernetes.

Understanding the data within your application

Microservices architecture improves the performance and resilience of your applications by distributing functionalities in different pieces, allowing us to scale them up or down and continue serving some functionalities even when some components fail. But this distribution of functionalities entails the distribution of the data associated with each component and somehow sharing it when more than one component needs it. It is very important to also understand that your application must allow scaling up without corrupting the data in case more than one replica is accessing the same data. Running application components as containers will help us distribute the processes, keeping the same data content in each replica, and starting and stopping processes quickly. The container runtime will attach defined volumes to the containers, but it doesn't manage your application's logic. That's why it is key to understand how data will be used when you are preparing your applications for running in container-orchestrated environments.

Container orchestrators will provide you with mechanisms for injecting configurations into your containers, maintaining these configurations synced within all container replicas. These configurations can be used as either files within the containers or environment variables. You must understand that if your application uses configurations in clear text, you will not be able to protect them from attackers if they get into your containers. This will always be the case, even if you encrypt your configurations before they are injected into the containers. If your code reads the configuration content in clear text, it will always be accessible because permissions will be adequate to allow your processes to read the configurations, and the container will use the main process user to attach any new processes (via `docker exec` or `kubectl exec`). If you use environment variables, they will be easily readable by any process running inside the container. But these things don't mean that your information is insecure inside containers. Different mechanisms, such as RBAC, allow us to limit access to containers from the orchestrated platform, but accessing cluster nodes will override the platform's security. You should never run commands from cluster nodes. Your container orchestrator administrators may provide you with a complete **continuous deployment** (**CD**) solution, or you may use your platform client (command line or graphical interface) to gain access.

Injecting data into an application's containers can be accomplished by using any of the following mechanisms:

- **Command-line arguments**: We can pass values to a container's main process by adding the `arguments` key to any Pod resource. You should never pass sensitive data using this method.

- **Environment variables**: It is usual to include information by using environment variables. Some coding languages even have standardized nomenclature for working directly with variables; in any case, your application must be prepared to include them. This method should also be avoided when including sensitive data unless it is combined with Secret resources, as we will learn about in the next section.

- **ConfigMaps**: These resources are the best option for adding configurations to our workloads. We can use them to add files inside containers, knowing that they will be available no matter which node runs the instance. The container orchestrator will manage and sync any change in its content. They can also be used to set up some variables with their values and use them as environment variables.

- **Secrets**: Secret resources are the appropriate method for managing sensitive data in either Kubernetes or Docker Swarm platforms. However, there's a big difference in how they are packaged inside each platform. Docker Swarm encrypts their content, and we aren't even allowed to retrieve the content, while Kubernetes uses the Base64 format for storing content within a cluster. Encryption for storing Secret resources at rest in etcd can be enabled, but it is not enabled by default. This only affects the etcd database, which shouldn't be accessible to normal users, but it may be useful to ask your Kubernetes administrators about such configuration if you are worried about the data you keep in your Secret resources. It is quite common to use Secret resources for either adding sensitive files, such as passwords, authentication tokens, certificates, or even container image registry connection strings, or to present variables to containers.

- **Volumes**: Volumes are not intended to be used for injecting data but for storing it during the execution of the containers. They can be either ephemeral or stateful, to persist data between executions. In Kubernetes, we consider volumes as those storage resources integrated into the Kubernetes platform's code. A lot of cloud storage solutions were integrated from the beginning of its development because it was part of the design, although host bind mounts, ephemeral directories, NFS, and other on-premises solutions are also available. ConfigMap and Secret resources are also Volumes, but we will treat them differently because of their content.

- **The downward API**: Although the downward API is considered a Volume resource, we can think of it as a completely different concept due to its usage. You can mount metadata information from the current namespace's resources to be used in your application by using the downward API, which automatically manages the required requests to the Kubernetes API to retrieve it.

- **PVs**: A PV is storage provisioned by the Kubernetes administrator to accommodate a **PersistentVolumeClaim** (**PVC**) resource, which is a request for storage for your application component. Whenever you create a PVC resource, it will be bound to an existing PV resource if there is one free with the required size and properties.

> Important note
>
> The concept of **Projected Volumes** also exists, which is a specific map of different volumes integrated into the same directory inside a container. This feature allows us to locate Secret and ConfigMap resources in the same directory.

We will learn how to inject data and use it inside containers by using ConfigMaps to include non-sensitive information.

Applying configurations using ConfigMaps

In this section, we are going to learn how to use ConfigMap resources, mount files inside containers or as environment variables, and present the information for our application's processes.

The content of a ConfigMap resource is stored in the Kubernetes etcd key-value store. Due to this, the content can't exceed 1 MB in size. The manifest of these resources doesn't have a spec section. Instead, we can have either data or binaryData (for Base64 content) keys for defining the content. The following screenshot shows an example of a ConfigMap manifest:

```
apiVersion: v1
kind: ConfigMap
metadata:
 name: settings
 labels:
   key1: value1
data:
 APP_VAR1: "200"
 APP_VAR2: "MY_CONFIG_STRING1"
 appsettings: |
   APP_VAR3: "MY_CONFIG_STRING2"
   APP_VAR4:
     APP_VAR5: "MY_CONFIG_STRING3"
```

Figure 10.1 – ConfigMap resource manifest

In the code in the presented screenshot, we have declared two types of configurations. While APP_VAR1 and APP_VAR2 are defined in key-value format, the appsettings section defines a complete configuration file that can be mounted. Notice the pipe symbol (|) used to define the appsettings key. This allows us to include all the subsequent content as the value for the key. You should be very careful with the indentation of the YAML file to avoid any issues with the file content.

Let's see now how we will use these configurations in a Pod:

```
apiVersion: v1

kind: Pod

metadata:

  name: demo-pod

spec:

  containers:

  - name: demo

    image: myimage

    env:

    - name: APP_VAR1

      valueFrom:

        configMapKeyRef:

        name: settings

        key: APP_VAR1
```

```
    - name: APP_VAR2

      valueFrom:

        configMapKeyRef:

        name: settings

        key: APP_VAR2

    volumeMounts:

    - name: mysettings

      mountPath: /app/config

      readOnly: true

  volumes:

  - name: mysettings

    configMap:

      name: settings

      items:

      - key: appsettings

        path: appsettings
```

Figure 10.2 – Pod resource manifest using a ConfigMap resource

In the Pod manifest shown in the preceding screenshot, we have presented two mechanisms for using the information declared in a ConfigMap resource. We used the key-value definitions in the settings ConfigMap as environment variables. But the content of the appsettings key, defined in the settings ConfigMap too, is presented as a volume in the demo container. In this case, a /app/config/appsettings file will be created, with the content of the appsettings key. Notice that we used the ReadOnly key to define that the mounted file will not be writable.

In this example, we didn't use the simplest mechanism for mounting configuration files. Let's see how we simply add a complete configuration file, created with kubectl create configmap <CONFIGMAP_NAME> --from-file=<CONFIGURATION_FILE>. We will use the appsettings.json file as an example, with any content, and created using kubectl create cm appsettings -from-file=appsettings.json:

```
apiVersion: v1
kind: Pod
metadata:
  name: mypod
spec:
```

```
containers:
- name: mypod
  image: myregistry/myimage
  volumeMounts:
  - name: config
    mountPath: "/app/config/appsettings.json"
    subPath: appsettings.json
    readOnly: true
volumes:
- name: config
  configMap:
    name: appsettings
```

In this case, we used the subPath key to set up the filename and complete path for the configuration file.

Configuration files can be updated at any time unless we have used the immutable key (which defaults to false), in which case we will need to recreate the resource. To modify the content or any of the allowed keys (use kubectl explain configmap to review them), we can use any of the following methods:

- Use kubectl edit to edit and modify its values online.

- Patch the file by using kubectl patch (https://kubernetes.io/docs/tasks/manage-kubernetes-objects/update-api-object-kubectl-patch).

- Replace the resource with a new manifest file by using kubectl replace -f <MANIFEST_FILE>. This is the preferred option as all changes can be followed by storing the manifest files (using **GitOps** methodology, as we will learn in *Chapter 13, Managing the Application Life Cycle*).

ConfigMap resources will be updated on your running workloads unless their values are used in your containers as environment variables or mounted using the subPath key, in which case this will not be done. It is very important to understand how updates will be managed by Kubernetes in your application's workloads. Even if your configuration is updated, it depends on how your application uses it, when it is loaded, and how these changes affect your container processes. If your processes only read configurations when they start, you will need to recreate the application's Pods. Hence, the only way you can ensure that a new configuration is applied is by recreating the containers. Depending on the workloads you used for your configuration, you will just need to remove or scale your resources down/up to make the configuration changes update.

> **Important note**
> We can use kubectl create cm <CONFIGMAP_NAME> --from-literal=KEY=VALUE to create ConfigMap resources with key-value resource types directly.

We can add annotations to our ConfigMap resources and update them to trigger the update of your workloads. This will ensure that your Pods will be recreated, hence the ConfigMap is updated immediately. By default, when a ConfigMap is updated, Kubernetes will update the content on the Pods using this configuration at regular intervals. This automatic update doesn't work if you use the `subPath` key to mount the content. However, notice that updating the content of the file does not include the update of your application; it depends on how your application works and how often the configuration is refreshed.

Kubernetes also allows us to include information in Pods at runtime. We will use the downward API to inject Kubernetes data into Pods, which we will learn about in the next section.

Using the downward API to inject configuration data

We will use the downward API mount endpoints to inject information about the current Pod resource. This way, we can import information such as the Pod's name (`metadata.name`), its annotations, labels, and service account, for example. Information can be passed as environment variables or mounted as a volume file.

Imagine a Pod with the following annotations:

```
...
metadata:
  annotations:
    environment: development
    cluster: cluster1
    location: Berlin
...
```

We can mount this information in a Pod's container with a volume definition and the `mountPath` parameter inside it:

```
...
  containers:
  ...
    volumeMounts:
        - name: podinfo
          mountPath: /etc/pod-annotations
  volumes:
    - name: podinfo
      downwardAPI:
        items:
          - path: "annotations"
```

```
        fieldRef:
          fieldPath: metadata.annotations
  ...
```

Notice that we are injecting the annotations inside the /etc/pod-annotations file. We can use either static data (added manually) or dynamic information, retrieved from Kubernetes.

Next, let's see how to include sensitive data by using Secrets.

Managing sensitive data using Secret resources

We should always avoid adding sensitive information to our application images. Neither passwords, connection strings, tokens, certificates, nor license information should be written inside container images; all this content must be included in the runtime. Therefore, instead of using ConfigMap resources, which are stored in clear text, we will use Secrets. Kubernetes Secret resource content is described in base64 format. They are not encrypted, and anyone with access to them can read their data. This includes any user who can create a Pod in the same namespace, as the Secret can be included and hence read. Only appropriate RBAC resource access can ensure Secrets' security. Therefore, it is important to understand that you should avoid access to your Secret resources using appropriate Kubernetes **Roles** and **RoleBindings** (Kubernetes RBAC). Also, by default, Kubernetes doesn't encrypt Secrets in etcd, hence access to the key-value data files at the filesystem level shouldn't be allowed. Secrets are namespaced resources, therefore we will be able to manage Kubernetes access at the namespace level. Your Kubernetes administrators should ensure the appropriate access at the cluster level.

We will use Secret resources as **volumes** (presenting files such as certificates or tokens), as **environment variables** (with their content hidden when you review the online Pod resource's manifest), or as **authentication** for accessing a remote registry (on-premise or cloud service).

We can use kubectl create secret generic <SECRET_NAME> --from-file=<SENSITIVE_DATA_FILE> or kubectl create secret generic <SECRET_NAME> --from-literal=SECRET_VARIABLE_NAME=SECRET_VALUE. Either the --from-file or --from-literal arguments can be used multiple times to add multiple data keys. The following Secret resource types can be created:

- generic: This is the most usual type and can be used to include any sensitive file or value.
- tls: This stands for **Transport Layer Security**. We will use this type to add certificates to our application. Notice that this will only add a Secret with a key and associated certificate that you must include in your application somehow; for example, by adding an SSL_CERT_FILE variable with its content or by using the associated .cert and .key files in your configuration.
- docker-registry: These resources can include dockercfg or dockerconfigjson content. They will be used to configure a profile for pulling images from a registry.

Kubernetes, by default, creates a Secret for each service account automatically. These Secret resources contain an associated token that can be used for interacting with the Kubernetes API from your application's components. This token is used to authenticate and authorize your processes with Kubernetes.

Ask your Kubernetes administrators if you have some **ResourceQuota** resources associated with your applications' namespaces because Secrets, as with many other resources, can be limited in their number. Hence, you may be limited when creating lots of Secrets, and you have to think about the information included. The size of Secret content is also limited to 1 MB, which will usually be more than enough for delivering sensitive configurations. If you need more, you may need to use appropriate Volumes or PVs.

Let's look at a quick example. We used `kubectl create secret settings --from-literal=user=test --from-literal=pass=testpass --from-file=mysecretfile --dry-run=client -o yaml` to generate the following Secret manifest:

```
apiVersion: v1
kind: Secret
metadata:
 name: settings
 labels:
   key1: value1
data:
  mysecretfile: IyB+Ly5iYXNocmM6...aW9uCiAgZmkKZmkK
  pass: dGVzdHBhc3M=
  user: dGVzdA==
```

Figure 10.3 – Secret manifest

We will now use the Secret values in a Pod as environment variables and mounted as a volume:

```
apiVersion: v1
kind: Pod
metadata:
 name: mypod
spec:
 containers:
 - name: mypod
   image: myimage
   env:
    - name: USERNAME
      valueFrom:
        secretKeyRef:
          name: settings
          key: user
    - name: PASSWORD
      valueFrom:
        secretKeyRef:
          name: settings
          key: pass
```

```
volumeMounts:
 - name: mysecret
   mountPath: "/etc/settings"
   readOnly: true
volumes:
- name: mysecret
  secret:
    secretName: settings
    optional: true
    defaultMode: 0400
```

Figure 10.4 – Example of the usage of a Secret resource

In the previous screenshot, we used `user` and `pass` two times. First, we added their values as environment variables, but we also used them as volumes, mounted inside `/etc/settings`. In this example, three files were created: `/etc/settings/user`, `/etc/settings/pass`, and `/etc/settings/mysecretfile`. Each file's content was defined in the Secret. Notice that we defined the file permissions using the `defaultMode` key, and we mounted the volumes in read-only mode. If we just need to mount a Secret as a file and we require this file in a specific path, we use the `subPath` key to define the name of the file. For example, if we used `kubectl create secret generic example --from-file=mysecretfile`, we could mount it as follows:

```
volumeMounts:
 - name: mysecret
   mountPath: "/etc/settings/mysecretfile"
   subPath: mysecretfile
```

We will never store Secret file manifests in clear text in our code repository. You can use any third-party tool to encrypt its content before uploading it. On the other hand, a better solution may be a solution such as **Bitnami's Sealed Secrets** (https://github.com/bitnami-labs/sealed-secrets) to create an intermediate encrypted Kubernetes **custom resource** (**CR**) entity, `SealedSecret`, which generates your Secret for you. In the `SealedSecret` manifests, the data is encrypted and you can manage it in your repositories without any problems. The `SealedSecret` entity works inside the Kubernetes cluster, hence your `SealedSecret` instance is the only software that can decrypt your data (encrypted by using certificate exchange). Your data will be safely encrypted, and it will be automatically decrypted when needed.

You can use more complex solutions such as **Hashicorp's Vault** to manage your Kubernetes Secret resources. This solution provides a lot of functionalities that can help you manage sensitive data for multiple platforms, not only Kubernetes, but it may require lots of hardware resources and management if you plan to have a highly available environment.

Cloud providers have their own software tools for deploying sensitive data on your Kubernetes cluster. They provide different access control integrations with your cloud **Identity and Access Management** (**IAM**) and may be a better solution if you plan to use a cloud platform for production. It is always interesting to ask your Kubernetes administrators for the best solution for deploying your Secrets in your environment.

Projected Volumes allow us to mount multiple resources (only Secrets, ConfigMaps, the downward API, or ServiceAccount tokens are allowed) into a container's directory. The following example shows how to mount a Secret, a container's specifications, and a ConfigMap resource:

```
...
...
spec:
  containers:
  - name: mycontainer
    image: myimage
    volumeMounts:
    - name: all-configs
      mountPath: "/etc/configs"
      readOnly: true
  volumes:
  - name: all-configs
    projected:
      sources:
      - secret:
          name: mysecret
          items:
          - key: username
            path: mygroup/myusername
      - downwardAPI:
          items:
          - path: "labels"
            fieldRef:
              fieldPath: metadata.labels
          - path: "cpu_limit"
            resourceFieldRef:
              containerName: mycontainer
              resource: limits.cpu
      - configMap:
          name: myconfigmap
          items:
          - key: config
            path: mygroup/myconfig
```

Figure 10.5 – Example of a Projected Volume

Now that we know how to inject configurations and sensitive data cluster-wide using Kubernetes resources, we will continue by reviewing how to store data from our applications. We will start with volumes, which are a storage solution included in Kubernetes' core.

Managing stateless and stateful data

When we think about storing an application's data, we must consider whether the data should persist or whether it's temporary. If the data must persist when a Pod is recreated, we must take into account that data should be available cluster-wide because containers may run on any worker host. Containers' state isn't stored by Kubernetes. If your application manages its state by using files, you may use volumes, but if this is not possible – for example, because multiple instances work at the same time – we should implement a mechanism such as a database to store its status and make it available to all instances.

Storage can be either filesystem-based, block-based, or object-based. This also applies to Kubernetes data volumes, hence before moving forward on how Kubernetes provides different solutions for volumes, let's have a quick review of these storage-type concepts:

- **Filesystem-based storage**: When we use filesystem-based storage (or file storage), we manage it by saving all the data in a hierarchal structure of folders and subfolders, provided by a local or remote operating system, where the files are kept. We use access control lists to decide whether a user is able to read or modify files' content.

  Filesystems are quite common and easy to use. We use them every day locally in our workstation or remotely via NAS, **Common Internet File System (CIFS)**, or even with cloud-specific solutions. Filesystem storage works fine for a limited number of files, with limited size, but it may be problematic with large files. In these solutions, files are indexed and this limits usage when we have an enormous number of files. It also doesn't scale well, and it's difficult to manage file locks when multiple processes are accessing the same file remotely, although it works very well for local storage.

- **Block storage**: Block storage (or block devices) is used by splitting your data into small blocks of a defined size that can be distributed in different physical local devices. Block devices can be used locally or remotely (SAN or even cloud-provided solutions), and they may be used directly or formatted using different filesystem options, depending on the underlying operating system. This storage solution is faster than filesystems and can include HA and resilience using distributed devices. But redundancy isn't cheap as it is based on the duplication of blocks. Applications must be prepared for working with block devices, and block storage isn't commonly used because it depends a lot on the infrastructure. It is very efficient for virtualization and database solutions prepared for it.

- **Object storage**: Object storage is a solution that divides the data into separate units that are stored in an underlying storage backend (block devices or filesystems). We can use distributed backends, which improves resilience and HA. Files are identified uniquely using IDs, and access to the data is easier to manage by using ACLs. It also provides redundancy and versioning. It was first developed for publishing storage on cloud providers but it is now very common in data centers. It's usually consumed via a REST API (HTTP/HTTPS), which makes it easy to include in our applications. Many backup solutions are prepared for object storage backends nowadays because they are suitable for storing large files and an enormous number of items.

Kubernetes includes some drivers that will allow us to use the reviewed storage solutions mounted as volumes, but we will use PVC resources for more advanced results.

Using volumes for storing data

Volumes in Kubernetes are used by adding their definitions to the `.spec.volumes` key. Kubernetes supports different types of volumes, and Pod resources can use any number of volume types at the same time. We have temporal (or ephemeral) volumes that will only exist during the Pod's execution, while data may persist using non-ephemeral volumes. All the volumes associated with a Pod can be mounted on any mount point (directories from containers' filesystems) on all the containers running inside. Using these volumes, we can persist an application's data.

While the `.spec.volumes` key allows us to define the volumes to be included in the Pod's mount namespace, we will use `.spec.containers[*].volumeMounts` to define in each container how and where the volumes should be used.

Let's review some of the most popular volume types:

- `emptyDir`: An `emptyDir` definition asks the kubelet component to create an empty temporary directory on your host that will follow the associated container's life cycle. When the Pod is deleted, the containers free up this storage, and kubelet removes it. By default, `emptyDir` type volumes are created on a host's filesystem but can use the `medium` subkey to define where the storage should be created. Size can also be limited by using the `sizeLimit` key. It is important to understand that these volumes are created on the hosts, hence you must be careful about their content and size. Never use them for storing unlimited logs, for example. And remember that they will be removed once the Pod is deleted/recreated.

- `hostPath`: These volumes allow us to include a defined host's storage inside the Pods. This is an important security breach that must be avoided unless your application needs to monitor or modify your host's files. Different types of `hostPath` volumes can be used by setting the `type` key. By default, a directory will be created if doesn't exist when the Pod starts, but we can create or use specific files, sockets, block devices, or even special types such as **char devices** (devices in which direct hardware access is required for interaction). Usage of `hostPath` volumes should be limited, and you must inform your Kubernetes administrators about their use in your workloads.

- `iscsi`: If you are already using **Internet Small Computer Systems Interface (iSCSI)** devices, you can expose them directly to your Pods. This allows us to include disks directly (similar to using block devices with `hostPath`). It is not very common to use `iscsi` volumes nowadays because it requires all the worker nodes to be completely equal (disk devices' names must be completely equal in all cluster hosts). You can use node labels to specify where the workloads should run, but this makes them too fixed to the underlying infrastructure.

- `nfs`: NFS volume types allow us to include a remote NFS filesystem in a Pod. The content of the mounted filesystem is maintained unless you remove it from your Pod's processes. We specify the server and the exposed path, and we can mount it in read-only mode by setting the `readOnly` key to `true`. The following screenshot shows a quick example displaying the required keys:

```
apiVersion: v1
kind: Pod
metadata:
  name: nfs-test-pod
spec:
  containers:
  - image: myimage
    name: test-container
    volumeMounts:
    - mountPath: /my-nfs-data
      name: nfs-volume
  volumes:
  - name: nfs-volume
    nfs:
      server: mynfsserver.example.com
      path: /mynfsvolume
      readOnly: true
```

Figure 10.6 – Manifest showing an NFS volume mounted in a Pod

- `PersistentVolumeClaim`: This is the most advanced volume definition and requires a full section to describe its usage. We will learn how to use them in the next subsection.

- `ephemeral`: These volumes may be considered similar to `emptyDir` because they are designed to provide ephemeral storage while a Pod is running, but they differ in that we can integrate PVC resources as ephemeral volumes or local storage. The volumes can be empty or they can already have some content when they are attached to the Pod.

A lot of cloud providers' volumes have moved from Kubernetes core-based storage to modern PVC management using external PV dynamic provisioners. Disassembling the storage provisioning from the Kubernetes code allows hardware storage manufacturers, cloud providers, and middleware software creators to prepare their own solutions and evolve them out of the Kubernetes development cycle. We will now learn how PV and PVC resources allow us to improve storage management on Kubernetes.

Enhancing storage management in Kubernetes with PV resources

A PV resource presents a unit of storage that can be used within Kubernetes. PVs can be created manually or dynamically by a storage provisioning backend.

While the volume definitions seen so far are declared namespace-wide, PV resources are defined cluster-wide, by Kubernetes administrators. We use them to define the storage capacity (size) using the `capacity.storage` key and the mode it will be consumed using the `accessMode` key. Let's quickly review these modes, because our applications may need specific access to the data, especially when we run multiple replicas of a process:

- `ReadWriteOnce`: This mode presents the storage in write mode only for the first Pod that attaches it. Other Pods (replicas or even defined in other different workloads) can only have read access to the data.

- `ReadOnlyMany`: In this mode, all the Pods will mount the volume in read-only mode.

- `ReadWriteMany`: This is the option to use when processes running in different Pods need to write on a PV at the same time. You must ensure your application manages the locks for writing your data without corrupting it.

You have to be aware that `accessMode` does not enforce write protection once the volume is mounted. If you need to ensure that it is mounted in read-only mode in some Pods, you must use the `readOnly` key.

Multiple access modes can be defined in a PV resource but it only uses one when it is mounted. This helps the cluster to bind PVs, but it will fit the specifications in each volume request.

> **Important note**
>
> The mode also affects how applications will be updated by issuing a rolling update. By default, Deployment resources will start a new Pod instance before the old one is stopped to maintain the application working. But the `ReadWriteOnce` access mode will only allow the old Pod to access the storage while the new one will wait forever to attach it. In such situations, it may be interesting to change the default rolling update behavior to ensure that the old processes stop completely and free the volume before the new ones start with the storage attached.

To use an available PV, we will use a PVC resource that may be considered a request for storage. These resources are defined namespace-wide because they will be used by our workloads. You, as a developer, will ask for storage for your application, defining a PVC with the required `capacity.storage` and `accessMode` keys. When both options match an already created and free PV resource, they are bound and the storage is attached to the Pod. In fact, the volumes are attached to the host that runs the Pod, and kubelet makes it available inside the Pod.

Labels can be used to fix PVCs with a subset of PVs by using `selector.matchLabels` and an appropriate label. The following screenshot shows an example of `PersistentVolume` created as a local `hostPath` variable:

```
apiVersion: v1                          apiVersion: v1

kind: PersistentVolume                  kind: PersistentVolumeClaim

metadata:                               metadata:

 name: task-pv-volume                    name: task-pv-claim

 labels:                                spec:

   type: local                            storageClassName: manual

spec:                                     accessModes:

   storageClassName: manual                - ReadWriteOnce

   capacity:                             resources:

     storage: 10Gi                         requests:

   accessModes:                             storage: 3Gi

    - ReadWriteOnce

   hostPath:

    path: "/mnt/data"
```

Figure 10.7 – PV and PVC manifests

If the PVC is created, but there isn't a PV available matching the PVC's defined requirements, it will stay unbound forever, waiting for a PV to be created or free in the Kubernetes cluster.

When a Pod is using a PVC and it is bound to a PV, the storage can't be removed until the Pod frees it when the Pod is removed. The following screenshot shows how a Pod uses a PVC:

```
apiVersion: v1

kind: Pod                               containers:

metadata:                                 - name: test-pv-container

 name: test-pvc-pod                         image: nginx

spec:                                       ports:

   volumes:                                   - containerPort: 80

    - name: task-pv-storage                     name: "http-server"

      persistentVolumeClaim:            volumeMounts:

        claimName: task-pv-claim           - mountPath: "/usr/share/nginx/html"

                                              name: task-pv-storage
```

Figure 10.8 – Pod manifest using a PVC associated with a PV

PVs can be resized if the underlying filesystem is either `xfs`, `ext3`, or `ext4`. Depending on the storage backend, we can clone the PV content or even create snapshots, which may be very interesting for debugging purposes when a problem is occurring in your application with a set of data or for backups.

Node affinity can be used to use specific cluster nodes with specific directories or even disk devices and mount the PVs wherever the directory is present. You, as a developer, should avoid node affinity unless your Kubernetes administrators ask you to use this feature as it makes your workloads infrastructure-dependent.

PVs can be provisioned either manually by your Kubernetes administrator or dynamically, using a container storage interface integrated solution that will interact with a storage backend to create the volumes for you. Multiple storage solutions can coexist in your cluster, which may provide different storage capabilities. No matter whether your platform uses dynamic provisioning or manually created volumes, we may access faster backends, more resilient ones, or some with specific housekeeping.

> **Important note**
>
> It is important to understand that matching a PVC to an appropriate PV can lead to bad use of disk space. When a PVC asks for 5 Gi and a PV with a size of 10 Gi is available, they will be bound despite their size not matching. This means we'll use only 50% of the available space. That's why dynamic provisioning is so important, because it creates PVs with the exact size required and then they are bound, hence the data fits perfectly in the provisioned storage.

We will use StorageClass resources to classify the PVs by their capabilities. Kubernetes administrators will associate created PV resources to any of the configured StorageClass resources. A default StorageClass may be defined to associate all PVs without a `spec.storageClassName` key. Whenever we create a PVC resource, we can either define a specific StorageClass or wait for the cluster to assign a PV from the default StorageClass to our claim.

Now that we know about the basic concepts of storage management, we can take a quick look at the provisioning and decommissioning of storage.

Provisioning and decommissioning storage

Kubernetes does not manage how PVs are created or destroyed in storage backends. It just interacts with their interfaces via a REST API to retrieve and follow the changes and states of the storage.

Dynamic provisioning requires some Kubernetes administration work. Kubernetes administrators will deploy **Container Storage Interface (CSI)** solutions and attach them to specific StorageClass resources, making them available for users. Each StorageClass resource includes its own parameters for invoking the provisioner, hence different PV resources will be created, depending on the mechanism used to create them.

The following screenshot shows two StorageClass resources using a **Google Compute Engine (GCE)** storage backend with different parameters (appropriate provisioner software must be installed in your Kubernetes cluster to use it, via the `kubernetes.io/gce-pd` API endpoint) for creating standard and SSD (fast hard drives) PVs:

```
apiVersion: storage.k8s.io/v1

kind: StorageClass

metadata:

 name: standard

reclaimPolicy: Retain

provisioner: kubernetes.io/gce-pd

parameters:

 type: pd-standard
```

```
apiVersion: storage.k8s.io/v1

kind: StorageClass

metadata:

 name: sssd

reclaimPolicy: Delete

provisioner: kubernetes.io/gce-pd

parameters:

 type: pd-ssd
```

Figure 10.9 – StorageClass manifests for standard and SSD storage

Notice the `reclaimPolicy` key, which manages the behavior of the provisioner when a Pod frees a PV and this is reused:

- A `Retain` policy will not modify the content of the PV, therefore we need to remove it from the provisioner backend manually when the data isn't needed anymore.

- A `Recycle` policy will basically delete all the content of the volume (such as issuing `rm -rf /volumedata/*` from a Pod mounting the PV). This is only supported in NFS and `hostPath` provisioners.

- A `Delete` policy will completely delete the PV and its content by asking the provisioner API to delete the associated volume.

> **Important note**
> Using PV and PVC resources allows us to prepare our application's workloads to work on any Kubernetes infrastructure. We will just modify `StorageClassName` to fit any of the StorageClasses present in the deployment platform.

We will now review some labs that will help us better understand some of the content of this chapter.

Labs

In this section, we will use some of the resources presented to improve our `simplestlab` application. We will include sensitive data in a Secret resource, NGINX configurations in a ConfigMap resource, and a simple StorageClass resource to implement a PV resource and a PVC resource to present storage for our StatefulSet resource.

The code for all the labs is available in this book's GitHub repository at `https://github.com/PacktPublishing/Containers-for-Developers-Handbook.git`. Ensure you have the latest revision available by simply executing `git clone https://github.com/PacktPublishing/Containers-for-Developers-Handbook.git` to download all its content or `git pull` if you have already downloaded the repository. All the manifests and the steps required for adding storage to the `simplestlab` application are located inside the `Containers-for-Developers-Handbook/Chapter10` directory.

This section will show you how to implement different volume solutions on the `simplestlab` tier-three application, prepared for Kubernetes in *Chapter 9, Implementing Architecture Patterns*.

These are the tasks you will find in the Chapter 10 GitHub repository:

- We will improve the security of the `simplestlab` application by adding Secret resources on any sensitive data required by the application. We will review each application component, create the required Secret resources, and modify the component's manifests to include the newly created resources. This includes some database initialization scripts required for the database component.

- We will also prepare different ConfigMap resources for the `lb` application's component and verify how these changes impact the application behavior.

You can use one of the following Kubernetes desktop environments:

- Docker Desktop

- Rancher Desktop

- Minikube

The labs will work on any of them, and of course, on any other Kubernetes environment. You may find issues with their default storage class, but there are some comments on the files that may be changed. We'll start with improving the `simplestlab` application on Kubernetes:

1. The `simplestlab` application is a very simple tier-3 application (a load balancer could present additional static content but it is not added for the purposes of the labs). To make it more secure, we're going to add Secret resources for defining all the required user authentications.

The application is composed of three components:

- A db component: Postgres database

- An app component: Application backend in Node.js

- An lb component: NGINX fronted for static content

We have included in their manifests some of the storage solutions learned in this chapter.

2. Let's look at the database component first. As we are already using a StatefulSet resource, a PVC resource is configured, but the postgres user password was presented in the database container in clear text. We added a Secret resource to include this password, and we also included a complete database initialization script that will allow us to prepare the database. In the previous labs, this component was initialized with the script included in the container image. In this case, we can manage how the database for the application will be created by modifying this script and replacing the Secret. You must know that we can't use this mechanism to modify a previously initialized database. That's why we expect to deploy this component from scratch.

These are two Secret manifests created:

- dbcredentials.secret.yaml, the content of which is shown here:

```
apiVersion: v1
data:
  POSTGRES_PASSWORD: Y2hhbmdlbWU=
kind: Secret
metadata:
  name: dbcredentials
```

- initdb.secret.yaml: This Secret is created by including the content of init-demo.sh script:

```
$ cat init-demo.sh
#!/bin/bash
set -e
psql -v ON_ERROR_STOP=1 --username "$POSTGRES_USER" <<-EOSQL
    CREATE USER demo with PASSWORD 'd3m0' ;
    CREATE DATABASE demo owner demo;
    GRANT ALL PRIVILEGES ON DATABASE demo TO demo;
    \connect demo;
    CREATE TABLE IF NOT EXISTS hits
    (
        hitid serial,
        serverip varchar(15) NOT NULL,
        clientip varchar(15) NOT NULL,
        date timestamp without time zone,
        PRIMARY KEY (hitid)
```

```
        );
      ALTER TABLE hits OWNER TO demo;
EOSQL
```

We used the following command line to create an `initdb.secret.yaml` Secret manifest:

```
$ kubectl create secret generic initdb \
--from-file=init-demo.sh --dry-run=client \
-o yaml |tee initdb.secret.yaml
apiVersion: v1
data:
  init-demo.sh: IyEvYmluL2Jhc2gKc2V0IC1lCgpwc3FsIC12IE9OX-
0VSUk9SX1NUT1A9MSAt....pZCkKICAgICk7CiAgICBBBTFRFRUiBUQUJMRSBoaXR-
zIE9XTkVSIFRPIGR1bW87CkVPU1FMCg==
kind: Secret
metadata:
  creationTimestamp: null
  name: initdb
```

3. In both cases, the values are encoded in Base64 format. They aren't encrypted, as we can verify:

```
$ kubectl apply -f dbcredentials.secret.yaml
$ kubectl get secret dbcredentials \
-ojsonpath="{.data.POSTGRES_PASSWORD}"|base64 -d
changeme
```

4. Let's see now how the `StatefulSet` manifest was modified:

```
apiVersion: apps/v1
kind: StatefulSet
metadata:
  name: db
  labels:
    component: db
    app: simplestlab
spec:
  replicas: 1
...
      volumes:
      - name: initdb-secret
        secret:
          secretName: initdb
          optional: true
      containers:
      - name: database
        image: docker.io/frjaraur/simplestdb:1.0
```

```
. . .
        env:
        - name: POSTGRES_PASSWORD
          valueFrom:
            secretKeyRef:
              name: dbcredentials
              key: POSTGRES_PASSWORD
        - name: PGDATA
          value: /data/postgres
        volumeMounts:
        - name: postgresdata
          mountPath: /data
        - name: initdb-secret
          mountPath: "/docker-entrypoint-initdb.d/"
          readOnly: true
. . .
  volumeClaimTemplates:
  - metadata:
      name: postgresdata
    spec:
      accessModes: [ "ReadWriteOnce" ]
      #storageClassName: "csi-hostpath-sc"
      resources:
        requests:
          storage: 1Gi
```

The full manifest file can be found in the Chapter10/db.statefulset.yaml file.

5. We create all the components for the database:

```
$ kubectl create -f dbcredentials.secret.yaml \
-f initdb.secret.yaml
secret/dbcredentials created
secret/initdb created
$ kubectl create -f db.statefulset.yaml
statefulset.apps/db created
$ kubectl create -f db.service.yaml
service/db created
```

The service wasn't modified at all.

6. We check if the user was created by connecting to the database server and showing its users:

```
$ kubectl exec -ti db-0 -- psql -U postgres
psql (15.3)
Type "help" for help.
```

```
postgres=# \du
                              List of roles
 Role name |                      Attributes
 | Member of
-----------+---------------------------------------------------
--------+-----------
 demo      |
 | {}
 postgres  | Superuser, Create role, Create DB, Replication,
Bypass RLS | {}
postgres=# \q
could not save history to file "//.psql_history": Permission
denied
```

Notice that we modified the content of the POSTGRES_PASSWORD variable, and now it's taken from the Secret we created.

We also included initdb-secret as a volume, and it's mounted in the /docker-entrypoint-initdb.d directory. Notice that we didn't use subPath because this directory is empty. You can change the content of the Secret and it will be synced inside the containers, but this will not change the authentication values in the database because it is an initialization script. You can modify it to enforce the change of the password via SQL.

7. We can now review the PVC, created by using a template (because it is a StatefulSet) and the associated PV:

```
$ kubectl get pvc
NAME                  STATUS    VOLUME
CAPACITY    ACCESS MODES    STORAGECLASS    AGE
postgresdata-db-0    Bound     pvc-4999f00b-deb3-4cec-97a0-
3a289c4457d9    1Gi          RWO             hostpath
168m
$ kubectl get pv pvc-4999f00b-deb3-4cec-97a0-3a289c4457d9
NAME                                          CAPACITY    ACCESS
MODES    RECLAIM POLICY    STATUS    CLAIM        STORAGE-
CLASS    REASON    AGE
pvc-4999f00b-deb3-4cec-
97a03a289c4457d9    1Gi          RWO Delete        Bound     default/
postgresdata-db-0    hostpath                      168m
```

8. A StorageClass resource is defined in Docker Desktop and we use it by default:

```
$ kubectl get sc
NAME                  PROVISIONER         RECLAIMPOL-
ICY    VOLUMEBINDINGMODE    ALLOWVOLUMEEXPANSION    AGE
hostpath (default)    docker.io/hostpath    Delete      Immedi-
ate           false                12d
```

We can now continue and review the changes in the app component.

9. For the application backend component, we used the imperative method to create an `appcre-dentials` Secret. This method does not generate a YAML manifest, which may be a problem because you will need to store your passwords somewhere. If you need to store all your manifest in your code repository, which is always recommended, you must always encrypt your Secret manifests:

```
$ kubectl create secret generic appcredentials \
--from-literal=dbhost=db \
--from-literal=dbname=demo \
--from-literal=dbuser=demo \
--from-literal=dbpasswd=d3m0
secret/appcredentials created
```

The values for these variables must be the ones used in the initialization script.

10. Let's review the changes included to load the database authentication in the application:

```
apiVersion: apps/v1
kind: Deployment
metadata:
  name: app
...
        containers:
        - name: app
          image: docker.io/frjaraur/simplestapp:1.0
          ports:
          - containerPort: 3000
          env:
          - name: dbhost
            valueFrom:
              secretKeyRef:
                name: appcredentials
                key: dbhost
          - name: dbname
            valueFrom:
              secretKeyRef:
                name: appcredentials
                key: dbname
    ...
```

The full manifest file can be found in the `Chapter10/app.deployment.yaml` file.

11. We have just included all the required environment variables from the Secret created before. We deploy the app manifests:

```
$ kubectl create -f app.deployment.yaml
-f app.service.yaml
```

```
deployment.apps/app created
service/app created
```

We now verify the content included in the containers:

```
$ kubectl exec -ti app-5f9797d755-2bgtt - env
...
dbhost=db
dbname=demo
dbuser=demo
dbpasswd=d3m0
...
$ kubectl get pods
NAME                        READY   STATUS    RESTARTS   AGE
app-5f9797d755-2bgtt        1/1     Running   0          100s
app-5f9797d755-gdpw7        1/1     Running   0          100s
app-5f9797d755-rzkqz        1/1     Running   0          100s
db-0                        1/1     Running   0          179m
```

12. We can now continue with the frontend component. In the previous versions of this application's deployment, we were already using a ConfigMap resource for configuring the NGINX load balancer.

This is the content of the ConfigMap resource with these special configurations for our NGINX load balancer:

```
apiVersion: v1
kind: ConfigMap
metadata:
  name: lb-config
  labels:
    component: lb
    app: simplestlab
data:
  nginx.conf: |
    user  nginx;
    worker_processes  auto;
    error_log  /tmp/nginx/error.log warn;
    pid        /tmp/nginx/nginx.pid;
    events {
      worker_connections  1024;
    }
    http {
      server {
        listen 8080; # specify a port higher than 1024 if
running as non-root user
        location /healthz {
```

```
                    add_header Content-Type text/plain;
                    return 200 'OK';
                }
              location / {
                proxy_pass http://app:3000;
              }
            }
          }
```

13. We deploy all the `lb` manifests:

```
$ kubectl create -f lb.daemonset.yaml \
-f lb.configmap.yaml -f  lb.service.yaml
daemonset.apps/lb created
configmap/lb-config created
service/lb create
```

14. We review the status of all the application's components and the configuration applied to the NGINX component:

```
$ kubectl get pods
NAME                        READY   STATUS     RESTARTS    AGE
app-5f9797d755-2bgtt        1/1     Running    0           5m52s
app-5f9797d755-gdpw7        1/1     Running    0           5m52s
app-5f9797d755-rzkqz        1/1     Running    0           5m52s
db-0                        1/1     Running    0           3h4m
lb-zcm6q                    1/1     Running    0           2m2s
$ kubectl exec lb-zcm6q -- cat /etc/nginx/nginx.conf
user  nginx;
worker_processes  auto;
...     location / {
          proxy_pass http://app:3000;
        }
      }
    }
$
```

15. And now, we can reach our application in any Kubernetes cluster host's port 32000. Your browser should access the application and show something like this (if you're using Docker Desktop, you will need to use `http://localhost:32000`):

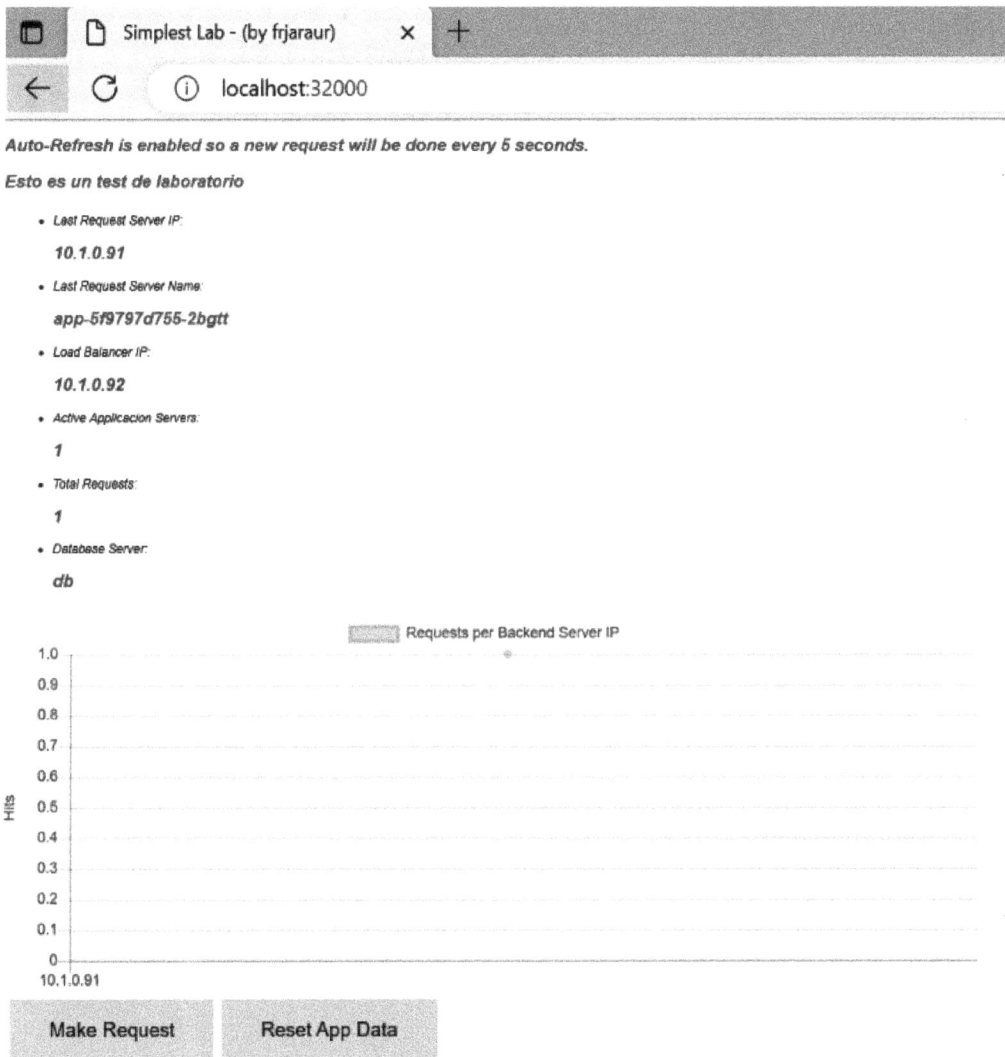

Figure 10.10 – simplestlab application web GUI

With this final step, you deployed completely the three components of the simplestlab application. You will find additional steps in the Chapter10 folder in the GitHub repository to add a new app component instance and modify the load balancer component to reach this new backend.

These labs will help you understand how Secret resources improve the security of an application and how to implement different volume types for different application needs.

Summary

In this chapter, we reviewed how we can manage data in a Kubernetes cluster, which is currently the most popular container orchestrator, and you will probably use it for most of your projects. We learned how to include configurations and sensitive data in our application's containers, and how we can manage stateless and stateful storage using different Kubernetes resources. At the end of the chapter, we learned how we can use dynamic provisioning of data volumes for our applications, which really fits in the microservices model, where automation is crucial for abstracting resources from the underlying infrastructure. This chapter is very important because it has taught you how to manage data in microservices and container-based environments.

We will follow this up in the next chapter, in which we will learn how to publish our applications using best security practices, isolating all backend components from users and other applications.

11

Publishing Applications

Running applications on Kubernetes adds resilience to all of an application's components by running its processes as containers. This helps us to provide stability and update these components without impacting our users. Although Kubernetes provides a lot of resources to simplify the cluster-wide management of the applications, we do need to understand how using these resources will affect the way our applications are reached by our users.

In this chapter, we will learn how to publish our applications to make them accessible to our users. This will involve publishing certain Pods or containers to provide services, but sometimes, we may also need to debug our applications to fix issues that arise.

By the end of this chapter, we will have learned how **NetworkPolicy resources** help us to isolate the workloads deployed in our cluster, and we will have reviewed the use of **service mesh** solutions to improve the overall security between our applications' components.

We will cover the following topics in this chapter:

- Understanding Kubernetes features for publishing applications cluster-wide
- Proxying and forwarding applications for debugging
- Using the host network namespace for publishing applications
- Publishing applications with Kubernetes' NodePort feature
- Providing access to your Services with LoadBalancer Services
- Understanding Ingress Controllers
- Improving our applications' security

We will start this chapter by reviewing the different options we have with Kubernetes out of the box for delivering our applications to our users.

Technical requirements

You can find the labs for this chapter at `https://github.com/PacktPublishing/ Containers-for-Developers-Handbook/tree/main/Chapter11`, where you will find some extended explanations, omitted in the chapter's content to make it easier to follow. The *Code In Action* video for this chapter can be found at `https://packt.link/JdOIY`.

Understanding Kubernetes features for publishing applications cluster-wide

Kubernetes is a container orchestrator that allows users to run their applications' workloads cluster-wide. We reviewed in *Chapter 9, Implementing Architecture Patterns*, the different patterns we can use to deploy our applications using different Kubernetes resources. Pods are the minimum deployment unit for our applications and have dynamic IP addresses, thus we can't use them for publishing our applications. Dynamism affects the exposure of all the components internally and externally – while Kubernetes successfully makes the creation and removal of containers simple, the IP addresses used will continuously change. Therefore, we need an intermediate component, the Service resource, to manage the interaction of any kind of client with the Pods (running on the backend) associated with an application component. We can also have Service resources pointing to external resources (for example, the `ExternalName` Service type).

> **Important note**
> It is crucial to understand that not all an application's components need to be accessible outside of the cluster or even namespace scopes. In this chapter, we are going to learn different options and mechanisms for publishing applications for access both inside and outside of the Kubernetes cluster. You as a developer must know and understand which of the application's components will act as frontends for your application and thus must be accessible and which should act as backends and be reachable, and employ the appropriate mechanisms in each case.

We will use Kubernetes Service resources to publish the application's Pods internally or externally as required. We will never connect to Pods' published ports directly. Pods' ports will be associated with a Service resource using labels. An intermediate resource is created to associate Services with Pods, EndpointSlices, and Endpoint resources. These resources are created automatically for you when you create a Service and the associated Pods are located. The EndpointSlices point to Endpoint resources, which are updated when the backend Pods (or external Services) change.

Let's see how this works with an example. We will create a Service resource before its actual Pods. The following code snippet shows an example of a Service resource manifest:

```
apiVersion: v1
kind: Service
metadata:
```

```
  name: myservice
spec:
  selector:
    myapp: test
  ports:
    - protocol: TCP
      port: 80
      targetPort: 8080
```

If we create the Service resource using the preceding YAML manifest and retrieve the created endpoints, we will see which Pods (with their IP addresses) are associated as backends. Let's see the currently associated endpoints:

```
$ kubectl get endpoints myservice
NAME          ENDPOINTS    AGE
myservice     <none>       29s
```

The list of endpoints is empty because we don't have any associated backend Pod (with the myapp=test label). Let's create a simple Pod with this label using kubectl run:

```
$ kubectl run mypod1 --labels myapp=test \
--image=nginx:alpine
pod/mypod1 created
```

We now review the associated Pods again:

```
$ kubectl get endpoints myservice
NAME          ENDPOINTS          AGE
myservice     10.1.0.113:8080    5m58s
```

Notice that we didn't specify any port for the Pod, hence the association may be wrong (in fact, the docker.io/nginx:alpine image defines port 80 for the process). Kubernetes does not verify this information; it just creates the required links between resources.

EndpointSlice resources are managed by Kubernetes and are dynamically updated whenever a new Pod is created or an old one fails (in fact, the backend Endpoint resources change and the update is propagated). If you are having problems with a Service not responding but your Pods are running, this is something you may need to check.

This is just an example of creating a simple **ClusterIP** Service, which is the default option. We already learned the different Service resource types in *Chapter 9*, *Implementing Architecture Patterns*, but it may be important to quickly review those types that allow us to publish applications:

- **ClusterIP**: This is the default type, used to publish a Service internally. An FQDN is created in Kubernetes' internal DNS (CoreDNS component) associated with the IP address of the Service resource (assigned by the internal IPAM from the Services' pool).

- **Headless**: These Services don't have an associated IP address, although they also have an FQDN. In this case, all IP addresses of the Pods associated with the Endpoint resource will be resolved.

- **NodePort**: This Service type allows us to publish an application's component externally, outside of the cluster network. It will create a NAT association between a port in all the cluster hosts and the port of the Service. When external network packets reach the host's associated port, the overlay network routes the traffic to the appropriate Pods. Services are a logical resource; they don't really have a physical device or associated software component. They just link the Service's IP address with the backend endpoints via DNS. The ports used by the NodePort Service resource can be fixed when we create it, or we can let Kubernetes choose a random one for us from the 30000-32767 port range. It is important to understand that NodePort Services have a ClusterIP address, associated via the internal Service's FQDN.

- **LoadBalancer**: This type of Service resource integrates with external cloud or on-premise software or hardware load balancers (or it creates them in your cloud infrastructure) from the underlying infrastructure to route user traffic to an application's Pods. In this case, a NodePort is created (along with its associated ClusterIP) to route the traffic from the external load balancer to the backend Endpoint resources. Kubernetes will use its own integration with the cloud infrastructure to create the required load balancers or apply specific configurations pointing to the associated NodePorts whenever a LoadBalancer Service resource is created.

> **Important note**
> We employ ClusterIP and Headless Services for internal use and NodePort and LoadBalancer Services whenever we are going to expose our applications. But this is not strictly true, as we can also use **Ingress Controllers** to publish applications without using either NodePort or LoadBalancer resources. This helps you to abstract your applications from the underlying infrastructure.

Let's continue exploring the different options provided by the Kubernetes platform for publishing applications by introducing the **Ingress Controller** concept. An Ingress Controller is a Kubernetes controller that we can add to our cluster to implement reverse proxy functionalities. This will allow us to use ClusterIP Service resources to expose our applications because the traffic coming from our users will be routed entirely internally from this proxy component to the Service and then reach the associated Pods. This proxy is configured dynamically by using **Ingress** resources. These resources allow us to define our applications' host headers and link them to our Service resources. Your work as a developer involves creating appropriate Ingress resources for your frontend application's components.

Finally, let's introduce the Kubernetes **proxy** publishing feature that directly proxifies the cluster API on a defined port. This will allow us to directly access the API and all the resources included in the Kubernetes cluster. For example, we can get all the Pods defined in the default namespace using the /api/v1/namespaces/default/pods path.

For debugging purposes, we can also use kubectl port-forward, which proxies specific Services to our desktop computer client. Note that neither method, proxy or port-forward, should be permitted in production because they directly expose important resources, bypassing our Kubernetes and load balancer infrastructure security.

In the next section, we will use the `kubectl proxy` feature to access a Service resource and reach our application.

Proxying and forwarding applications for debugging

In this section, we will learn how to publish the Kubernetes API directly on our desktop computer and reach any Service created in the cluster (if we have the appropriate permissions), and how to forward a Service directly to our client computer using the `port-forward` feature.

Kubernetes client proxy feature

We use `kubectl proxy` to enable the Kubernetes proxy feature. Some important options help us manage how and where the Kubernetes API will be accessible. We use the following options to define where the Kubernetes API will be published:

- `--address`: This option allows us to define the IP address of our client host used for publishing the Kubernetes API. By default, `127.0.0.1` is used.

- `--port` or `-p`: This option is used to set the specific port where the Kubernetes API will be available. The default value is `8001`, and although we can let Kubernetes use a random port by using `-p=0`, it is recommended to always define a specific port.

- `--unix-socket` or `-u`: This option is used to define a Unix socket instead of a TCP port, which is more secure if you limit access to the socket at the filesystem level.

The following options are used to secure the Kubernetes API access:

- `--accept-hosts` and `--accept-paths`: These options allow us to ensure that only specific host headers and API paths will be allowed. For example, we can ensure local access only using the following regex pattern, `'^localhost$,^127\.0\.0\.1$,^\[::1\]$'`, with the `--accept-hosts` argument.

- `--reject-methods`: We can block specific API methods by rejecting them. For example, we can disable the patching of any Kubernetes resource by using `kubectl proxy --reject-methods='PATCH'`.

- `--reject-paths`: We can specify certain paths to be denied by using this option. We can, for example, disable the attachment of a new process to a Pod resource (`kubectl exec` equivalent) by using `--reject-paths='^/api/.*/pods/.*/exec,'`.

It is important to understand that, although we have seen some options for ensuring security, the Kubernetes proxy feature shouldn't be used in production environments because it may be possible to bypass the RBAC system if someone gets access to the API via the proxied port. The user authentication used for creating the proxy will allow access to anyone via the exposed API.

This method should only be used for debugging in either your own Docker Desktop, Rancher Desktop, or Minikube for exposing a Kubernetes remote development environment. Your Kubernetes administrators must enable this method for you if you are not using your own Kubernetes environment. You must ensure that your operating system allows access to the specified port by reviewing your firewall settings if you still aren't able to reach the proxied Kubernetes API.

Now we have reviewed how we can employ this method to publish Kubernetes APIs, let's use it to access a created Service resource with a quick example:

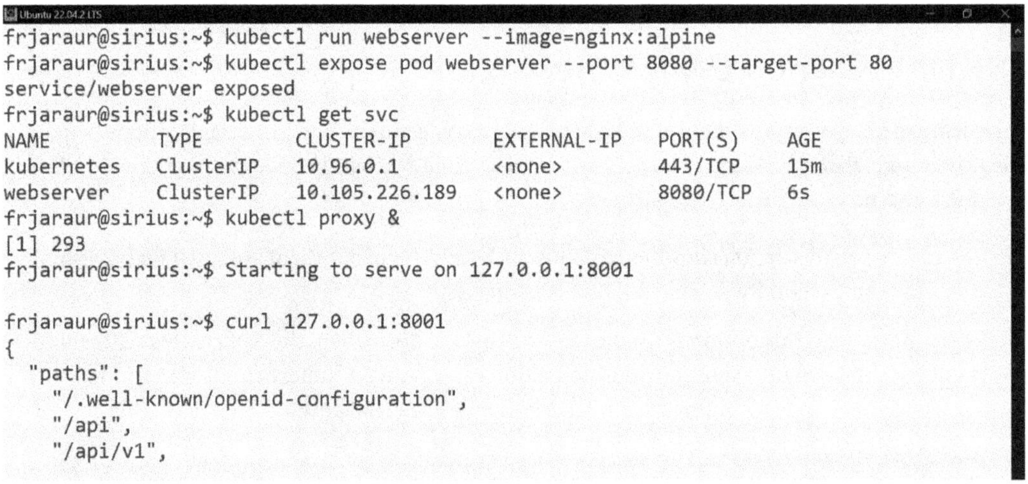

```
frjaraur@sirius:~$ kubectl run webserver --image=nginx:alpine
frjaraur@sirius:~$ kubectl expose pod webserver --port 8080 --target-port 80
service/webserver exposed
frjaraur@sirius:~$ kubectl get svc
NAME          TYPE        CLUSTER-IP       EXTERNAL-IP    PORT(S)    AGE
kubernetes    ClusterIP   10.96.0.1        <none>         443/TCP    15m
webserver     ClusterIP   10.105.226.189   <none>         8080/TCP   6s
frjaraur@sirius:~$ kubectl proxy &
[1] 293
frjaraur@sirius:~$ Starting to serve on 127.0.0.1:8001

frjaraur@sirius:~$ curl 127.0.0.1:8001
{
  "paths": [
    "/.well-known/openid-configuration",
    "/api",
    "/api/v1",
```

Figure 11.1 – Creating a simple webserver Service with NGINX and exposing the Kubernetes API

Important note

Notice that we executed `kubectl proxy` in the background using &. We did this to be able to continue in the current terminal. The `kubectl proxy` action runs in the foreground, and it will keep running until we issue *Ctrl + C* to terminate the process. To end the background execution, we can use the following steps:

```
$ jobs
[1]+  Running                 kubectl proxy &
$ kill %1
```

Now that we have access to Kubernetes API, we can access the ClusterIP Service resource directly using the proxied port, but first, let's review the Service resource:

```
Ubuntu 22.04.2 LTS                                                              — σ  ×
frjaraur@sirius:~$ curl 127.0.0.1:8001/api/v1/namespaces/default/services/webserver/
{
  "kind": "Service",
  "apiVersion": "v1",
  "metadata": {
    "name": "webserver",
    "namespace": "default",
    "uid": "01a675ab-c5ae-4f26-9c51-45f44cc85f80",
    "resourceVersion": "1577",
    "creationTimestamp": "2023-07-15T08:40:15Z",
    "labels": {
      "run": "webserver"
    },
```

Figure 11.2 – Accessing the webserver Service resource using the Kubernetes proxy

We configured port `8080` for the `webserver` Service resource. The Kubernetes proxy will publish the Service resources using the following URI format (Kubernetes API):

`/api/v1/namespaces/<NAMESPACE>/services/<SERVICE_NAME>:<SERVICE_PORT>/proxy/`

Therefore, the `webserver` Service is accessible in `/api/v1/namespaces/default/services/webserver:8080/proxy/`, and we can reach NGINX's default `index.html` page, as we can see in the following screenshot:

```
Ubuntu 22.04.2 LTS                                                              — σ  ×
frjaraur@sirius:~$ curl 127.0.0.1:8001/api/v1/namespaces/default/services/webserver:8080/proxy/
<!DOCTYPE html>
<html>
<head>
<title>Welcome to nginx!</title>
<style>
html { color-scheme: light dark; }
body { width: 35em; margin: 0 auto;
font-family: Tahoma, Verdana, Arial, sans-serif; }
</style>
</head>
<body>
<h1>Welcome to nginx!</h1>
<p>If you see this page, the nginx web server is successfully installed and
working. Further configuration is required.</p>

<p>For online documentation and support please refer to
<a href="http://nginx.org/">nginx.org</a>.<br/>
Commercial support is available at
<a href="http://nginx.com/">nginx.com</a>.</p>

<p><em>Thank you for using nginx.</em></p>
</body>
</html>
```

Figure 11.3 – Accessing the webserver Service using the kubectl proxy feature

The Service is accessible and we reached the `webserver` Service's default page. Let's now review how we can forward the Service's port to our desktop computer without having to implement a complex routing infrastructure.

Kubernetes client port-forward feature

Instead of accessing the full Kubernetes API, we can use `kubectl port-forward` to forward ports from a Service, Deployment, ReplicaSet, StatefulSet, or even a Pod resource directly. In this case, a transparent NAT is used to forward a backend port to a port defined on our desktop computer by executing the `kubectl` command-line client.

Let's see how this works, using the `webserver` Service defined in the previous section as an example:

```
frjaraur@sirius:~$ kubectl get svc
NAME         TYPE        CLUSTER-IP       EXTERNAL-IP    PORT(S)    AGE
kubernetes   ClusterIP   10.96.0.1        <none>         443/TCP    131m
webserver    ClusterIP   10.105.226.189   <none>         8080/TCP   115m
frjaraur@sirius:~$ kubectl port-forward svc/webserver 8080 &
[1] 368
frjaraur@sirius:~$ Forwarding from 127.0.0.1:8080 -> 80
Forwarding from [::1]:8080 -> 80

frjaraur@sirius:~$ curl 127.0.0.1:8080
Handling connection for 8080
<!DOCTYPE html>
<html>
<head>
<title>Welcome to nginx!</title>
<style>
html { color-scheme: light dark; }
body { width: 35em; margin: 0 auto;
font-family: Tahoma, Verdana, Arial, sans-serif; }
</style>
</head>
<body>
<h1>Welcome to nginx!</h1>
```

Figure 11.4 – Using port-forward to publish the webserver Service resource example

As you can see in this example, it is quite simple to forward any application's Kubernetes resource listening on a defined port. We can specify the port attached to our application in our local client by using `[LOCAL_PORT:]RESOURCE_PORT`. Note that it is important to choose the local IP address when working on a multihomed host with multiple IP addresses using the `--address` argument. This will improve the overall security by attaching an interface if we define the appropriate firewall rules to only allow our host. By default, `localhost` is used, which means that it will remain secure as long as we are the only user with access to our desktop computer.

In the next section, we will discuss the direct use of the host's kernel network namespace for publishing Pod resources.

Using the host network namespace for publishing applications

So far, we have seen different methods for accessing ClusterIP Service resources or Pods (created using different workload types) by either proxying or forwarding their ports to our desktop computers. Sometimes, however, the applications require a direct connection to the host's interfaces, without the bridge interface created by the container runtime. In this case, the containers in the Pod will use the network namespace of the host, which allows the processes inside to control the host because they will have access to all the host's interfaces and network traffic. This can be dangerous and must only be used to manage and monitor the host's interfaces.

Using the hostNetwork key

To use the host's network namespace, we set the `hostNetwork` key to `true`. The Pod will now get all the IP addresses associated with the host, including those of all the virtual interfaces associated with the containers running in that host. But what is particularly important in terms of publishing our applications is that they will be accessible through any of the host's IP addresses, waiting for requests on the ports defined by the `ports` keys in the Pod `spec` section. Let's see how this works by executing an NGINX Pod with the aforementioned `hostNetwork` key. We will use `cat` (redirected to `kubectl`) to create a Pod resource on the fly using the `nginx/nginx-unprivileged:stable-alpine3.18` image (which uses the unprivileged port `8080`):

```
frjaraun@sirius:~$ cat <<EOF|kubectl create -f -
apiVersion: v1
kind: Pod
metadata:
  labels:
    run: webserver
  name: webserver
spec:
  hostNetwork: true
  containers:
  - image: nginxinc/nginx-unprivileged:stable-alpine3.18
    name: webserver
EOF
pod/webserver created
frjaraun@sirius:~$ kubectl get pods -o wide
NAME        READY   STATUS    RESTARTS   AGE   IP             NODE             NOMINATED NODE   READINESS GATES
webserver   1/1     Running   0          10s   192.168.65.4   docker-desktop   <none>           <none>
frjaraun@sirius:~$ kubectl exec webserver -- curl -I http://192.168.65.4:8080
  % Total    % Received % Xferd  Average Speed   Time    Time     Time  Current
                                 Dload  Upload   Total   Spent    Left  Speed
  0   615    0     0    0     0      0      0 --:--:-- --:--:-- --:--:--     0
HTTP/1.1 200 OK
Server: nginx/1.24.0
Date: Sat, 15 Jul 2023 14:42:44 GMT
Content-Type: text/html
Content-Length: 615
Last-Modified: Sat, 17 Jun 2023 01:44:36 GMT
```

Figure 11.5 – Exposing a Pod using hostNetwork

This way, your NGINX web server will be accessible in the host's IP address where it is running (in this example, on the IP address 192.168.65.4, which is the address of our Docker Desktop worker and master host). The following code snippet shows the creation of the webserver application using the host's interfaces, and how we get the content of the NGINX process:

```
frjaraur@sirius:~$ kubectl run nettools --image=frjaraur/nettools:small-1.0 -- curl -I http://192.168.65.4:8080
pod/nettools created
frjaraur@sirius:~$ kubectl get pods
NAME        READY   STATUS      RESTARTS     AGE
nettools    0/1     Completed   1 (2s ago)   4s
webserver   1/1     Running     0            12m
frjaraur@sirius:~$ kubectl logs nettools
  % Total    % Received % Xferd  Average Speed   Time    Time     Time  Current
                                 Dload  Upload   Total   Spent    Left  Speed
HTTP/1.1 200 OK
  0   615    0     0    0     0       0      0 --:--:-- --:--:-- --:--:--     0
Server: nginx/1.24.0
Date: Sat, 15 Jul 2023 14:53:47 GMT
Content-Type: text/html
Content-Length: 615
Last-Modified: Sat, 17 Jun 2023 01:44:36 GMT
Connection: keep-alive
ETag: "648d1004-267"
Accept-Ranges: bytes
```

Figure 11.6 – Accessing the webserver application using the host's IP address

Notice that we executed the curl binary inside the webserver Pod. In this example, we are using Docker Desktop with **Windows Subsystem for Linux 2** (**WSL 2**), which doesn't have direct routing from another WSL client; but the important thing here is the accessibility to the NGINX process using the host's IP address. We can use another Pod, running the frjaraur/nettools image (developed and maintained by me), to verify that the application is accessible.

In this case, we are using just one port on our Pod; in fact, we didn't even declare the ports on our Pod's container, hence all the ports defined in the container image will be used. Using hostNetwork, all the ports defined in the image will be exposed, which may be a problem if you don't want to expose some specific Pods externally (for example, if your application has an internal API or administration interface that you will not be able to access). If you manage the platform yourself, you can manage access by modifying the host's firewall, but this can be tricky. In such situations, we can use the hostPort key at container level, instead of using hostNetwork at the Pod resource level. Let's explore this in the next section.

Using hostPort

The hostPort key is used inside the containers section of the Pod, where we define the ports to be exposed either internally or externally. With hostPort, we can expose only those ports that are required, while the remainder can stay internal. Let's see an example involving defining two containers within the webserver Pod:

```
[ Ubuntu 22.04.2 LTS
frjaraur@sirius:~$ cat <<EOF|kubectl create -f -
> apiVersion: v1
> kind: Pod
> metadata:
>   labels:
>     run: webserver
>   name: webserver
> spec:
>   containers:
>   - image: nginxinc/nginx-unprivileged:stable-alpine3.18
  name:>       name: webserver1
>     ports:
   - co>     - containerPort: 8080
>       hostPort: 8080
image: >
>   - image: nginx:alpine
>     name: webserver2
>     ports:
>     - name: http
>       containerPort: 80
> EOF
pod/webserver created
frjaraur@sirius:~$ kubectl get pods -o wide
NAME        READY   STATUS    RESTARTS   AGE   IP           NODE             NOMINATED NODE   READINESS GATES
webserver   2/2     Running   0          26s   10.1.0.137   docker-desktop   <none>           <none>
frjaraur@sirius:~$
```

Figure 11.7 – Example with two containers but only one exposed at the host level

In the preceding screenshot, we have two containers. Let's verify whether they are reachable using the frjaraur/nettools image again, trying to access both ports via curl:

```
[ Ubuntu 22.04.2 LTS
frjaraur@sirius:~$ kubectl run nettools --restart=Never --image=frjaraur/nettools:small-1.0 -ti --rm  -- /bin/sh
If you don't see a command prompt, try pressing enter.
/ $ curl -I http://192.168.65.4:8080
HTTP/1.1 200 OK
Server: nginx/1.24.0
Date: Sat, 15 Jul 2023 17:10:35 GMT
Content-Type: text/html
Content-Length: 615
Last-Modified: Sat, 17 Jun 2023 01:44:36 GMT
Connection: keep-alive
ETag: "648d1004-267"
Accept-Ranges: bytes

/ $ curl -I http://192.168.65.4:80
curl: (7) Failed to connect to 192.168.65.4 port 80 after 0 ms: Connection refused
/ $ curl -I http://10.1.0.137:80
HTTP/1.1 200 OK
Server: nginx/1.25.1
Date: Sat, 15 Jul 2023 17:11:09 GMT
Content-Type: text/html
Content-Length: 615
Last-Modified: Tue, 13 Jun 2023 17:34:28 GMT
Connection: keep-alive
ETag: "6488a8a4-267"
Accept-Ranges: bytes

/ $ exit
pod "nettools" deleted
frjaraur@sirius:~$
```

Figure 11.8 – Access to the webserver Service's ports 8080 and 80

In the preceding screenshot, we can see that only port 8080 is accessible on the host's IP address. Port 80 is accessible locally within the Kubernetes cluster.

Neither hostNetwork nor hostPort should be used without a Kubernetes administrator's supervision. Both represent a security breach and should be avoided unless strictly necessary for our application. They are commonly used for monitoring or administrative workloads when we need to manage or monitor the hosts' IP addresses.

Now that we have learned the different options we have at the host level, let's continue reviewing the NodePort mechanism associated with Service resources.

Publishing applications with Kubernetes' NodePort feature

As we mentioned at the beginning of this chapter, in the *Understanding the Kubernetes features for publishing applications cluster-wide* section, every NodePort Service resource has an associated ClusterIP IP address. This IP address is used to internally load balance all the client requests (from the Kubernetes cluster, internal, and external clients). Kubernetes provides this internal load to all available Pod replicas. All replicas will have the same weight, hence they will receive the same amount of requests. The ClusterIP IP address makes the applications running within Pods accessible internally. To make them available externally, the NodePort Service type attaches the defined port on all cluster nodes using NAT. The following schema represents the route taken by a request to an application running inside a Kubernetes cluster:

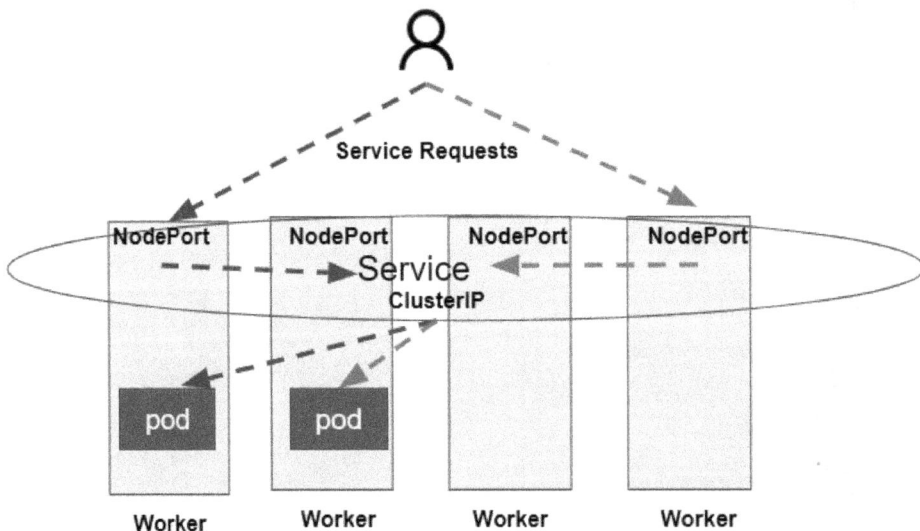

Figure 11.9 – NodePort simplified communications schema

The Endpoint resource is used to map the Pods' backends with the Service's ClusterIP. This resource is dynamically configured using label selectors in the Service's YAML manifest. Here is a simple example:

```
apiVersion: v1
kind: Pod
metadata:
 name: mypod
 labels:
   app.kubernetes.io/name: myapp
spec:
 containers:
 - name: nginx
   image: nginx:stable
   ports:
     - containerPort: 80
       name: http-web-svc
```

```
apiVersion: v1
kind: Service
metadata:
 name: myservice
spec:
 selector:
   app.kubernetes.io/name: myapp
 ports:
   - protocol: TCP
     port: 8080
     targetPort: 80
```

Figure 11.10 – Simple Pod and NodePort YAML manifests

The preceding screenshot shows the most common Service resource usage. With this manifest, an EndpointSlice resource will be created, associating the application's Pods with the Service by using the labels defined in the selector section. Notice that using these label selectors will create these EndpointSlice resources pointing to backend Pod resources running in the same namespace. But we can create Service resources without dynamic Pods attachment. This scenario could be useful, for example, to link external Services running outside of Kubernetes with an internal Service resource (this is how the ExternalName Service resource type works), or to access a Service from another namespace as if it were deployed on your current namespace. The internal Pods are made accessible thanks to the kube-proxy component, which will inject the traffic to the Pod's containers. This only happens in those nodes where the actual Pods are running, although the Service is accessible cluster-wide.

EndpointSlice resources using label selectors will create Endpoint resources, and thus, their status updates are propagated. Failed Pod resources will be deprecated from the actual Service and requests will not be routed to those backends, hence Kubernetes will only route to healthy Pods. This is the most popular and recommended method for using Service resources because this way, your resources are infrastructure agnostic.

Let's see a quick example of how Endpoint resource creation works by creating a webserver Pod and publishing the web process in NodePort mode by using kubectl expose:

```
Ubuntu 22.04.2 LTS
frjaraur@sirius:~$ kubectl run webserver --image=nginx:alpine --port=80
pod/webserver created
frjaraur@sirius:~$ kubectl expose pod webserver --port=8080 --type=NodePort --target-port=80
service/webserver exposed
frjaraur@sirius:~$ kubectl get svc
NAME         TYPE        CLUSTER-IP      EXTERNAL-IP    PORT(S)          AGE
kubernetes   ClusterIP   10.96.0.1       <none>         443/TCP          30h
webserver    NodePort    10.106.47.153   <none>         8080:31526/TCP   7s
frjaraur@sirius:~$ kubectl get endpointslices
NAME             ADDRESSTYPE   PORTS   ENDPOINTS        AGE
kubernetes       IPv4          6443    192.168.65.4     30h
webserver-zbr9g  IPv4          80      10.1.0.139       37s
frjaraur@sirius:~$ kubectl get endpoints
NAME         ENDPOINTS             AGE
kubernetes   192.168.65.4:6443     30h
webserver    10.1.0.139:80         41s
frjaraur@sirius:~$ kubectl get pods -o wide
NAME        READY   STATUS    RESTARTS   AGE    IP           NODE             NOMINATED NODE   READINESS GATES
webserver   1/1     Running   0          107s   10.1.0.139   docker-desktop   <none>           <none>
frjaraur@sirius:~$
```

Figure 11.11 – Exposing a Pod using the imperative format

In the preceding example, we created a Pod and then exposed it by using **imperative mode**. We didn't specify a NodePort port, hence Kubernetes assigned one from the 30000-32767 range to the port of the Service resource. We also retrieved a list of endpoints and the dynamic configurations created. We used the kubectl expose <WORKLOAD_TYPE> <WORKLOAD_NAME> format syntax to create the Service. This uses label selectors for the creation of the Service resource, taking the labels from the actual workload, hence an EndpointSlice resource was created to attach the available Pods to the Service.

In this example, the webserver application will be accessible using the docker-desktop node IP address, which may require additional configuration if you use WSL2 for execution. This is because in this infrastructure we will need to declare an NAT IP address to forward to your desktop computer. This will not be required if Hyper-V or Minikube are used as a Kubernetes environment on your PC. In a remote Kubernetes cluster, you must ensure that the IP addresses of the hosts and the ports are reachable from your computer.

> **Important note**
>
> Because the NodePort Services use the host's ports, these ports must be allowed in each node's firewall. Your Kubernetes administrator may have configured multiple interfaces on your Kubernetes platform nodes and should inform you about which IP addresses to use to make your applications accessible.

If your workloads run on a cloud infrastructure, additional steps may be required to allow access to your Service resources, and thus this is often not a good option for publishing your applications.

In the next section, we will review the LoadBalancer Service type, which was created specifically for cloud environments but is now also available for on-premises infrastructure thanks to software load balancers such as MetalLB.

Providing access to your Services with LoadBalancer Services

A LoadBalancer-type Service requires an external device to integrate your application. This type of Service resource includes a NodePort and its ClusterIP IP address. The external device provides a LoadBalancer IP address that will be load-balanced to the IP addresses of the cluster nodes and associated NodePorts. This configuration is completely managed for you by Kubernetes, but you must define an appropriate `spec` section for your infrastructure. This type of resource depends on the actual infrastructure because it will use the APIs from software-defined networking infrastructure to route and publish the applications' Services. Loadbalancer Service resources were prepared primarily for Kubernetes cloud platforms but are now more commonly encountered in modern local data centers with software-defined networks and API-managed devices, although they require a good knowledge of the underlying platform to work. As mentioned before, each LoadBalancer Service resource is assigned an IP address dynamically, which may require additional management on your cloud infrastructure and even additional costs.

The cloud provider decides how the Service is to be load-balanced. Depending on the cloud platform used, the NodePort part can sometimes be omitted as direct routing may be available if defined by the platform vendor.

On-premises virtual cloud infrastructures such as OpenStack can be integrated into our Kubernetes platforms to manage this type of Service resource because they are also part of the Kubernetes core. But if you are not using OpenStack or any other on-premise virtual cloud infrastructure, there are solutions such as MetalLB (`https://metallb.org/`) that make it possible to run a Kubernetes-compatible and dynamically configurable load balancer on any bare-metal infrastructure.

This type of Service resource is not recommended if you are looking for maximum compatibility and want to avoid vendor-specific resources. It really has a lot of dependencies on the underlying infrastructure and may require additional configurations to be done on the platform.

If you as a developer have to implement a Service of type LoadBalancer (or you're simply curious about their definition), you can use Minikube as it implements this functionality on your desktop computer without any external requirements to negotiate. Docker Desktop will report the LoadBalancer IP address as `localhost`, hence you will be able to connect to the given Services directly using the `127.0.0.1` IP address.

Let's see how this works with a simple example. We will first start a new Minikube cluster environment (ensure your Docker Desktop or Rancher Desktop instances are stopped before starting Minikube), and then we will create a **Minikube tunnel**, which is a Minikube feature that will create a tunnel between your host and the Minikube node. We will use an administrator console for executing the `minikube start` and `minikube tunnel` commands:

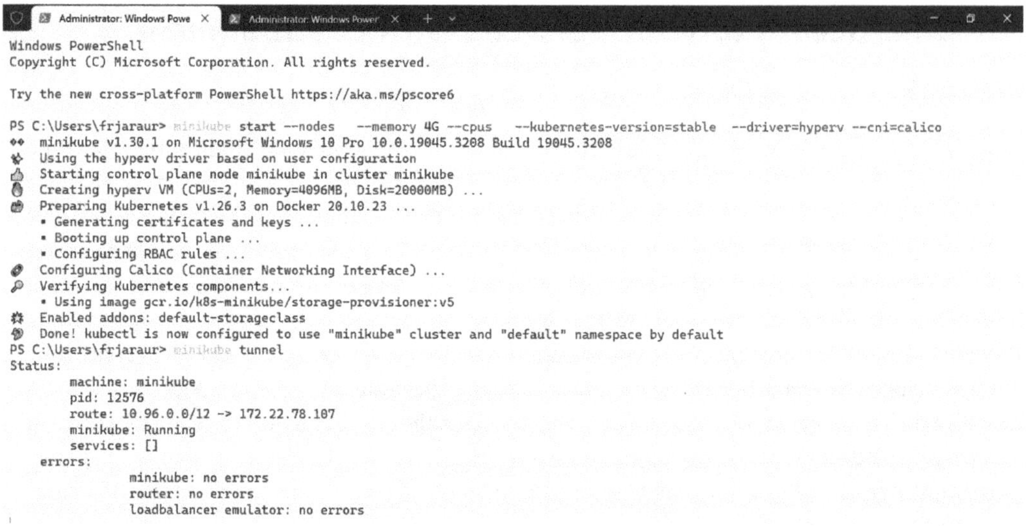

```
Windows PowerShell
Copyright (C) Microsoft Corporation. All rights reserved.

Try the new cross-platform PowerShell https://aka.ms/pscore6

PS C:\Users\frjaraur> minikube start --nodes   --memory 4G --cpus   --kubernetes-version=stable  --driver=hyperv --cni=calico
  minikube v1.30.1 on Microsoft Windows 10 Pro 10.0.19045.3208 Build 19045.3208
  Using the hyperv driver based on user configuration
  Starting control plane node minikube in cluster minikube
  Creating hyperv VM (CPUs=2, Memory=4096MB, Disk=20000MB) ...
  Preparing Kubernetes v1.26.3 on Docker 20.10.23 ...
    ▪ Generating certificates and keys ...
    ▪ Booting up control plane ...
    ▪ Configuring RBAC rules ...
  Configuring Calico (Container Networking Interface) ...
  Verifying Kubernetes components...
    ▪ Using image gcr.io/k8s-minikube/storage-provisioner:v5
  Enabled addons: default-storageclass
  Done! kubectl is now configured to use "minikube" cluster and "default" namespace by default
PS C:\Users\frjaraur> minikube tunnel
Status:
        machine: minikube
        pid: 12576
        route: 10.96.0.0/12 -> 172.22.78.107
        minikube: Running
        services: []
    errors:
                minikube: no errors
                router: no errors
                loadbalancer emulator: no errors
```

Figure 11.12 – Execution of a Minikube cluster from an administrator console

We will open another console, but this time we will connect to the Kubernetes cluster, so we don't need to execute the commands as an administrator. We create a Pod and then expose it using imperative mode with the LoadBalancer type:

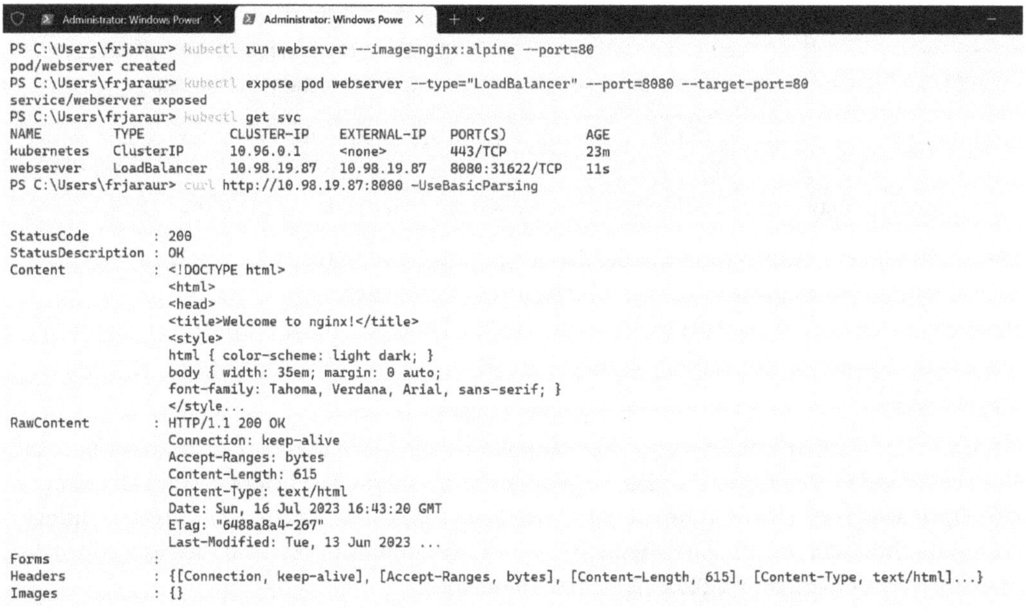

```
PS C:\Users\frjaraur> kubectl run webserver --image=nginx:alpine --port=80
pod/webserver created
PS C:\Users\frjaraur> kubectl expose pod webserver --type="LoadBalancer" --port=8080 --target-port=80
service/webserver exposed
PS C:\Users\frjaraur> kubectl get svc
NAME         TYPE           CLUSTER-IP     EXTERNAL-IP    PORT(S)          AGE
kubernetes   ClusterIP      10.96.0.1      <none>         443/TCP          23m
webserver    LoadBalancer   10.98.19.87    10.98.19.87    8080:31622/TCP   11s
PS C:\Users\frjaraur> curl http://10.98.19.87:8080 -UseBasicParsing


StatusCode        : 200
StatusDescription : OK
Content           : <!DOCTYPE html>
                    <html>
                    <head>
                    <title>Welcome to nginx!</title>
                    <style>
                    html { color-scheme: light dark; }
                    body { width: 35em; margin: 0 auto;
                    font-family: Tahoma, Verdana, Arial, sans-serif; }
                    </style...
RawContent        : HTTP/1.1 200 OK
                    Connection: keep-alive
                    Accept-Ranges: bytes
                    Content-Length: 615
                    Content-Type: text/html
                    Date: Sun, 16 Jul 2023 16:43:20 GMT
                    ETag: "6488a8a4-267"
                    Last-Modified: Tue, 13 Jun 2023 ...
Forms             :
Headers           : {[Connection, keep-alive], [Accept-Ranges, bytes], [Content-Length, 615], [Content-Type, text/html]...}
Images            : {}
```

Figure 11.13 – Creating a LoadBalancer Service in Minikube

Notice that we have a new column with the external IP addresses. In this case, it is an emulation of a real external load balancer device that provides a specific IP address for the new Service manifest. Minikube, in fact, creates a tunnel from the Kubernetes node to your desktop computer, making the Pod accessible over the assigned load-balanced IP address (`EXTERNAL-IP`) and the Service's port. In this case, we will reach the NGINX web server at `http://10.98.19.87:8080`. We then test the accessibility of the application with `curl` (which is an alias for Windows PowerShell's `Invoke-WebRequest` command).

The dependency of the LoadBalancer Service type on the platform infrastructure makes this type too specific for day-to-day usage and may not be available in all Kubernetes clusters. Therefore, the best solution for compatibility is to use Ingress Controllers, as we will learn in the following section.

Understanding Ingress Controllers

An **Ingress Controller** is a piece of software that provides load balancing, SSL termination, and host-based virtual routing. It is a reverse proxy that runs in the Kubernetes cluster, which manages a reverse proxy network component that can run inside the Kubernetes cluster or externally, just like any other network infrastructure device. An Ingress Controller acts just like any other controller deployed in a Kubernetes cluster, although it is not managed by the cluster itself. We must deploy this controller manually as it is not part of the Kubernetes core. If required, we can deploy multiple Ingress Controllers in a cluster and define which one is to be used by default if none is specified.

Ingress Controllers work very well with HTTP/HTTPS applications (OSI Layer 7, the application layer), but we can publish TCP and UDP applications too (OSI Layer 4, the transport layer), although this does require more configuration and may not be the best option. In such cases, it may be better to use an external load balancer and route traffic to NodePort Service resources because TCP and UDP Ingress resources will need additional ports to distribute incoming traffic.

Kubernetes administrators use **IngressClass resources** to declare the different Ingress Controllers available on a platform. Each Ingress Controller is associated with an IngressClass resource. You as the developer must create Ingress resources, which are the definitions required for reverse-proxying your application's workloads.

There are multiple options for deploying an Ingress Controller: cloud providers and many software vendors have developed their own solutions, and you can include any of them in your own Kubernetes setup, but you must understand their specific features and particularities. You can review the available solutions in the Kubernetes documentation at `https://kubernetes.io/docs/concepts/services-networking/ingress-controllers/#additional-controllers` and ask your Kubernetes administrator about the Ingress Controllers available on your platform before preparing your applications. Small tweaks may be necessary on your side in your Ingress resources. In the following section, we will examine the most frequently encountered option, the **Kubernetes NGINX Ingress Controller**, included by default in some Kubernetes solutions.

Deploying an Ingress Controller

To deploy an Ingress Controller, we simply follow the specific instructions for the chosen solution. There may be different approaches for installing the given software in your cluster, but we will follow the easiest one: deploying a YAML file containing all the required resources in one file (`https://raw.githubusercontent.com/kubernetes/ingress-nginx/controller-v1.8.1/deploy/static/provider/cloud/deploy.yaml`). Make sure to check the latest available release before using the URL just provided and consult the specific instructions provided for it. At the time of writing this book, NGINX Controller release `1.8.1` was the latest release available. In this example, we use the cloud YAML file, although you can use the bare-metal option if you have a fully functional Kubernetes environment installed (this version uses NodePort instead of the LoadBalancer type). Let's work through our simple example:

1. We start by running `kubectl apply` on a Docker Desktop environment:

```
🔵 Ubuntu 22.04.2 LTS        ×    + ∨                                                                    –  ⤢  ×

frjaraur@sirius:~$ kubectl apply -f https://raw.githubusercontent.com/kubernetes/ingress-nginx/controller-v1.8.1/deploy/stat
ic/provider/cloud/deploy.yaml
namespace/ingress-nginx created
serviceaccount/ingress-nginx created
serviceaccount/ingress-nginx-admission created
role.rbac.authorization.k8s.io/ingress-nginx created
role.rbac.authorization.k8s.io/ingress-nginx-admission created
clusterrole.rbac.authorization.k8s.io/ingress-nginx created
clusterrole.rbac.authorization.k8s.io/ingress-nginx-admission created
rolebinding.rbac.authorization.k8s.io/ingress-nginx created
rolebinding.rbac.authorization.k8s.io/ingress-nginx-admission created
clusterrolebinding.rbac.authorization.k8s.io/ingress-nginx created
clusterrolebinding.rbac.authorization.k8s.io/ingress-nginx-admission created
configmap/ingress-nginx-controller created
service/ingress-nginx-controller created
service/ingress-nginx-controller-admission created
deployment.apps/ingress-nginx-controller created
job.batch/ingress-nginx-admission-create created
job.batch/ingress-nginx-admission-patch created
ingressclass.networking.k8s.io/nginx created
validatingwebhookconfiguration.admissionregistration.k8s.io/ingress-nginx-admission created
friaraur@sirius:~$ █
```

Figure 11.14 – Deployment of the popular Kubernetes NGINX Ingress Controller

As you can see in the previous screenshot, many resources are created for the Ingress Controller to work. A new namespace was created, `ingress-nginx`, and some Pods are now running there:

```
🔵 Ubuntu 22.04.2 LTS        ×    + ∨

frjaraur@sirius:~$ kubectl get deploy -n ingress-nginx
NAME                      READY   UP-TO-DATE   AVAILABLE   AGE
ingress-nginx-controller  1/1     1            1           163m
frjaraur@sirius:~$ kubectl get pods -n ingress-nginx
NAME                                    READY   STATUS      RESTARTS   AGE
ingress-nginx-admission-create-c5pq9    0/1     Completed   0          163m
ingress-nginx-admission-patch-2gscp     0/1     Completed   0          163m
ingress-nginx-controller-74469fd44c-gfv7l  1/1  Running     0          163m
frjaraur@sirius:~$ kubectl get ingressclass
NAME    CONTROLLER            PARAMETERS   AGE
nginx   k8s.io/ingress-nginx  <none>       164m
frjaraur@sirius:~$ █
```

Figure 11.15 – Deployment, Pods, and IngressClass resources created

In the preceding screenshot, we can see an `IngressClass` resource created. We may need to configure it as default.

2. Let's check the deployed Ingress Controller. We first check the Service resource created to reach the Deployment resource using `kubectl get svc`:

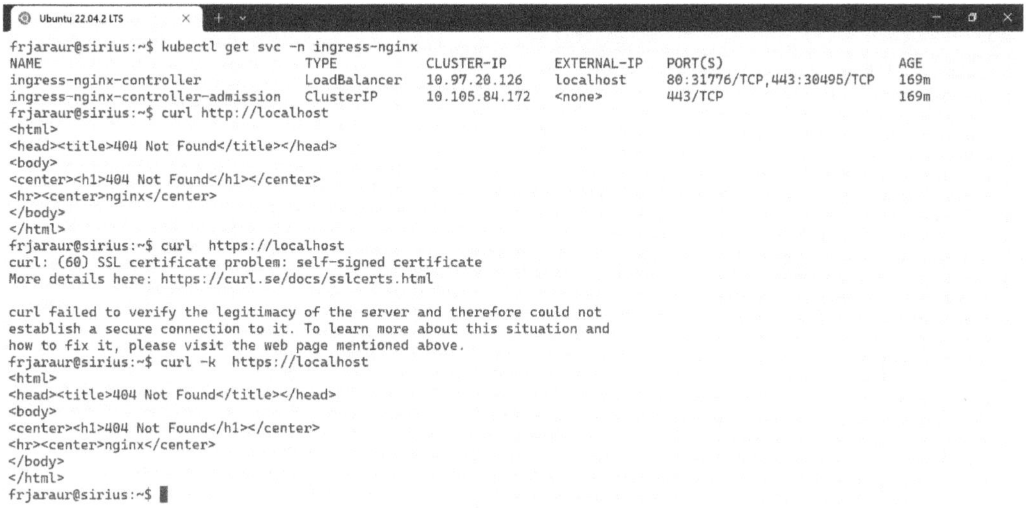

```
frjaraur@sirius:~$ kubectl get svc -n ingress-nginx
NAME                              TYPE           CLUSTER-IP      EXTERNAL-IP   PORT(S)                      AGE
ingress-nginx-controller          LoadBalancer   10.97.20.126    localhost     80:31776/TCP,443:30495/TCP   169m
ingress-nginx-controller-admission ClusterIP     10.105.84.172   <none>        443/TCP                      169m
frjaraur@sirius:~$ curl http://localhost
<html>
<head><title>404 Not Found</title></head>
<body>
<center><h1>404 Not Found</h1></center>
<hr><center>nginx</center>
</body>
</html>
frjaraur@sirius:~$ curl  https://localhost
curl: (60) SSL certificate problem: self-signed certificate
More details here: https://curl.se/docs/sslcerts.html

curl failed to verify the legitimacy of the server and therefore could not
establish a secure connection to it. To learn more about this situation and
how to fix it, please visit the web page mentioned above.
frjaraur@sirius:~$ curl -k  https://localhost
<html>
<head><title>404 Not Found</title></head>
<body>
<center><h1>404 Not Found</h1></center>
<hr><center>nginx</center>
</body>
</html>
frjaraur@sirius:~$
```

Figure 11.16 – Verification of the deployed Ingress Controller

Note in the preceding screenshot how the Service resource was created as a `LoadBalancer` type. It acquired the `localhost` IP address (we are using Docker Desktop in this example), which means that we should be able to reach the NGINX Ingress Controller backend directly with `curl` using `localhost`. The Service is listening on ports 80 and 443, and we were able to reach both (we passed the `-k` argument to `curl` to avoid having to verify the associated auto-signed and untrusted SSL certificate).

The use of Ingress Controllers improves security when we add SSL certificates to implement SSL tunnels between our applications' exposed components and our users, or even between different components that use the Ingress URL associated with the Service resource.

Let's go ahead now and learn how to manage the behavior of our applications using Ingress resources.

Ingress resources

As with any other resource, we need to define `apiVersion`, `kind`, `metadata`, and `spec` keys and sections. The most important section is `.spec.rules`, which defines a list of host rules that dynamically configure the reverse proxy deployed by the Ingress Controller. Let's see a basic example:

```
apiVersion: networking.k8s.io/v1
kind: Ingress
metadata:
  name: webserver
  annotations:
    nginx.ingress.kubernetes.io/rewrite-target: /
spec:
  ingressClassName: myingresscontroller
  tls:
  - hosts:
      - www.webserver.com
    secretName: webserver-tls-secret
  rules:
  - host: www.webserver.com
    http:
      paths:
      - path: /test
        pathType: Prefix
        backend:
          service:
            name: webserver
            port:
              number: 80
```

Figure 11.17 – Ingress resource example

In the preceding screenshot, we can see the `ingressClassName` key, which indicates the Ingress Controller to be used. The `rules` section defines a list of host headers and the paths associated with the different backends. In our example, the `www.webserver.com` host header is required; if requests do not include it, they will be redirected to the default backend (if defined) or be shown a `404` error (page not found). The `backend` section describes the Service resource that will receive the application's requests.

Let's run a quick example using the `webserver` Service resource created in the previous section. It will listen on port `8080`, hence we create an Ingress resource with a fake hostname and validate its accessibility with `curl -H "host: <FAKE_HOST>" http://localhost` (we use `localhost` because its IP address is the one associated with the `LoadBalancer` Service):

```
Ubuntu 22.04.2 LTS          ×    +  ∨
frjaraur@sirius:~$ cat <<EOF|kubectl create -f -
apiVersion: networking.k8s.io/v1
kind: Ingress
metadata:
  name: webserver
spec:
  ingressClassName: nginx
  rules:
  - host: webserver.example.local
    http:
      paths:
      - path: /
        pathType: Prefix
        backend:
          service:
            name: webserver
            port:
              number: 8080

EOF
ingress.networking.k8s.io/webserver created
frjaraur@sirius:~$ curl -H "host: webserver.example.local" http://127.0.0.1:80/
<!DOCTYPE html>
<html>
<head>
<title>Welcome to nginx!</title>
<style>
html { color-scheme: light dark; }
body { width: 35em; margin: 0 auto;
font-family: Tahoma, Verdana, Arial, sans-serif; }
</style>
</head>
```

Figure 11.18 – Ingress webserver resource for the webserver Service example

Security features are implemented in the `.spec.tls` section, where we link the hosts with their keys and certificates, integrated into a Secret resource. This Secret must be included in the namespace in which you defined the Ingress resource, and it is of the `tls` type. The `data` sections in these Secrets must include the key for the generation of the certificate along with the generated certificate itself. We will learn how to create this via an example in the *Labs* section.

We can have an Ingress resource with rules for multiple hosts and each host with multiple paths, although it is more common to separate each host on a different Ingress resource for easier management and include multiple paths for reaching different backend Service resources. This combination represents a typical microservice architecture where each application functionality is provided by different backend Services.

The `annotations` section can be used to instruct the Ingress Controller regarding special configurations. Here is a list of some of the most important configurations we can manage via annotations for the Kubernetes NGINX Ingress Controller:

- `nginx.ingress.kubernetes.io/rewrite-target`: It is usual to integrate some rewrite rules in our Ingress resource for rewriting the application's URI paths. There are also options for redirecting URLs.

- `nginx.ingress.kubernetes.io/auth-type` and `nginx.ingress.kubernetes.io/auth-secret`: These will allow us to use basic authentication at the Ingress level for our applications.

- `nginx.ingress.kubernetes.io/proxy-ssl-verify`: If our Service resource backends use TLS, there are many annotations available to manage how NGINX connects to them.

- `nginx.ingress.kubernetes.io/enable-cors`: We may need to enable **Cross-Origin Resource Sharing** (**CORS**) in our application to allow some external routes and URLs. There are also other interesting options here for managing and securing CORS behavior.

- `nginx.ingress.kubernetes.io/client-body-buffer-size`: It's quite common to limit the size of client requests to avoid overall performance issues, but your application may require larger responses.

There are many options available beyond these, and you may need to ask your Kubernetes and infrastructure administrators for advice. The range on offer includes integrating external authentication backends, limiting the rate of requests to avoid **distributed denial-of-service** (**DDoS**) attacks, redirecting and rewriting URLs, enabling SSL passthrough, and even managing canary application deployments, routing some requests to a newer release of your workload backends. Some of the options can be defined at the Ingress Controller level, which will affect all Ingress resources at once. For a full list of available annotations, please refer to the following page: `https://kubernetes.github.io/ingress-nginx/user-guide/nginx-configuration/annotations/`.

It is very important to understand that the options mentioned here may differ from those available for other Ingress Controllers (at least, they will use other annotation keys, for sure). Some Ingress Controllers such as Kong also implement API management for your backend Services, which can be very useful if they are involved in many interactions. Ask your Kubernetes administrators about the Ingress Controllers deployed on your platform.

We covered the basics here, but as always, note that your Ingress resource may need some small tweaks to fully implement your platform requirements. In OpenShift, for example, Ingress Controllers can be enabled, but by default, Kubernetes will use OpenShift Route, which is the Red Hat implementation of a L7 reverse proxy for publishing applications in Kubernetes. Ingress Controllers and OpenShift Route are quite similar (even their resources look alike) but you should review further specific information about it if your application needs to run on an OpenShift cluster. The following link may help you decide which one to use if both implementations are available for your application: `https://cloud.redhat.com/blog/kubernetes-ingress-vs-openshift-route`.

By default, Kubernetes implements a flat network, without any access boundaries between applications. This applies no restrictions to any lateral movement (East-West traffic), a configuration that could cause critical security issues. In the next section, we will review some security improvements to help us publish our applications securely.

Improving our applications' security

In Kubernetes, application traffic flows freely by default. A flat network is deployed to cover Pod-to-Pod and Service-to-Pod communications – remember that containers within a Pod have a common, shared IP address. Pods running within a Kubernetes cluster will see each other, and it will require some extra work to protect one application from another, even if they run on different nodes and in different namespaces.

It may be strange to hear, but applications running in different namespaces can see each other. In fact, if they have an associated Service resource, it would be easy to use the internal DNS to resolve its associated IP address and access its processes.

In the next section, we will learn how NetworkPolicy resources can be used to define our applications' communications and have Kubernetes block any unwanted connectivity for us.

Network policies

NetworkPolicy resources (also referred to as `netpol`) allow us to manage OSI Layer 3 and 4 communications (IP and port access, respectively). Kubernetes provides the NetworkPolicy resource as part of its core, but its implementation depends on the CNI deployed in your cluster. Therefore, it is vital to use a CNI (such as Calico, Canal, or Cilium, among others) that implements this feature.

NetworkPolicy resources define all aspects of Pod communications: egress (output traffic) and ingress (input traffic). As we will see in a few moments, NetworkPolicies are applied to specific sets of Pods by using the `.spec.podSelector` section. NetworkPolicy resources are namespaced, hence `podSelector` allows us to decide which Pods are to be affected by our rule definitions. Multiple rules can be applied to a Pod, and your Kubernetes administrators may have included some `GlobalNetworkPolicy` resources that affect the entire cluster, thus you should inquire whether any cluster default rules require allowing some egress or ingress traffic. It is quite common to allow only DNS traffic by default, disallowing all additional egress traffic. If this is the case in your cluster, you will need to declare all egress (as well as ingress) communications in your applications' manifests. Let's see a quick example of a NetworkPolicy in which we declare both ingress and egress communications:

```
apiVersion: networking.k8s.io/v1          ingress:
kind: NetworkPolicy                         - from:
metadata:                                     - ipBlock:
 name: mynetworkpolicy                           cidr: 172.17.0.0/16
spec:                                            except:
 podSelector:                                      - 172.17.1.0/24
   matchLabels:                               - namespaceSelector:
     app: myapp                                  matchLabels:
 policyTypes:                                     project: myproject
   - Egress                                    - podSelector:
   - Ingress                                     matchLabels:
 egress:                                           role: frontend
   - to:                                     ports:
     - ipBlock:                                 - protocol: TCP
         cidr: 10.0.0.0/24                       port: 6379
     ports:                                  - from:
       - protocol: TCP                         - ipBlock:
         port: 5978                              cidr: 192.168.200.0/24
   - to:                                     ports:
     - ipBlock:                                 - protocol: TCP
         cidr: 0.0.0.0/0                         port: 80
     ports:
       - protocol: UDP
         port: 53
```

Figure 11.19 – NetworkPolicy resource manifest with both egress and ingress rules

Let's review some of the most important keys and sections available in the preceding code snippet. The `.spec.podSelector` section declares which Pods in the current namespace (if none are declared under the `metadata` section) will be affected by this policy. Under the `policyTypes` key, we can see a list of policy types defined. We should clarify here that egress communications are those initiated from a Pod, while ingress communications are those that go into the Pod. If you declare both types, egress and ingress, and then only declare a section for one of them (either an `egress` or `ingress` section, as in the preceding example), the omitted one is declared as empty, meaning that that type of communication will not be allowed *at all*. The `egress` section in this example is a list of rules to be applied. Let's have a closer look at this:

- The first rule allows egress communications from selected Pods (those with the `app=myapp` label) to port `5978` on any host in the `10.0.0.0/24` subnet

- The second rule allows egress communications to UDP port `53` on any host (Kubernetes internal and external DNS)

In the `ingress` section, two rules are also declared:

- The first rule allows access to port `6379` on the selected Pods (those containing the `app=myapp` label) for any communication coming from the `172.17.0.0/16` subnet (except those hosted on the `172.17.1.0/24` subnet), from Pods running in namespaces with the `project=myproject` label, and from Pods in the current namespace with the `role=frontend` label. We can say that *Kubernetes is all about labels*.

- The second ingress rule allows access to selected Pods on port 80 from hosts on the 192.168.200.0/24 subnet.

These rules may seem complex but are quite easy to implement in practice.

If you are planning to deploy all your application's components in a specific namespace, it could be worthwhile to allow all egress and ingress communications between your Pods. This isn't a good idea for production because only attackers' lateral movements to other namespaces will be blocked, but a security issue in one of your Pods could affect others in the same namespace. While preparing your NetworkPolicy resources or debugging your application, allowing all namespace East-West traffic may also be necessary. The following YAML manifests allow all internal communication and expose only the frontend component externally:

```
apiVersion: networking.k8s.io/v1
kind: NetworkPolicy
metadata:
 name: allow-namespaced-networking
spec:
 podSelector: {}
 policyTypes:
   - Egress
   - Ingress
 egress:
 -  to:
     -  podSelector: {}
 ingress:
 -  from:
     -  podSelector: {}
```

```
apiVersion: networking.k8s.io/v1
kind: NetworkPolicy
metadata:
 name: allow-access-to-frontend
spec:
 podSelector:
   matchLabels:
     appcomponent: frontend
 policyTypes:
   - Ingress
 ingress:
   - from:
     - ipBlock:
         cidr: 192.168.200.0/24
     ports:
       - protocol: TCP
         port: 80
```

Figure 11.20 – Example manifests for allowing all namespaced
communications as well as access to port 80 on a specific Pod

The preceding screenshot shows two manifests:

- The one on the left declares a NetworkPolicy resource that allows all communications between all Pods deployed in the current namespace. The rule applies to all Pods in the namespace because podSelector is empty.

- The manifest on the right allows access to the Pod with the appcomponent=frontend label (podSelector applies on this Pod) on port 80, but only from hosts on the 192.168.200.0/24 subnet.

> **Important note**
>
> NetworkPolicy resources apply at the connection level and don't leave any connectivity traces by default, which may be inconvenient when trying to fix some connectivity issues between components. Some CNI plugins such as Calico enable you to log connections between Pods. This requires additional permissions on your Kubernetes environment. Ask your Kubernetes administrators if they can provide some connectivity traces if required for debugging your applications. In some cases, it is best to start with a NetworkPolicy that allows and logs all the connections made in the namespace.

You as a developer are responsible for creating and maintaining your application's resource manifests and thus, the NetworkPolicy resources required by your application. It is up to you how to organize them, but it's recommended to use descriptive names and group multiple rules in a manifest per each application component. This way, you will be able to fine-tune each component's configuration. In the *Labs* section, we have prepared for you a specific exercise where you will protect an application by allowing only trusted access.

NetworkPolicies allow you to thoroughly isolate all your application's components, and although they may be hard to implement, this solution does provide great granularity and does not depend on the underlying infrastructure. You only require a Kubernetes CNI that supports this feature.

In the next section, we will review how service mesh solutions can provide more complex security functionality by injecting small, lightweight proxies on all your application's Pods.

Service mesh

By implementing NetworkPolicies, we enforce some firewall-like connectivity rules between our applications' workloads, but this may not be enough. A service mesh is considered an infrastructure layer that interconnects services and manages how they will interact with each other. A service mesh is used to manage East-West and North-South traffic to background Services, in some cases even substituting the Ingress Controller if the service mesh solution is deployed in Kubernetes.

The most popular service mesh solution is **Istio** (an open source solution that is part of the **Cloud Native Computing Foundation** (**CNCF**)), although other options worth mentioning include Linkerd, Consul Connect, and Traefik Mesh. If you are running a cloud Kubernetes platform, you may have your own cloud provider solution available.

Service mesh solutions are capable of adding TLS communications, traffic management, and observability to your applications without having to modify their code. If you are looking for a transparent security and management layer, using a service mesh solution may be perfect for you, but it also adds a high level of complexity and some platform overhead.

Service mesh solutions deploy a small proxy on all your application workloads. These proxies intercept all your application's network traffic and apply rules to allow or disallow your application processes' communications.

The implementation and use of service meshes are out of the scope of this book but it is worth investigating whether your Kubernetes administrators have deployed a service mesh solution on your platform that may necessitate the implementation of service mesh-specific resources.

As mentioned earlier in this section, NetworkPolicy resources isolate your application's workloads by disabling unauthorized communications, which may provide sufficient security for a production environment. These resources are highly configurable, and you are responsible for defining the required communications between your application's components and preparing the required YAML manifests to fully implement all your application communications. In the following *Labs* section, we will see some of the content learned in this chapter in action as we try publishing the `simplestlab` application used in previous chapters.

Labs

In this section, we show you how to work through the implementation of the Ingress resources for the `simplestlab` Tier-3 application, prepared for Kubernetes in *Chapter 9, Implementing Architecture Patterns*, and improved upon in *Chapter 10, Leveraging Application Data Management in Kubernetes*. Manifests for all the resources have been prepared for you in this book's GitHub repository at `https://github.com/PacktPublishing/Containers-for-Developers-Handbook.git` and can be found in the `Chapter11` folder. Ensure you have the latest revision available by simply executing `git clone` to download all its content, or use `git pull` if you have already downloaded the repository before. All the manifests and steps required for running the `simplestlab` application are located inside the `Containers-for-Developers-Handbook/Chapter11/simplestlab` directory, while all the manifests for Ingress and NetworkPolicy resources can be found directly in the `Chapter11` folder.

These labs will help you learn and understand how Ingress and NetworkPolicy resources work in Kubernetes. You will deploy an Ingress Controller, publish the `simplestlab` example application using the HTTP and HTTPS protocols, and create some NetworkPolicy resources to allow only appropriate connectivity. The Ingress Controller lab will work on Docker Desktop, Minikube, and Rancher, but for the NetworkPolicy resources part, you will need to use an appropriate Kubernetes CNI with support for such resources, such as Calico. Each Kubernetes desktop or platform implementation manages and presents its own networking infrastructure to users in a different way.

These are the tasks you will find in this chapter's GitHub repository:

1. We will first deploy the Kubernetes NGINX Ingress Controller (if you don't have your own Ingress Controller in your labs platform).

2. We will deploy all the manifests prepared for the `simplestlab` application, located inside the `simplestlab` folder. We will use `kubectl create -f simplestlab`.

3. Once all the components are ready, we will create an Ingress resource using the manifest prepared for this task.

4. In the GitHub repository, you will find instructions for deploying a more advanced Ingress manifest with a self-signed certificate and encrypting the client communications.

5. There is also a NetworkPolicy lab in the GitHub repository that will help you understand how to secure your applications using this feature with a compatible CNI (Calico).

In the first task, we will deploy our own Ingress Controller.

Improving application access by deploying your own Ingress Controller

For this task, we will use Docker Desktop, which provides a good LoadBalancer service implementation. These Service resources will attach the localhost IP address, which will make it easy to connect to the published services. We will use the cloud deployment of Kubernetes NGINX Ingress Controller (`https://kubernetes.github.io`) based on the LoadBalancer Service type, described in the following manifest: `https://raw.githubusercontent.com/kubernetes/ingress-nginx/controller-v1.8.1/deploy/static/provider/cloud/deploy.yaml`. If you are using a completely bare-metal infrastructure, you can use the bare-metal YAML (`https://raw.githubusercontent.com/kubernetes/ingress-nginx/controller-v1.8.1/deploy/static/provider/baremetal/deploy.yaml`) and follow the additional instructions at `https://kubernetes.github.io/ingress-nginx/deploy/baremetal/` for NodePort routing.

> **Important note**
>
> Local copies of both YAML files are provided in the repository as `kubernetes-nginx-ingress-controller-full-install-cloud.yaml` and `kubernetes-nginx-ingress-controller-full-install-baremetal.yaml`.

Once this is done, follow these steps:

1. We will just deploy the cloud version, provided in the YAML as a series of concatenated manifests. We just use `kubectl apply` to deploy the controller:

```
Chapter11$ kubectl apply -f https://raw.githubusercontent.com/
kubernetes/ingress-
nginx/controller-v1.8.1/deploy/static/provider/cloud/deploy.yaml
namespace/ingress-nginx created
serviceaccount/ingress-nginx created
serviceaccount/ingress-nginx-admission created
role.rbac.authorization.k8s.io/ingress-nginx created
role.rbac.authorization.k8s.io/ingress-nginx-admission created
clusterrole.rbac.authorization.k8s.io/ingress-nginx created
```

```
clusterrole.rbac.authorization.k8s.io/ingress-nginx-admission
created
rolebinding.rbac.authorization.k8s.io/ingress-nginx created
rolebinding.rbac.authorization.k8s.io/ingress-nginx-admission
created
clusterrolebinding.rbac.authorization.k8s.io/ingress-nginx
created
clusterrolebinding.rbac.authorization.k8s.io/ingress-nginx-
admission created
configmap/ingress-nginx-controller created
service/ingress-nginx-controller created
service/ingress-nginx-controller-admission created
deployment.apps/ingress-nginx-controller created
job.batch/ingress-nginx-admission-create created
job.batch/ingress-nginx-admission-patch created
ingressclass.networking.k8s.io/nginx created
validatingwebhookconfiguration.admissionregistration.k8s.io/
ingress-nginx-admission created
```

2. We can review the workload resources created:

```
Chapter11$ kubectl get all -n ingress-nginx
NAME                                        READY    STATUS
      RESTARTS    AGE
pod/ingress-nginx-admission-create-
9cpnb           0/1       Completed   0         13m
pod/ingress-nginx-admission-patch-
6gq2c           0/1       Completed   1         13m
pod/ingress-nginx-controller-74469fd44c-
h6nlc   1/1      Running      0          13m
NAME                                        TYPE
CLUSTER-IP        EXTERNAL-IP   PORT(S)                AGE
service/ingress-nginx-
controller             LoadBalancer   10.100.162.170    localhost
    80:31901/TCP,443:30080/TCP    13m
service/ingress-nginx-controller-
admission   ClusterIP      10.100.197.210    <none>        443/
TCP                    13m
NAME                                        READY    UP-TO-
DATE    AVAILABLE    AGE
deployment.apps/ingress-nginx-
controller   1/1       1                1          13m
NAME                                            DESIRED
   CURRENT    READY    AGE
replicaset.apps/ingress-nginx-controller-
74469fd44c   1          1          1         13m
```

```
NAME                                          COMPLETIONS
DURATION    AGE
job.batch/ingress-nginx-admission-create      1/1              7s
        13m
job.batch/ingress-nginx-admission-patch       1/1              8s
        13m
```

3. The `ingress-nginx-controller` Service is attached to the `localhost` IP address, so we can check its availability at `http://localhost:80` and `https://localhost:443` (exposed ports):

```
Chapter11$ curl http://localhost
<html>
<head><title>404 Not Found</title></head>
<body>
<center><h1>404 Not Found</h1></center>
<hr><center>nginx</center>
</body>
</html>
```

It also works with HTTPS (we add the −k argument to avoid certificate validation):

```
Chapter11$ curl https://localhost
curl: (60) SSL certificate problem: self-signed certificate
More details here: https://curl.se/docs/sslcerts.html
curl failed to verify the legitimacy of the server and therefore
could not
establish a secure connection to it. To learn more about this
situation and
how to fix it, please visit the web page mentioned above.
Chapter11$ curl -k https://localhost
<html>
<head><title>404 Not Found</title></head>
<body>
<center><h1>404 Not Found</h1></center>
<hr><center>nginx</center>
</body>
</html>
```

The Ingress Controller is now deployed and listening, and the `404` error indicates that there isn't an associated Ingress resource with the `localhost` host (in fact there isn't even a default one configured, but the Ingress Controller responds correctly).

Publishing the simplestlab application on Kubernetes using an Ingress Controller

In this lab, we will deploy `simplestlab`, a very simplified tier-3 application, located in the `simplestlab` directory, and we'll publish its frontend, the `lb` component, without TLS encryption. You can follow these steps:

1. The manifests for the application are already written for you; we will just have to use `kubectl` to create an appropriate namespace for the application and then deploy all its resources:

```
Chapter11$ kubectl create ns simplestlab
namespace/simplestlab created
Chapter11$ kubectl create -n simplestlab \
-f simplestlab/
deployment.apps/app created
service/app created
secret/appcredentials created
service/db created
statefulset.apps/db created
secret/dbcredentials created
secret/initdb created
configmap/lb-config created
daemonset.apps/lb created
service/lb created
```

2. We can now verify the resources created in the `simplestlab` namespace:

```
Chapter11$ kubectl get all -n simplestlab
NAME                          READY   STATUS    RESTARTS   AGE
pod/app-5f9797d755-5t4nz      1/1     Running   0          81s
pod/app-5f9797d755-9rzlh      1/1     Running   0          81s
pod/app-5f9797d755-nv58j      1/1     Running   0          81s
pod/db-0                      1/1     Running   0          80s
pod/lb-5wl7c                  1/1     Running   0          80s
NAME            TYPE        CLUSTER-IP      EXTERNAL-IP   PORT(S)
     AGE
service/app     ClusterIP   10.99.29.167    <none>        3000/
TCP    81s
service/db      ClusterIP   None            <none>        5432/
TCP    81s
service/lb      ClusterIP   10.105.219.69   <none>        80/
TCP       80s
NAME                 DESIRED   CURRENT   READY   UP-TO-
DATE    AVAILABLE   NODE SELECTOR   AGE
```

```
daemonset.apps/lb    1           1           1           1           1
             <none>          80s
NAME                      READY    UP-TO-DATE    AVAILABLE    AGE
deployment.apps/app      3/3      3             3            81s
NAME                              DESIRED    CURRENT    READY    AGE
replicaset.apps/app-5f9797d755    3          3          3        81s
NAME                      READY    AGE
statefulset.apps/db      1/1      80s
```

Our application is ready but inaccessible; the lb component isn't exposed. It is listening on port 80, but ClusterIP is used, hence the Service is only available internally, cluster-wide.

3. We will now create an Ingress resource. There are two manifests in the ingress directory. We will use simplestlab.ingress.yaml, which will be deployed without custom TLS encryption:

```
Chapter11$ cat ingress/simplestlab.ingress.yaml
apiVersion: networking.k8s.io/v1
kind: Ingress
metadata:
  name: simplestlab
  annotations:
    # nginx.ingress.kubernetes.io/rewrite-target: /
spec:
  ingressClassName: nginx
  rules:
  - host: simplestlab.local.lab
    http:
      paths:
      - path: /
        pathType: Prefix
        backend:
          service:
            name: lb
            port:
              number: 80
```

4. We will just deploy the previously created manifest:

```
Chapter11$ kubectl create \
-f ingress/simplestlab.ingress.yaml -n simplestlab
ingress.networking.k8s.io/simplestlab created
Chapter11$ kubectl get ingress -n simplestlab
NAME CLASS HOSTS ADDRESS PORTS AGE
simplestlab nginx simplestlab.local.lab 80 16s
```

5. We can check the defined host URL with `curl`:

```
Chapter11$ curl -H "host: simplestlab.local.lab" http://
localhost/
<!DOCTYPE html>
<html>
<head>
  ...
</head>
<body>
  ...
</body>
</html>
</body>
</html>
```

The `simplestlab` application is now available and accessible.

6. We can change our `/etc/hosts` file (or equivalent MS Windows `c:\system32\drivers\etc\hosts` file). Add the following line and open the web browser to access the `simplestlab` application:

```
127.0.0.1 simplestlab.local.lab
```

This requires root or Administrator access, hence it may be more interesting to use `curl` with the `-H` or `--header` arguments to check the application.

> **Important note**
>
> You can use an extension on your web browser that allows you to modify the headers of your requests or an FQDN including `nip.io`, which will be used in *Chapter 13, Managing the Application Life Cycle*. For example, you can simply add the `simple-modify-headers` extension if you are using MS Edge (you will find equivalent ones for other web browsers and operating systems). Additional information for configuring this extension is discussed in the GitHub `Readme.md` file for this chapter.

The application will be available at `http://localhost` (notice that we defined the URL pattern as `http://locahost/*` in the `simple-modify-headers` extension configuration):

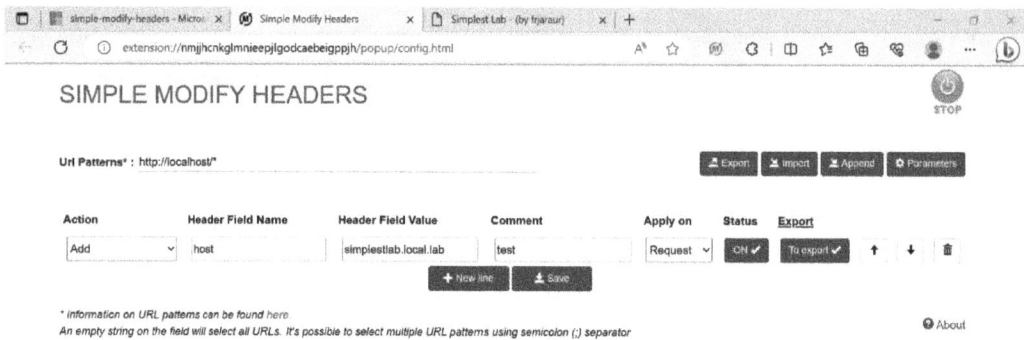

Figure 11.21 – SIMPLE MODIFY HEADERS Edge extension configuration

7. Once the extension is configured, we can reach the `simplestlab` application using `http://localhost`:

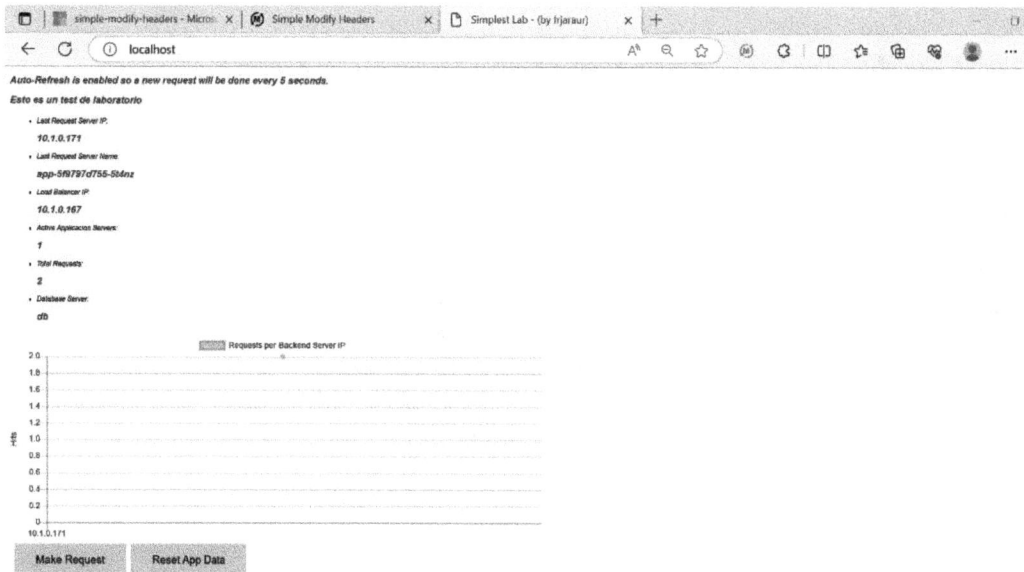

Figure 11.22 – The simplestlab application is accessible thanks to the Ingress Controller

In the GitHub repository, you will find instructions to add TLS to the Ingress resource to improve our application security and how to implement NetworkPolicy resources using Calico as a CNI with Minikube.

These labs have helped you understand how to improve the security of your applications by isolating their components and exposing and publishing only those required by the users and other applications' components.

Summary

In this chapter, we learned how to publish our applications in Kubernetes for our users and for other components deployed either internally in the same cluster or externally. Different mechanisms for this were examined, but ultimately, it is up to you to determine which of your applications' components should be exposed and accessible.

Throughout this chapter, we reviewed some quick solutions for debugging and publishing Service resources directly on our desktop computers with the `kubectl` client. We also examined different Service types that could be useful for locally accessing our remote applications on remote Kubernetes development clusters. We discussed how LoadBalancer Services are part of the Kubernetes core and were prepared for cloud platforms, due to which they may be difficult to implement on-premises, and this is why the recommended option for delivering applications is to create your own Ingress resource manifest. Ingress Controllers will help you to publish applications on any Kubernetes platform. You will use Ingress resources to define how applications will be published, and you may need to tweak their syntax according to the Ingress Controller deployed in your Kubernetes platform.

Toward the end of the chapter, we introduced the NetworkPolicy resource and the service mesh concept, which offers the means to improve the security of our applications by dropping any untrusted and undefined communications. This was followed by some labs to test what we learned.

In the next chapter, we will review some useful mechanisms and tools for monitoring and gathering performance data from our applications.

12

Gaining Application Insights

So far in this book, we have seen how to implement our applications using software containers and how Kubernetes helps us run them in production with security and high availability. We can run and manage our own Kubernetes environments to prepare our applications for any Kubernetes environment; few changes will be necessary to customize our deployments for specific platforms. In this chapter, we will learn how to gain access to our application's Kubernetes resources and the different tools that can be used to identify problems in our applications. We will review **Prometheus** as a popular tool in the Kubernetes world for monitoring an application's component health, interactions, and resources. We will also explore **Loki**, which is an open source logging platform that's highly extensible, configurable, and easy to integrate with Kubernetes. By the end of this chapter, we will have taken a look at some **instrumentation** options for our applications.

Here is a summary of the content in this chapter:

- Understanding your application's behavior
- Obtaining access to your Kubernetes resources
- Monitoring your application's metrics
- Logging your application's important information
- Load testing your application
- Adding instrumentation to your application's code

Technical requirements

You can find the labs for this chapter at `https://github.com/PacktPublishing/Containers-for-Developers-Handbook/tree/main/Chapter12`, where you will find some extended explanations that have been omitted from this chapter's content to make it easier to follow. The *Code In Action* video for this chapter can be found at `https://packt.link/JdOIY`.

Understanding your application's behavior

Understanding how your application works can be very hard if you don't know how your application works. This may sound obvious, but the better you know your application, the better you can implement different mechanisms to verify its status at every moment.

You, as a developer, have to ask yourself which is the best place in your code to add monitoring endpoints or flags. But your application should also be monitored using external third-party tools, which leads us to the following list of monitoring mechanisms:

- **Internal application metrics**: Implementing monitoring points in our code may be difficult at the end of the project, but if you introduce them from the beginning and measure the time between transactions, you will have a great overall performance view of your application.

- **Internal health checks**: Health checks are crucial for identifying when your application fails, but we can go further. We can have some simple error/OK quick tests that can be executed very often and help Kubernetes keep the application up and running.

- **External application metrics**: Some components such as databases may allow you to query certain metrics that can be exported and used by an external component.

- **External health checks**: These health checks are used to provide a good overview of the application's behavior. They can be complex and include multiple components, and they may trigger some alarms or create events that will help us manage the application's status.

We haven't talked about how any of these mechanisms can be implemented yet. While including some metrics in our application may require us to make some changes in the code, add some entries with metric values, or create the sources to be retrieved for such metrics, other points can be achieved by using external tools, outside of the application's code.

Running your applications in containers can help you implement any of the models described here. But *never* include a parallel check process inside your application container. This is not how containers should be used. Remember that we introduced the concept of a container as the main process that runs isolated in a host, sharing its kernel among all other containers, in *Chapter 1, Modern Infrastructure and Applications with Docker*. Running more than one process is not a good practice because you will need to ensure that all processes receive SIGTERM or SIGKILL signals whenever the container needs to stop, and this may be tricky if you fork the processes and they run separately in the container's namespace (the same as a PID namespace but without dependencies).

In this chapter, we will learn how to implement different models for monitoring, logging, and even tracing our application's processes. But let's get started by reviewing how to retrieve and manage our application's Kubernetes resources from our own application's workloads, which is key for some of the different tools we are going to use later.

Obtaining access to your Kubernetes resources

Sometimes, your applications need to manage certain Kubernetes resources. Let's think about the following situations:

- The default Kubernetes autoscaling doesn't cover our requirements
- We need to create some resources triggered by an event – for example, when our application starts

In these scenarios, our application's processes will need to retrieve information from the Kubernetes API and create some resources. If we think of this workflow, at least our Pods will require network access to the Kubernetes API server's IP address and appropriate permissions for the required actions and resources. In this section, we will learn how to create and manage Kubernetes objects from our application's processes.

First, we need to remember a few concepts from *Chapter 8*, *Deploying Applications with the Kubernetes Orchestrator*. In that chapter, we talked about how Kubernetes improves our application's security by using different authentication and authorization strategies. You may need to ask your Kubernetes administrators about some Kubernetes platform insights, but you will probably be using **role-based access control** (**RBAC**) in your environment because, currently, all Kubernetes platforms work with such a mechanism. This means that all Kubernetes API requests must be authorized by a **role authorization system**. Kubernetes will use either client certificates, bearer tokens, or an authenticating proxy to authenticate API requests through authentication plugins, and when client requests are authenticated, the authorization system will allow or deny them.

By default, all Pods running in a namespace will inherit a service account and its token to authenticate the application processes within the Kubernetes cluster. This behavior is not secure and that's why it can be avoided by your Kubernetes administrators. But in any case, we will use a service account included in the Pod definition and its associated token to identify the processes and validate access to the specified resources and actions.

Within the Kubernetes cluster, the RBAC API declares roles and their bindings for pairing Kubernetes resources with the actions allowed for them. This role system will include namespaced authorizations (Role and RoleBinding resources) and cluster-wide authorizations (ClusterRole and ClusterRoleBinding resources), which help us provide fine-grained access.

Role and ClusterRole resources define rules that represent additive permissions; so, if no rule exists for a permission, it is denied. We will match resources and the verbs allowed for them in their manifest definitions. We will use ClusterRole resources to define cluster-wide permissions or define namespace-scoped permissions from a higher level, which allows us to reuse them in several namespaces.

Let's see an example of a Role and how we associate it with a ServiceAccount resource using a RoleBinding:

```
apiVersion: rbac.authorization.k8s.io/v1
kind: Role
metadata:
  namespace: default
  name: pod-reader
rules:
- apiGroups: [""]
  resources:
    - pods
  verbs:
    - get
    - watch
    - list
```

```
apiVersion: rbac.authorization.k8s.io/v1
kind: RoleBinding
metadata:
  name: read-pods
  namespace: default
subjects:
- kind: ServiceAccount
  name: myserviceaccount
  namespace: default
roleRef:
  kind: Role
  name: pod-reader
  apiGroup: rbac.authorization.k8s.io
```

Figure 12.1 – Role and RoleBinding resources allowing us to list and read Pods in a namespace

Notice that in both resources, we defined namespace because we are using namespace-scoped resources (they can be omitted if we deploy them on the current namespace). In the Role resource, we defined the list of verbs or actions allowed and resources in which the actions will be applied (you can use kubectl api-resources -o wide to retrieve the verbs available for each Kubernetes resource). The apiGroups key is used when we have different resources with the same name but belonging to different APIs. Let's see this configuration in action with a quick example. We will create both Role and RoleBinding resources in the default namespace:

```
Ubuntu 22.04.2 LTS          ×    +  ˅

frjaraur@sirius:~$ cat <<EOF|kubectl create -f -
apiVersion: rbac.authorization.k8s.io/v1
kind: Role
metadata:
  namespace: default
  name: pod-reader
rules:
- apiGroups: [""]
  resources:
  - pods
  verbs:
  - get
  - watch
  - list
EOF
role.rbac.authorization.k8s.io/pod-reader created
frjaraur@sirius:~$ cat <<EOF|kubectl create -f -
apiVersion: rbac.authorization.k8s.io/v1
kind: RoleBinding
metadata:
  name: read-pods
  namespace: default
subjects:
- kind: ServiceAccount
  name: myserviceaccount
  namespace: default
roleRef:
  kind: Role
  name: pod-reader
  apiGroup: rbac.authorization.k8s.io
EOF
rolebinding.rbac.authorization.k8s.io/read-pods created
frjaraur@sirius:~$ █
```

Figure 12.2 – Creating resources so that Pods can be listed and read in a namespace

---

Important note

We can use either **declarative** (YAML manifests) or **imperative** modes to create RBAC resources. Therefore, we could have used `kubectl create role pod-reader --verb=get,list,watch --resource=Pods` to create the `pod-reader` Role resource from the preceding code snippet.

We have applied the RoleBinding resource to a specific service account, but it doesn't exist yet. Let's create it before moving on:

```
Ubuntu 22.04.2 LTS          ×    +  ∨
frjaraur@sirius:~$ kubectl create sa myserviceaccount
serviceaccount/myserviceaccount created
frjaraur@sirius:~$ kubectl get rolebindings,roles,sa
NAME                                                 ROLE              AGE
rolebinding.rbac.authorization.k8s.io/read-pods      Role/pod-reader   3m52s

NAME                                            CREATED AT
role.rbac.authorization.k8s.io/pod-reader       2023-07-26T11:39:57Z

NAME                              SECRETS    AGE
serviceaccount/default            0          7d19h
serviceaccount/myserviceaccount   0          4s
frjaraur@sirius:~$ █
```

Figure 12.3 – Creating the myserviceaccount ServiceAccount resource

We have a Role that allows us to list, watch, and get Pods within any namespace (but applied to the current default namespace), and a RoleBinding associating this Role resource with a defined service account, myserviceaccount, in the default namespace. We will now run a Pod with this service account and get the list of current Pods. We will run a Pod with the kubectl command line included in the container image. We included the myserviceaccount service account resource and used get pod as arguments for the image:

```
Ubuntu 22.04.2 LTS          ×    +  ∨
frjaraur@sirius:~$ cat <<EOF|kubectl create -f -
apiVersion: v1
kind: Pod
metadata:
  creationTimestamp: null
  labels:
    run: kubectl
  name: kubectl
spec:
  serviceAccountName: myserviceaccount
  containers:
  - args:
    - get
    - pods
    image: bitnami/kubectl:latest
    name: kubectl
    resources: {}
  dnsPolicy: ClusterFirst
  restartPolicy: Always
EOF
pod/kubectl created
frjaraur@sirius:~$ kubectl get pods
NAME        READY    STATUS       RESTARTS    AGE
kubectl     0/1      Completed    0           4s
webserver   1/1      Running      0           36m
frjaraur@sirius:~$ kubectl logs kubectl
NAME        READY    STATUS       RESTARTS    AGE
kubectl     0/1      Completed    0           4s
webserver   1/1      Running      0           36m
frjaraur@sirius:~$ █
```

Figure 12.4 – Creating a Pod that lists all the Pods in the current namespace

In the preceding code snippet, we created a Pod whose container will use the `myserviceaccount` service account to interact with the Kubernetes API. Notice that we just retrieved the logs from the Pod and we get the output from the `kubectl get pods` command line that's executed. All the Pods running at the time of execution were listed. Let's try the same with a new Pod that doesn't use this service account:

```
frjaraur@sirius:~$ kubectl run kubectl2 \
--image=bitnami/kubectl:latest -- get pods
pod/kubectl2 created
```

Now, we can retrieve the logs from this new Pod:

```
$ kubectl logs kubectl2
Error from server (Forbidden): pods is forbidden: User
"system:serviceaccount:default:default" cannot list resource "pods" in
API group "" in the namespace "default"
```

As you can see, this time, the new Pod does not have access to the Kubernetes API (the `default` service account was used by default, which doesn't have permission to list the Pods in the namespace).

Accessing the Kubernetes API may become complex when you create custom resources and need fine-grained access, but be sure that the principles for managing RBAC access are the same.

> **Important note**
>
> You can implement your own `Kubeconfig` file to use with your application. Kubernetes offers out-of-the-box integration with a service account's tokens, but you can use a Secret resource to include your authentication file and use it in your application. This is especially useful when you're managing different Kubernetes clusters from a unified control plane cluster.

In these examples, we are not using a NetworkPolicy resource, but you will need to ensure your Pods have access to Kubernetes API server Pods. These Pods will use control plane hosts' IP addresses and port `6443`, although this may vary between Kubernetes platforms (confirm these requirements with your Kubernetes administrators). If your platform administrators configured GlobalNetworkPolicy resources, you may need to add some **egress** rules for your deployments.

In this example, we used the kubectl Kubernetes client, but you will find client libraries and modules for different code languages, which makes integration more secure. Attackers will not be able to exploit erroneous RBAC configurations if your code only manages required resources. On the other hand, adding new functionalities may require you to recompile your code, but it is worth it. Here is a link to the current documentation, where you will find more information about the client libraries: `https://kubernetes.io/docs/reference/using-api/client-libraries/`. At the time of writing this book, C, .NET, Go, Perl, Python, Java, JavaScript, and Ruby, among other languages, are officially supported. If you require a different language, such as Rust, there are community projects developing libraries for accessing Kubernetes.

Now, let's introduce the concept of Kubernetes operators, which will help us deploy and operate applications in Kubernetes. We will use some of them in this chapter and it is interesting to learn the basics before we do so.

Understanding Kubernetes operators

A **Kubernetes operator** is a piece of software that runs and integrates with Kubernetes to manage applications and their components. They use the concepts we've learned about so far in this chapter to monitor and create resources as needed by your applications.

Kubernetes operators are designed to automate most of the repetitive tasks a human operator must do to manage an application. There are many well-documented examples of Kubernetes operators that will help you perform tasks such as deploying databases with high availability, managing complex applications with multiple components with just one YAML manifest, and so on. These are some of the tasks you may expect from a Kubernetes operator designed for a certain application:

- Automated deployment of the application and all its components

- Management and creation of Kubernetes **CustomResourceDefinitions** (**CRDs**) required for the application

- Backup and restore features that will help recover the application with easy steps

- Fully managed application upgrades, with all the internal application components such as database schema migrations

- Choosing a leader when your application requires a distributed control plane or must manage the master-worker (or master-slave) relations between application components

As you can see, operators are key for managing complex application deployments, and their integration into Kubernetes makes things more simple. There are good examples of managing databases, created by the most important software database vendors, and for many other software categories. You may find what you need at `https://operatorhub.io`. In the following section, we will use the **Prometheus Operator** to deploy and manage the Prometheus monitoring tool.

In case you don't find a Kubernetes operator that meets your application requirements, you can code your own operator using any of the **software development kits** (**SDKs**) available for different languages (`https://kubernetes.io/docs/concepts/extend-kubernetes/operator/#writing-operator`).

So far, we have learned how Kubernetes can be queried for some resources' statuses and how to manage them within our applications. In the following section, we will learn how **third-party applications** can be used to monitor our applications.

Monitoring your application's metrics

Analyzing your application's metrics is key to understanding how you are serving your users or other applications. In this section, we will learn how to monitor our applications using **Prometheus**, a monitoring solution that fits very well in the Kubernetes ecosystem.

Prometheus is an open source monitoring solution that's been hosted by the **Cloud Native Computing Foundation** (**CNCF**) since 2016. It is used to collect, store, and represent metrics and alert users using thresholds. It can be integrated into any infrastructure, although it is known to work great with Kubernetes. It comes included within some Kubernetes platform deployments and that's why it is considered standard within the Kubernetes community.

Prometheus includes a data model that is easy to query, using its own **Prometheus query language** (**PromQL**), which stores metrics recorded over time from different sources identified by key-value pairs. By default, Prometheus will pull different endpoints via HTTP requests, although pushing data is also available, but less common. These endpoints, usually identified as **targets**, can be auto-discovered or manually configured, which makes Prometheus fit perfectly in Kubernetes clusters where dynamism is a must.

Although Prometheus provides a graphing tool, it is very common to integrate it as a data source in more advanced dashboard tools such as **Grafana** or to directly consume its data via an API (using PromQL queries).

We are not going to cover this tool in depth in this book as it would be out of its scope, but we will review some of its components, quick installation, and how to implement some monitoring endpoints for your applications.

Exploring Prometheus architecture

Prometheus is based on at least five different components:

- **Prometheus server**: This is the core component. It retrieves metrics data from Prometheus exporters and adds data from the **Pushgateway**. All this data is stored in its own **time series database** (**TSDB**) and is available via the HTTP API, also managed by the Prometheus server component. It also checks the different configured thresholds.

- **Pushgateway**: There are some devices or components that can't be scraped. The Pushgateway component allows you to directly push the data into Prometheus, instead of waiting for it to be pulled. Different libraries exist for common languages such as Java, Go, Python, and Ruby, among others supported by the community.

- **Alertmanager**: Alertmanager handles all the alerts generated from the Prometheus server. Different notification backends can be used, such as email or webhooks.

- **Prometheus web UI**: The provided web UI allows us to query stored metrics, quickly graph data, and review the status of the different targets configured.

- **Prometheus exporters**: These components are the key to the extensibility of the platform. Many client libraries are officially supported for different code languages (and others are unofficially supported) that allow us to create metrics for our applications. When Prometheus scrapes your endpoints, the client library presents the data, and it will be stored in the server for later access or threshold validation. You can find officially supported Prometheus exporters inside GitHub's Prometheus organization (`https://github.com/orgs/prometheus/repositories?q=exporter&type=all`).

Prometheus can monitor your applications running in Kubernetes by either running inside your Kubernetes cluster (in its own namespace or even within your application's namespace, which is not recommended) or externally, in a different infrastructure, such as virtual machines. It is recommended to run Prometheus inside your Kubernetes cluster because you will be able to use internal communications, instead of having to publish your exporters externally to be pulled from outside of the cluster. This will improve security, even if you expose your exporters internally using HTTP instead of HTTPS protocol. Running in Kubernetes will allow us to deploy Prometheus as a Kubernetes operator, which will help us implement the auto-discovery of targets as well as an easy-to-manage complete monitoring platform. All the components will be installed for us, and we will just have to configure how they should be deployed.

Installing Prometheus

To install the Prometheus monitoring platform, we will use **Helm**, which is a tool that allows us to easily customize and deploy a set of manifests. We will deep dive into using Helm to package applications in *Chapter 13, Managing the Application Life Cycle*. In this case, these manifests include the **kube-prometheus** platform components (`https://github.com/prometheus-operator/kube-prometheus`). The kube-prometheus community project installs a cluster-ready monitoring platform with the following components:

- The **Prometheus Operator**, which will create its own CRDs and manage the Service discovery integration.

- **Prometheus** and **Alertmanager** with high availability, both deployed as StatefulSets. A set of default alerts is included, which will help you start monitoring your Kubernetes environment.

- **Prometheus node-exporter**, deployed as a DaemonSet to all the Kubernetes cluster nodes. This exporter will have host-related metrics such as CPU, memory, and disk space available.

- **Prometheus Adapter for Kubernetes Metrics APIs**, which integrates all the Kubernetes Metrics Server metrics into Prometheus automatically.

- **kube-state-metrics**, which connects with the Kubernetes API server and retrieves the status of different resources such as Pods, Deployments, and so on, and delivers them as metrics for Prometheus. By default, important metrics are created and configured for you.

- **Grafana**, deployed as part of the platform to enhance Prometheus graphs in Grafana dashboards. A set of default dashboards is included to show you how your platform works.

Important note

The default alerts and dashboards included in the kube-prometheus project are taken from the **kubernetes-mixin** project (`https://github.com/kubernetes-monitoring/kubernetes-mixin`), which provides a set of well-documented rules and simple dashboards for monitoring Kubernetes.

You, as a developer, will probably not use many of the dashboards and metrics provided by this platform, but it will help you understand how to implement your metrics and rules and create dashboards with your data.

Installing the `kube-prometheus-stack` is easy. A Helm Chart (Helm-specific package) is ready for us. We will just add the Prometheus community Helm Charts repository, update the repositories cache, and install a Helm Chart release in our cluster.

Important note

If you are using Rancher Desktop, Helm comes preinstalled with your command-line tools, but on other platforms, it may be necessary to install it before using it. Helm is available for Windows, macOS, and Linux and there are different methods for installing it. We recommend that you use the binary directly. That way, you can update it whenever you need it and use a different release if required. If you are using Windows, you can use `Get-Content` to download it and then add it to your PATH:

```
PS C:\Users\frjaraur> Invoke-WebRequest -URI https://get.helm.sh/helm-v3.12.2-windows-amd64.zip -OutFile $env:TEMP\helm.zip -UseBasicParsing
PS C:\Users\frjaraur> Expand-Archive $env:TEMP\helm.zip -DestinationPath $env:TEMP\helm_install
PS C:\Users\frjaraur> md $HOME\bin


    Directory: C:\Users\frjaraur


Mode                 LastWriteTime         Length Name
----                 -------------         ------ ----
d-----          7/28/2023     12:02                bin


PS C:\Users\frjaraur> mv $env:TEMP\helm_install\windows-amd64\helm.exe $HOME\bin
PS C:\Users\frjaraur> $env:PATH += ";$HOME\bin"
PS C:\Users\frjaraur> helm version
version.BuildInfo{Version:"v3.12.1", GitCommit:"f32a527a060157990e2aa86bf45810dfb3cc8b8d", GitTreeState:"clean", GoVersion:"go1.20.4"}
PS C:\Users\frjaraur>
```

Figure 12.5 – Installing the Helm binary using gc to download the required package

Once Helm has been installed, we will just use `helm install` with `--create-namespace` to tell Helm to create a new namespace for us. In this example, we are using Minikube as the Kubernetes environment:

```
PS C:\Users\frjaraur> helm repo add prometheus-community https://prometheus-community.github.io/helm-charts
"prometheus-community" has been added to your repositories
PS C:\Users\frjaraur> helm install prometheus -n monitoring --create-namespace prometheus-community/kube-prometheus-stack
NAME: prometheus
LAST DEPLOYED: Fri Jul 28 20:48:43 2023
NAMESPACE: monitoring
STATUS: deployed
REVISION: 1
NOTES:
kube-prometheus-stack has been installed. Check its status by running:
  kubectl --namespace monitoring get pods -l "release=prometheus"

Visit https://github.com/prometheus-operator/kube-prometheus for instructions on how to create & configure Alertmanager and Prometheu
s instances using the Operator.
PS C:\Users\frjaraur> kubectl get pod,svc -n monitoring
NAME                                                         READY   STATUS             RESTARTS   AGE
pod/alertmanager-prometheus-kube-prometheus-alertmanager-0   0/2     Init:0/1           0          3s
pod/prometheus-grafana-7cb6877764-hc7dl                      0/3     ContainerCreating  0          17s
pod/prometheus-kube-prometheus-operator-6dd4cfff56-7wl2n     1/1     Running            0          17s
pod/prometheus-kube-state-metrics-6df4697c45-hjbjk           0/1     ContainerCreating  0          17s
pod/prometheus-prometheus-kube-prometheus-prometheus-0       0/2     Init:0/1           0          3s
pod/prometheus-prometheus-node-exporter-7sq2f                1/1     Running            0          17s

NAME                                                TYPE        CLUSTER-IP       EXTERNAL-IP   PORT(S)                      AGE
service/alertmanager-operated                       ClusterIP   None             <none>        9093/TCP,9094/TCP,9094/UDP   3s
service/prometheus-grafana                          ClusterIP   10.103.228.129   <none>        80/TCP                       18s
service/prometheus-kube-prometheus-alertmanager     ClusterIP   10.102.251.220   <none>        9093/TCP,8080/TCP            18s
service/prometheus-kube-prometheus-operator         ClusterIP   10.97.128.218    <none>        443/TCP                      18s
service/prometheus-kube-prometheus-prometheus       ClusterIP   10.108.154.247   <none>        9090/TCP,8080/TCP            18s
service/prometheus-kube-state-metrics               ClusterIP   10.101.5.41      <none>        8080/TCP                     18s
service/prometheus-operated                         ClusterIP   None             <none>        9090/TCP                     3s
service/prometheus-prometheus-node-exporter         ClusterIP   10.99.60.11      <none>        9100/TCP                     18s
PS C:\Users\frjaraur>
```

Figure 12.6 – Installing the Prometheus stack using Helm

After a few seconds, the Prometheus stack will be up and running. At this point, we can check the platform's Pod and Service resources:

```
PS C:\Users\frjaraur> kubectl get pod,svc -n monitoring
NAME                                                         READY   STATUS    RESTARTS   AGE
pod/alertmanager-prometheus-kube-prometheus-alertmanager-0   2/2     Running   0          72s
pod/prometheus-grafana-7cb6877764-hc7dl                      2/3     Running   0          86s
pod/prometheus-kube-prometheus-operator-6dd4cfff56-7wl2n     1/1     Running   0          86s
pod/prometheus-kube-state-metrics-6df4697c45-hjbjk           1/1     Running   0          86s
pod/prometheus-prometheus-kube-prometheus-prometheus-0       1/2     Running   0          72s
pod/prometheus-prometheus-node-exporter-7sq2f                1/1     Running   0          86s

NAME                                              TYPE        CLUSTER-IP       EXTERNAL-IP   PORT(S)                      AGE
service/alertmanager-operated                     ClusterIP   None             <none>        9093/TCP,9094/TCP,9094/UDP   72s
service/prometheus-grafana                        ClusterIP   10.103.228.129   <none>        80/TCP                       87s
service/prometheus-kube-prometheus-alertmanager   ClusterIP   10.102.251.220   <none>        9093/TCP,8080/TCP            87s
service/prometheus-kube-prometheus-operator       ClusterIP   10.97.128.218    <none>        443/TCP                      87s
service/prometheus-kube-prometheus-prometheus     ClusterIP   10.108.154.247   <none>        9090/TCP,8080/TCP            87s
service/prometheus-kube-state-metrics             ClusterIP   10.101.5.41      <none>        8080/TCP                     87s
service/prometheus-operated                       ClusterIP   None             <none>        9090/TCP                     72s
service/prometheus-prometheus-node-exporter       ClusterIP   10.99.60.11      <none>        9100/TCP                     87s
```

Figure 12.7 – Prometheus stack Pod and Service resources

This small installation is intended only for local usage – so that you can develop your own monitors for your application on your desktop computer. Notice that we didn't even include a PersistentVolume, so data will be lost every time you restart your environment. A lot of customizations can be done at the installation level to cover all your specific needs, but you should read the documentation before configuring your own Helm values YAML file (`https://github.com/prometheus-community/helm-charts/blob/main/charts/kube-prometheus-stack/values.yaml`).

Reviewing the Prometheus environment

In this section, we will learn about the GUI and features of Prometheus. We can get into Prometheus by using the `prometheus-stack-grafana` Service:

1. For a quick review, we will use `port-foward` to access the Grafana web UI:

    ```
    PS C:\Users\frjaraur> kubectl -n monitoring `
    port-forward service/prometheus-operated 9090:9090
    Forwarding from 127.0.0.1:9090 -> 9090
    Forwarding from [::1]:9090 -> 9090
    Handling connection for 9090
    ```

2. You can now open `http://localhost:8080` in your browser to access Prometheus. If you click on **Status**, you will see the available pages related to the monitored endpoints:

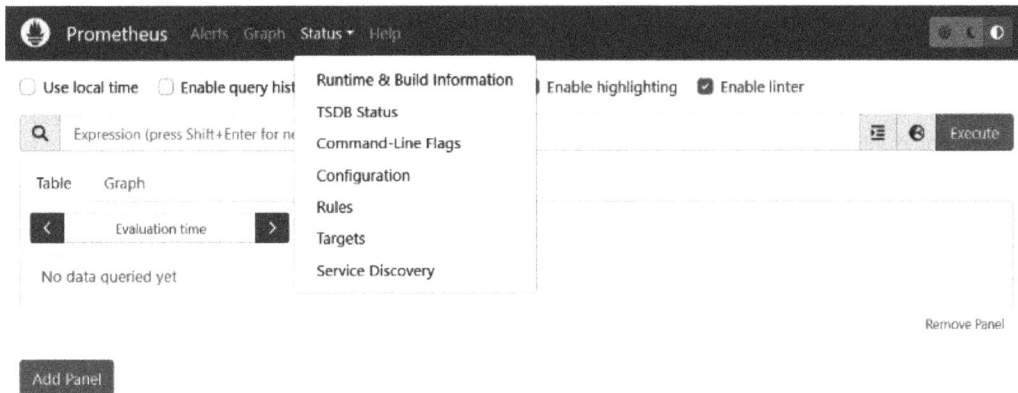

Figure 12.8 – Prometheus GUI showing the Status section

3. Now, we can go to the **Targets** section and review which targets are currently monitored by the Prometheus platform:

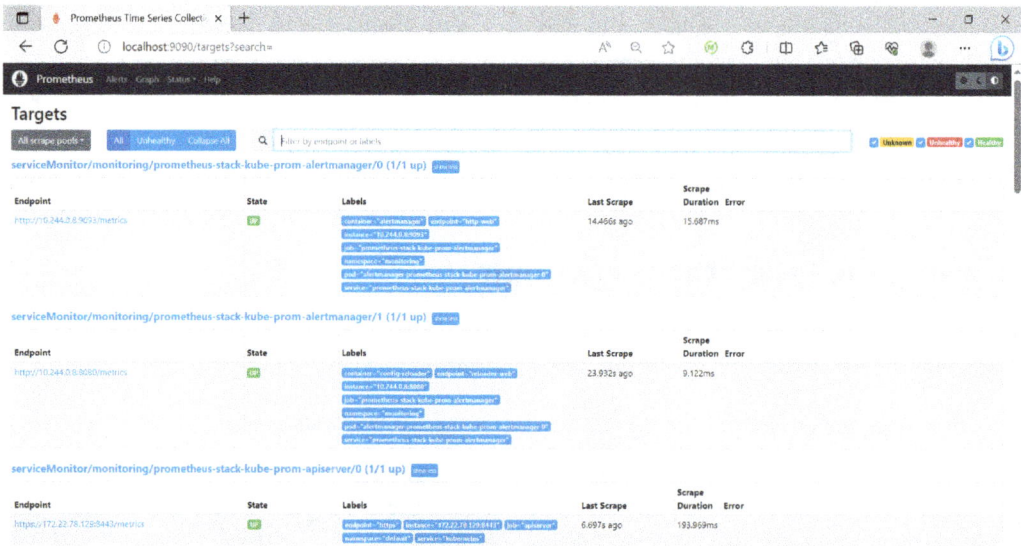

Figure 12.9 – Prometheus GUI showing the Targets section

4. We can use the filter on the right-hand side to uncheck the **Healthy** targets and view which targets are currently down:

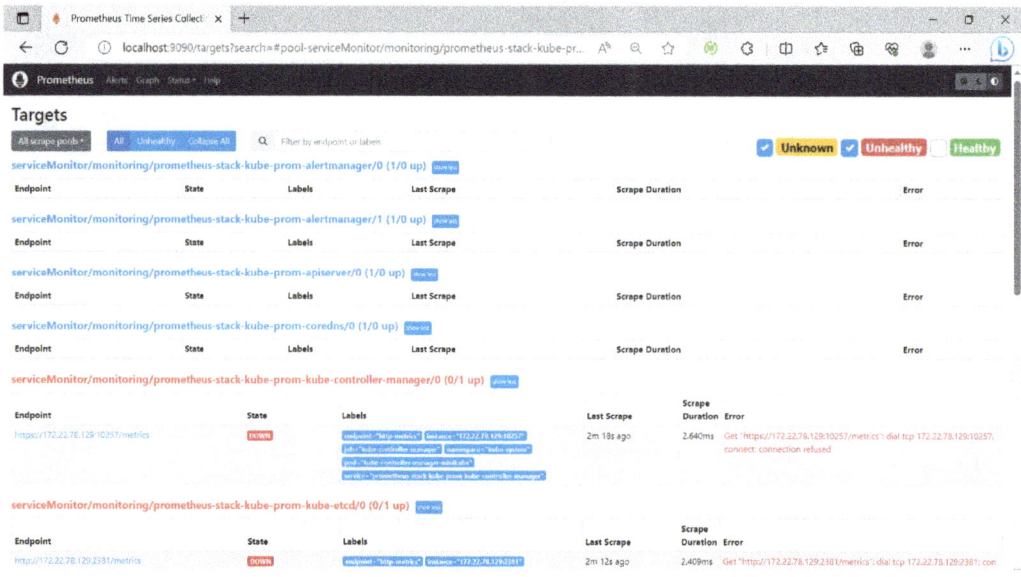

Figure 12.10 – Prometheus GUI showing the Unhealthy targets in the Targets section

Don't worry – this is normal. Minikube does not expose all the Kubernetes metrics; therefore, some monitoring endpoints will not be available.

5. Let's review some of the metrics that are currently retrieved by clicking on **Graph**, on top of the Prometheus GUI, and then write a simple PromQL query to retrieve the seconds of CPU in use per Pod (the `container_cpu_usage_seconds_total` metric):

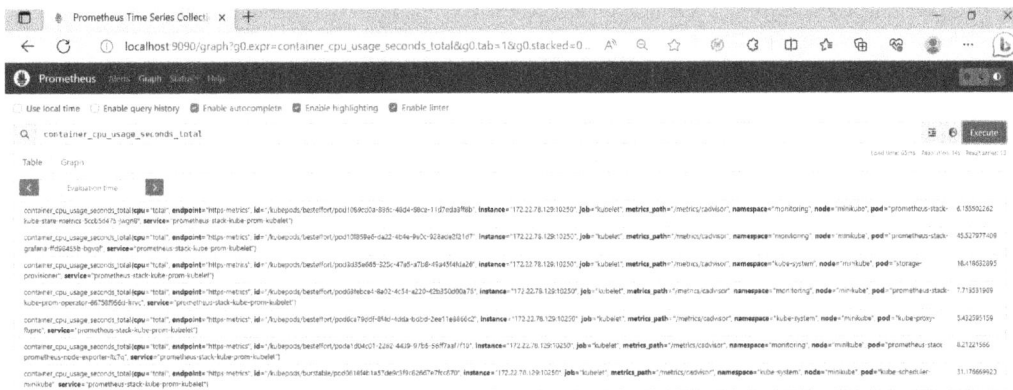

Figure 12.11 – Prometheus GUI showing the Graph section with a metric as a query

Notice that we only have metrics from objects running on the **kube-system** and **monitoring** (created for the stack itself) namespaces.

6. Let's create a quick web server deployment on the default namespace and verify that it appears in the monitoring platform:

```
PS C:\Users\frjaraur> kubectl create deployment `
webserver --image=nginx:alpine --port=80
deployment.apps/webserver created
```

In a few seconds, the new Pod for the web server deployment will appear in the Prometheus **Graph** section:

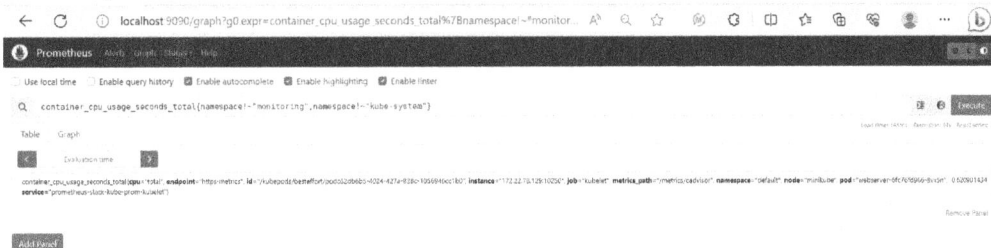

Figure 12.12 – Filtered list of Kubernetes containers using CPU

Notice that in this case, we used a new PromQL query, `container_cpu_usage_seconds_total{namespace!~"monitoring",namespace!~"kube-system"}`, in which we removed any metrics from the `monitoring` and `kube-system` namespaces. The metrics were automatically included thanks to the Prometheus Adapter for the Kubernetes Metrics APIs component.

> **Important note**
>
> Knowledge of Kubernetes metrics or Pod metrics is outside the scope of this chapter. We are using Prometheus to show you how you can deliver your application metrics. Each exporter or integration mentioned in this section has its documentation, where you will find information about the metrics available. The use of PromQL and Prometheus itself is also outside the scope of this book. You can find very useful documentation at `https://prometheus.io/docs`. To monitor our applications and retrieve their active hardware resource consumption, we don't need to deploy the Alertmanager component, which will reduce the requirements of your desktop environment.

Prometheus uses labels to filter resources. Choosing good metrics and label conventions will help you design your application monitoring. Take a close look at `https://prometheus.io/docs/practices/naming`, where the documentation explains a good naming and labeling strategy.

Now that we know about the basics of the Prometheus interface, we can move on and review how the Prometheus server gets infrastructure and application data.

Understanding how Prometheus manages metrics data

Let's review how targets are configured by the Prometheus Operator:

1. We will retrieve the new CRDs created by the Prometheus stack deployment:

```
PS C:\Users\frjaraur> kubectl api-resources |select-string "monitoring.coreos"

alertmanagerconfigs      amcfg       monitoring.coreos.com/v1alpha1    true      AlertmanagerConfig
alertmanagers            am          monitoring.coreos.com/v1          true      Alertmanager
podmonitors              pmon        monitoring.coreos.com/v1          true      PodMonitor
probes                   prb         monitoring.coreos.com/v1          true      Probe
prometheusagents         promagent   monitoring.coreos.com/v1alpha1    true      PrometheusAgent
prometheuses             prom        monitoring.coreos.com/v1          true      Prometheus
prometheusrules          promrule    monitoring.coreos.com/v1          true      PrometheusRule
scrapeconfigs            scfg        monitoring.coreos.com/v1alpha1    true      ScrapeConfig
servicemonitors          smon        monitoring.coreos.com/v1          true      ServiceMonitor
thanosrulers             ruler       monitoring.coreos.com/v1          true      ThanosRuler
```

Figure 12.13 – Filtered list of available Kubernetes API resources

The Prometheus Operator will use **PodMonitor** and **ServiceMonitor** resources to query the associated endpoints in time intervals. Therefore, to monitor our application, we will need to create a custom metric exporter, to provide the application metrics, and a PodMonitor or ServiceMonitor to expose them for Prometheus.

2. Let's take a quick look at some of the metrics that have been exposed. We will review the Node Exporter component here. The Service resource associated with this monitoring component can be easily retrieved using the following command:

```
PS C:\Users\frjaraur> kubectl get svc `
-n monitoring prometheus-prometheus-node-exporter `
-o jsonpath='{.spec}'
{"clusterIP":"10.108.206.5","clusterIPs":["10.108.206.5"],"in-
```

```
ternalTrafficPolicy":"Cluster","ipFamilies":["IPv4"],"ip-
FamilyPolicy":"SingleStack","ports":[{"name":"http-met-
rics","port":9100,"protocol":"TCP","targetPort":9100}],"se-
lector":{"app.kubernetes.io/instance":"prometheus-stack","app.
kubernetes.io/name":"prometheus-node-exporter"},"sessionAffini-
ty":"None","type":"ClusterIP"}
```

As you can see, this component is exposing data from Pods with the `app.kubernetes.io/name=prometheus-node-exporter` label.

3. Let's expose that Pod and review the presented data:

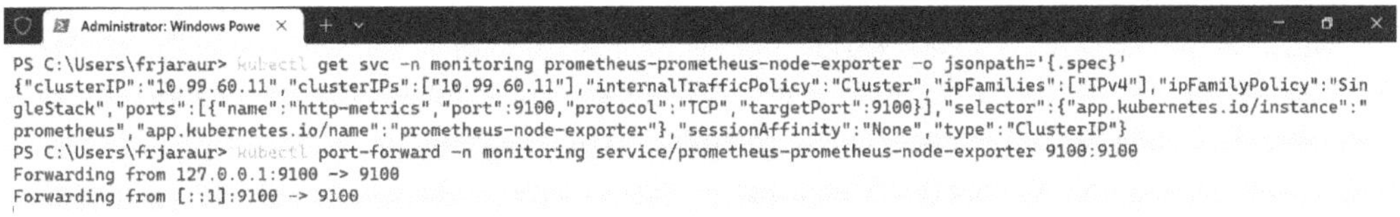

Figure 12.14 – Exposing Prometheus Node Exporter via the port forwarding feature

4. We can now open a new PowerShell terminal and use `Invoke-WebRequest` to retrieve the data or any web browser (you will obtain an intermediate web page indicating that the metrics will be found in the `/metrics` path):

Figure 12.15 – Metrics available from Node Exporter, exposed via port forwarding in local port 9100

The metrics for the Node Exporter component are published internally by an associated Service resource. This is how we should create our monitoring endpoint. We can use two different architectures here:

- Integrate our monitoring component inside our application Pod using a new container and sidecar pattern
- Run a separate Pod with the monitor component and retrieve the data internally using the Kubernetes overlay network

You must remember that the containers running inside a Pod share a common IP address and will always be scheduled together. This may be the main difference and why you will probably choose the first option from the preceding list. Running a different Pod will also require some dependency tracking between both Pods and node affinity patterns (if we want them to run together on the same host). In either of these situations, we will need a PodMonitor or ServiceMonitor resource.

Next, we'll take a quick look at how Prometheus will automatically scrape metrics from a new exporter when we create exporters for our applications.

Scraping metrics with Prometheus

Prometheus installation creates a new resource, `Prometheus`, which represents a Prometheus instance. We can list Prometheus instances in our cluster (we used `-A` to include all the namespaces in the search):

```
PS C:\Users\frjaraur> kubectl get Prometheus -A
NAMESPACE       NAME                 VERSION    DESIRED    READY    RECON-
CILED    AVAILABLE    AGE
monitoring    prometheus-kube-prom-pro-
metheus    v2.45.0    1              1         True                True        17h
```

If we take a look at this resource, we will realize that two interesting keys will decide which resources to monitor. Let's get the `Prometheus` resource YAML manifest and review some interesting keys:

```
PS C:\Users\frjaraur> kubectl get Prometheus -n monitoring prometheus-
kube-prom-prometheus  -o yaml
apiVersion: monitoring.coreos.com/v1
kind: Prometheus
...
spec:
  podMonitorNamespaceSelector: {}
  podMonitorSelector:
    matchLabels:
      release: prometheus-stack
...
```

```
serviceMonitorNamespaceSelector: {}
serviceMonitorSelector:
  matchLabels:
    release: prometheus-stack
```

This `Prometheus` resource includes important keys to modify the behavior of Prometheus itself (such as `scrapeInterval`, data `retention`, and `evaluationInterval`). From the preceding code snippet, we can see that all the ServiceMonitor resources from all namespaces will be monitored if they include the `release=prometheus-stack` label. The same is required for PodMonitors, so we will just create a ServiceMonitor resource for our new application monitor. Here is an example:

```
apiVersion:
monitoring.coreos.com/v1                    ...

kind: ServiceMonitor                         namespaceSelector:

metadata:                                      matchNames:

  labels:                                        - simplestlab

    release: prometheus-stack                selector:

  name: myappmonitor                           matchLabels:

spec:                                            component: lb

  endpoints:

  - interval: 30s

    targetPort: 8080

    path: /metrics
```

Figure 12.16 – ServiceMonitor example manifest

In the *Labs* section, we will add some open source monitoring endpoints for our application (the **Postgres exporter** from `https://grafana.com/oss/prometheus/exporters/`).

Now, let's review some quick concepts for configuring a good logging strategy for our applications.

Logging your application's important information

In this section, we will get a general overview of different logging strategies and how to implement an open source Kubernetes-ready solution such as **Grafana Loki**.

In *Chapter 4, Running Docker Containers*, we talked about which strategies were available for our applications. As a rule of thumb, processes running in containers should always log standard and error outputs. This makes it easier or at least prepares a good solution for all the processes at the same time. However, your application may need a different **logging strategy** for different use cases. For example, you shouldn't log any sensitive information to the standard output. While local logging can be helpful when you are developing your application, it may be very tricky (or even only available for Kubernetes administrators) in a development or production environment. Therefore, we should use either volumes or an external logging ingestion platform. Using volumes may require additional access so that you can recover your logs from a storage backend, so an external platform would be a better approach for covering your logging needs. The fact is that using an external platform can make your life easier if your application runs in Kubernetes. There are plenty of Kubernetes-ready logging platforms that will allow you to push all your container logs to a backend in which you will be able to manage and add appropriate views for different users. This will solve the problem of logging sensitive data because it may be necessary for debugging purposes but only visible to certain trusty users. You may need to ask your Kubernetes administrators because you will probably have a logging solution already working on your Kubernetes platform.

In this chapter, we'll discuss how you can use Grafana Loki in your Kubernetes environment to read and forward your application's containers and send them to a unified backend, in which we will be able to use additional tools such as Grafana to review the application's data.

Grafana Loki can be deployed using different modes, depending on the size of your platform and the number of logs expected. To develop and prepare the logs of your applications, we will use a minimal installation, as we already did with Prometheus. We will use the **monolithic** mode (`https://grafana.com/docs/loki/latest/fundamentals/architecture/deployment-modes/`), in which all of Loki's microservices will run together in a single container image. Loki is capable of managing a large number of logs and it uses object storage backends. For our needs as developers, it won't be necessary, and we will just use local storage (filesystem mode) provided by our own Kubernetes platform.

While Grafana Loki provides the server-side part, **Grafana Promtail** will work as an agent, reading, preparing, and sending logs to Loki's backend.

We are not interested in how Grafana Loki or Prometheus work or can be customized. The purpose of this chapter is to learn how we can use them to monitor and log our application's processes, so we will install Loki and configure Promtail to retrieve the logs from Kubernetes deployed applications.

Installing and configuring Loki and Promtail

Let's move forward by installing Grafana Loki in our Kubernetes cluster so that we can manage all the platform logs. After that, we will be ready to install Promtail to retrieve and push the logs to the Loki server:

1. We will use `helm` again to install Grafana Loki in a different namespace. The simplest installation is the single binary chart method with filesystem storage (`https://grafana.com/docs/loki/latest/installation/helm/install-monolithic`):

```
PS C:\Users\frjaraur> helm repo add grafana https://grafana.github.io/helm-charts
"grafana" has been added to your repositories
PS C:\Users\frjaraur> helm repo update
Hang tight while we grab the latest from your chart repositories...
...Successfully got an update from the "grafana" chart repository
...Successfully got an update from the "prometheus-community" chart repository
Update Complete. ⎈Happy Helming!⎈
PS C:\Users\frjaraur> helm install loki -n logging --create-namespace grafana/loki --set loki.commonConfig.replication_factor=1 --set
 loki.storage.type=filesystem --set singleBinary.replicas=1 --set loki.auth_enabled=false --set monitoring.lokiCanary.enabled=false -
-set test.enabled=false --set monitoring.selfMonitoring.enabled=false
NAME: loki
LAST DEPLOYED: Fri Jul 28 20:56:03 2023
NAMESPACE: logging
STATUS: deployed
REVISION: 1
NOTES:
***********************************************************************
 Welcome to Grafana Loki
 Chart version: 5.9.2
 Loki version: 2.8.3
***********************************************************************

Installed components:
* loki
PS C:\Users\frjaraur> kubectl get pods -n logging
NAME                                           READY   STATUS             RESTARTS   AGE
loki-0                                         0/1     ContainerCreating   0          3s
loki-gateway-66f55cfb99-z8hmw                  0/1     ContainerCreating   0          4s
loki-grafana-agent-operator-864d5f9d87-fsbfq   0/1     ContainerCreating   0          4s
```

Figure 12.17 – Grafana Loki installation using Helm

2. We used the following settings to apply the monolithic mode and remove API authentication:

    ```
    --set loki.commonConfig.replication_factor=1 --set loki.
    commonConfig.storage.type=filesystem --set singleBinary.
    replicas=1 --set loki.auth_enabled=false --set monitoring.
    lokiCanary.enabled=false --set test.enabled=false --set
    monitoring.selfMonitoring.enabled=false
    ```

    All these flags will help you reduce the hardware resources required for a testing environment.

3. It is now up and running. Let's take a quick look at the Service resources before installing the Promtail component:

```
PS C:\Users\frjaraur> kubectl get svc -n logging
NAME              TYPE        CLUSTER-IP      EXTERNAL-IP   PORT(S)             AGE
loki              ClusterIP   10.102.2.255    <none>        3100/TCP,9095/TCP   5m59s
loki-canary       ClusterIP   10.98.23.63     <none>        3500/TCP            5m59s
loki-gateway      ClusterIP   10.103.143.161  <none>        80/TCP              5m59s
loki-headless     ClusterIP   None            <none>        3100/TCP            5m59s
loki-memberlist   ClusterIP   None            <none>        7946/TCP            5m59s
PS C:\Users\frjaraur>
```

Figure 12.18 – Grafana Loki Service resources

We've reviewed the Loki Services because we are going to configure Promtail to send all the logging information retrieved to the `loki-gateway` Service, available in the `logging` namespace on port `80`.

4. Helm can show us the default values that will be used to install a Helm Chart if we execute `helm show values <CHART>`. So, we can retrieve the default values for the `grafana/promtail` chart by issuing `helm show values grafana/promtail`. The output is huge, with all the default values shown, but we only need to review the client configuration. This configuration applies to Promtail and defines where to send the logs that are read from the different sources:

```
PS C:\Users\frjaraur> helm show values `
grafana/promtail
...
config:
...
  clients:
    - url: http://loki-gateway/loki/api/v1/push
...
```

5. Promtail will read several log files from our cluster hosts (deployed as a DaemonSet), mounted as volumes (the `defaultVolumeMounts` key in the chart values YAML file). All the files included will be read and managed by the Promtail agent and the data extracted will be sent to the URL defined in `config.clients[].url`. This is the basic configuration we need to review because, by default, Kubernetes logs will be included in the `config.snippets` section. Prometheus, Grafana Loki, and Promtail are quite configurable applications, and their customization can be very tricky. In this chapter, we are reviewing the basics for monitoring and logging your applications with them. It may be very useful for you to review the documentation of each mentioned tool to extend these configurations.

6. By default, Promtail will send all the data to `http://loki-gateway`, which we have seen exists inside the Kubernetes cluster if we run this tool in the logging namespace, alongside Grafana Loki. We'll proceed to install Promtail using the logging namespace, using the default values:

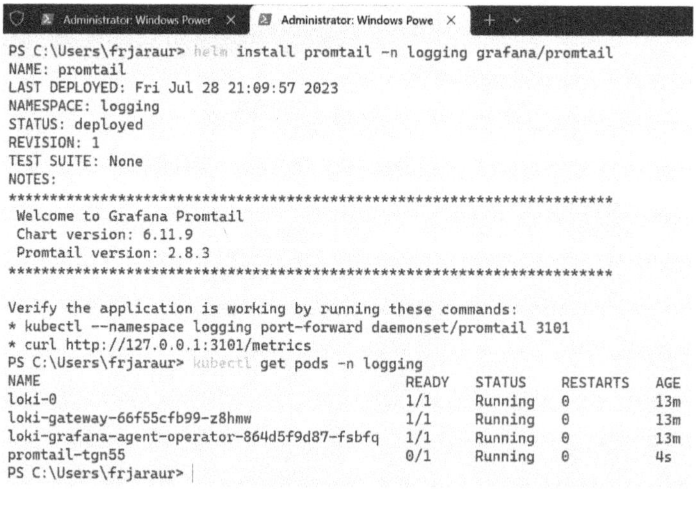

Figure 12.19 – Promtail installation using a Helm Chart

7. Once installed, we can review the logs of all Kubernetes clusters in Grafana. But first, we will need to configure Prometheus (monitoring) and Loki (logging) data sources in Grafana. We will use port forwarding to expose the Grafana Service:

```
PS C:\Users\frjaraur> kubectl port-forward `
-n monitoring service/prometheus-grafana 8080:80
Forwarding from 127.0.0.1:8080 -> 3000
Forwarding from [::1]:8080 -> 3000
```

8. And now, in the Grafana web UI (the admin username with the prom-operator password), accessible at http://localhost:8080, we can configure the data sources by navigating to **Home | Adminsitration | Datasources**:

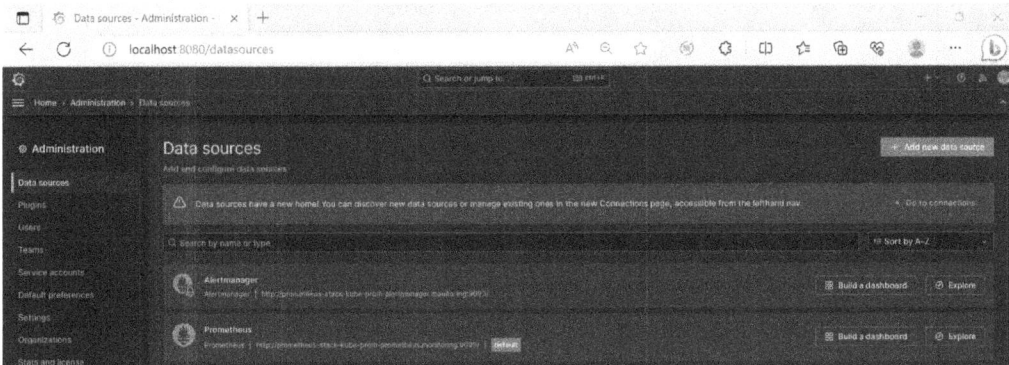

Figure 12.20 – Grafana – Data sources configuration

9. Notice that the deployment of Grafana using the kube-prometheus-stack Chart already configured the Prometheus and Alertmanager data sources for us. We will configure a new data source for Loki:

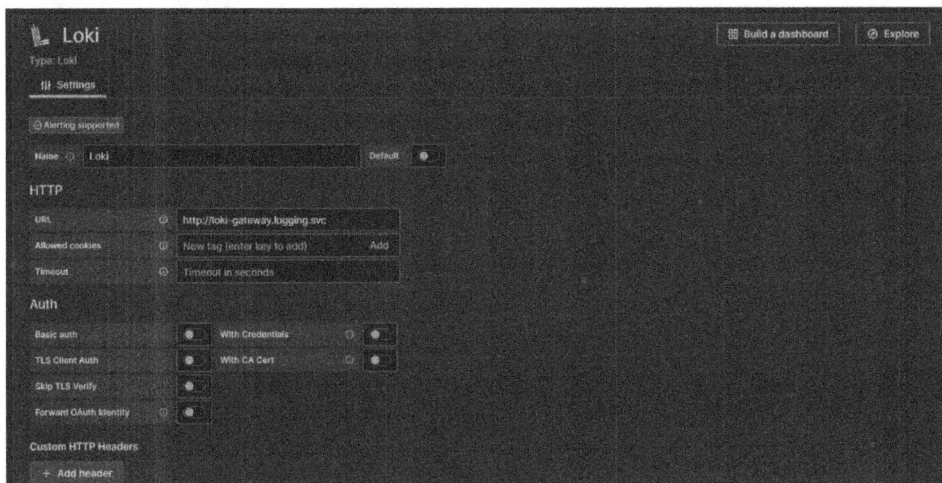

Figure 12.21 – Grafana Loki data source configuration

10. Click on **Save & Test** – and that's it! You may receive an error stating "**Data source connected, but no labels were received. Verify that Loki and Promtail are correctly configured**". This indicates that we don't have labels available yet for indexing data; it occurs when you have just installed the Promtail component. If you wait a few minutes, labels will be available, and everything will work correctly. Anyway, we can verify the Loki data by clicking the **Explore** button, at the beginning of the **Data sources | Settings** page. The **Explore** section allows us to retrieve data directly from any Loki-type source. In our example, we used Loki as the name of the source, and we can select the available labels generated by Promtail with the information from our containers:

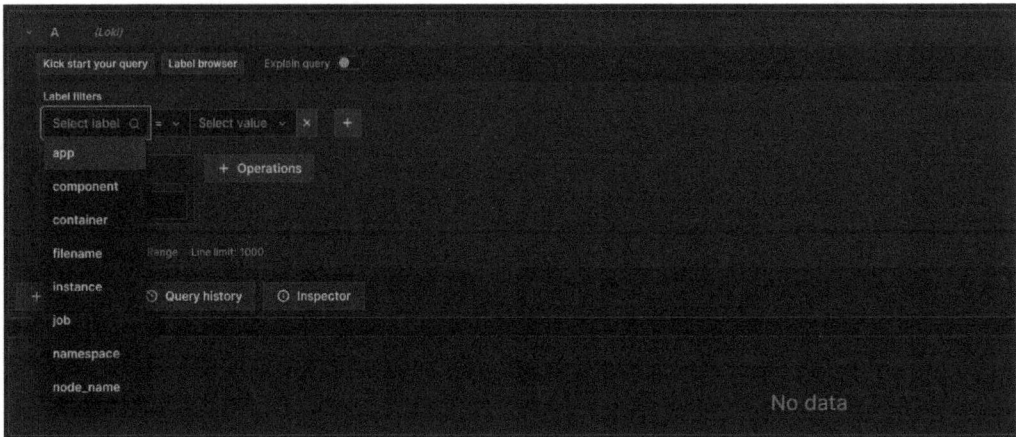

Figure 12.22 – Exploring Loki data sources

11. We can retrieve all the logs from the kube-system namespace by simply selecting the namespace label and the appropriate value from the list:

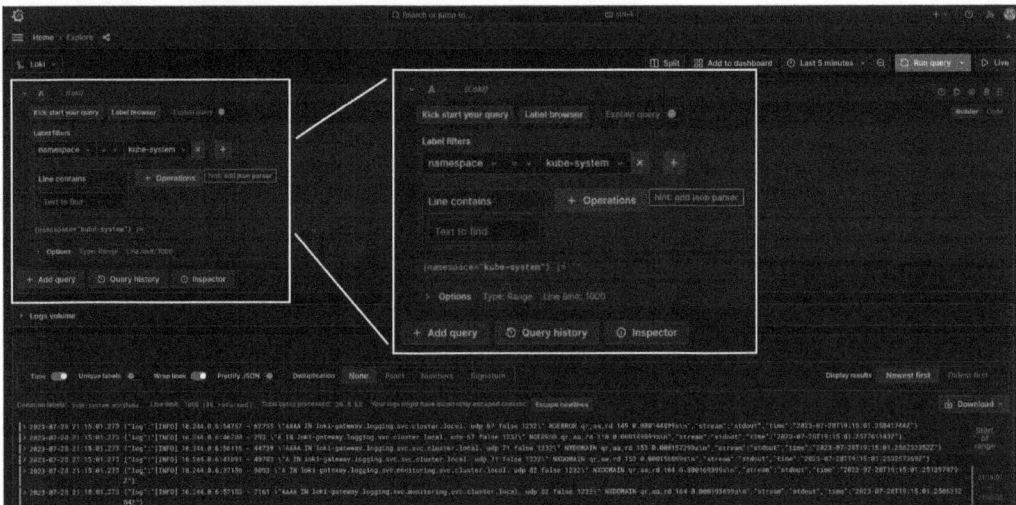

Figure 12.23 – Exploring the logs from all the Pod resources from the kube-system namespace

You will be able to filter by any of the current labels and add very useful operations such as grouping, counting the number of appearances of certain strings, searching for specific regex patterns, and so on. Lots of options are available and they will be very useful for you as a developer. You can have all the logs from different applications' components in one unified dashboard, or even prepare your own application dashboard with metrics and logs, mixing different sources.

Prometheus, Loki, and Grafana are very powerful tools and we can only cover the basics here. It is up to you to create dashboards using the Grafana documentation (`https://grafana.com/docs/grafana/latest/dashboards/use-dashboards/`). The Grafana community provides many dashboard examples (`https://grafana.com/grafana/dashboards/`). We will create a fully functional dashboard for the `simplestlab` application in the *Labs* section.

The next section will introduce some load-testing mechanisms and review how to use the **Grafana k6** open source tool.

Load testing your applications

Load testing is the task in which we review how our application works by measuring its behavior under different stressful circumstances. You, as a developer, always have to think about how your application will manage these stressful situations and try to answer some of the following questions:

- Will my application work under high user loads?

- How will my application's components be impacted in such a case?

- Will scaling up some components maintain the overall performance or might this cause problems with other components?

Testing the application before it goes to production will help us predict how the application is going to work. **Automation** is key to simulating thousands of requests at a time.

With load testing or **performance testing**, we try to put pressure on our applications and increase their workloads. We can test our applications for different reasons:

- To understand how our application behaves with an expected load

- To ascertain the maximum load under which our application will work

- To try to find possible memory or performance issues that could appear over time (memory leaks, fulfillment of certain storage resources, cache, and so on)

- To test how our application auto-scales in certain circumstances and how the different components will be affected

- To confirm some configuration changes in development before they are done in production

- To test the application performance from different locations that have different network speeds

All these test points can be delivered with some scripting techniques and automation. Depending on the application we are testing, we can use some well-known, simple but effective tools such as **Apache JMeter** (`https://jmeter.apache.org`) or – even simpler – **Apache Bench** (`https://httpd.apache.org/docs/2.4/programs/ab.html`). These tools can emulate application requests but they never behave like a real web browser, but **Selenium** (`https://www.selenium.dev`) does. This tool includes a **WebDriver** component that emulates a **World Wide Web Consortium (W3C)** browsing experience, but it may be complex to integrate into automated processes (different releases and integration with different languages can be time-consuming). Grafana provides k6, which is an open source tool that was created by Load Impact some years ago and is now part of the Grafana tools ecosystem. It is a very small tool, written in Go, and can be configured via JavaScript, which makes it very customized.

> **Important note**
>
> Grafana k6 functionality can be extended by using extensions, although you will need to recompile your k6 binary to include them (`https://k6.io/docs/extensions`).

Grafana k6 tests can be integrated into the **Grafana SaaS platform** (**Grafana Cloud**) or your own Grafana environment, although some features are only available in the cloud platform at the time of writing this book.

Installing this tool is very simple; it can be run on Linux, macOS, and Windows and is suitable for running within containers, which makes it perfect to run in Kubernetes as a Job resource.

To install this tool on Windows, open a PowerShell console with administrator privileges and execute `winget install k6`:

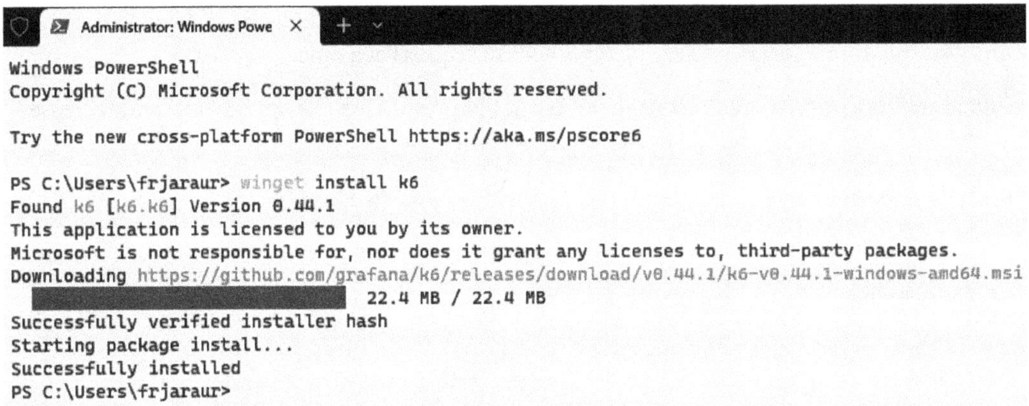

Figure 12.24 – Grafana k6 installation on Windows

Let's see a quick example of its usage by writing a simple JavaScript `check` script:

```
import { check } from "k6";
import http from "k6/http";
export default function() {
  let res = http.get("https://www.example.com/");
  check(res, {
    "is status 200": (r) => r.status === 200
  });
};
```

We can now test this example script for 10 seconds, simulating five virtual users with the k6 command line:

Figure 12.25 – Executing a k6 test script

In the preceding example, we are verifying a 200 code that was returned from the page. We can now use 5,000 virtual users for the same amount of time, which may create performance problems in the backend. The results can be integrated into different analysis tools such as Prometheus, Datadog, and so on. If you are already familiar with Prometheus and Grafana, you will probably like k6.

There is a bunch of good usage examples (https://k6.io/docs/examples) in which Grafana documents different use cases:

- Single and complex API requests
- HTTP and OAuth authentication with authorization
- The correlation of tokens, dynamic data, and cookie management
- POST data parameterization and HTML forms and the parsing of results
- Data uploads or scraping websites
- Load testing of HTTP2, WebSockets, and SOAP

We can create a more complex example by adding some content parsing and increasing and decreasing the number of virtual users during the execution of the test:

```
import http from 'k6/http';
import { sleep, check } from 'k6';
import {parseHTML} from "k6/html";
export default function() {
    const res = http.get("https://k6.io/docs/");
    const doc = parseHTML(res.body);
    sleep(1);
    doc.find("link").toArray().forEach(function (item) {
        console.log(item.attr("href"));
    });
}
```

In the preceding code snippet, we are looking for all the link strings and retrieving their href values:

```
PS C:\Users\frjaraur> gc k6.io-links-check.js
import http from 'k6/http';
import { sleep, check } from 'k6';
import {parseHTML} from "k6/html";
export default function() {
    const res = http.get("https://k6.io/docs/");
    const doc = parseHTML(res.body);
    sleep(1);
    doc.find("link").toArray().forEach(function (item) {
        console.log(item.attr("href"));
    });
}
PS C:\Users\frjaraur> k6 run -u 1 -d 2s k6.io-links-check.js --quiet
INFO[0001] https://k6.io/docs                    source=console
INFO[0001] /docs/favicon-32x32.png?v=b39a6f3e0dc925c8ec4f77e0a65490e9  source=console
INFO[0001] /docs/manifest.webmanifest            source=console
INFO[0001] /docs/icons/icon-48x48.png?v=b39a6f3e0dc925c8ec4f77e0a65490e9   source=console
INFO[0001] /docs/icons/icon-72x72.png?v=b39a6f3e0dc925c8ec4f77e0a65490e9   source=console
INFO[0001] /docs/icons/icon-96x96.png?v=b39a6f3e0dc925c8ec4f77e0a65490e9   source=console
INFO[0001] /docs/icons/icon-144x144.png?v=b39a6f3e0dc925c8ec4f77e0a65490e9   source=console
INFO[0001] /docs/icons/icon-192x192.png?v=b39a6f3e0dc925c8ec4f77e0a65490e9   source=console
INFO[0001] /docs/icons/icon-256x256.png?v=b39a6f3e0dc925c8ec4f77e0a65490e9   source=console
INFO[0001] /docs/icons/icon-384x384.png?v=b39a6f3e0dc925c8ec4f77e0a65490e9   source=console
INFO[0001] /docs/icons/icon-512x512.png?v=b39a6f3e0dc925c8ec4f77e0a65490e9   source=console
INFO[0001] https://fonts.googleapis.com/css?family=Roboto+Mono:300,400  source=console
INFO[0001] /docs/sitemap/sitemap-index.xml       source=console
INFO[0002] https://k6.io/docs                    source=console
INFO[0002] /docs/favicon-32x32.png?v=b39a6f3e0dc925c8ec4f77e0a65490e9  source=console
INFO[0002] /docs/manifest.webmanifest            source=console
INFO[0002] /docs/icons/icon-48x48.png?v=b39a6f3e0dc925c8ec4f77e0a65490e9   source=console
INFO[0002] /docs/icons/icon-72x72.png?v=b39a6f3e0dc925c8ec4f77e0a65490e9   source=console
INFO[0002] /docs/icons/icon-96x96.png?v=b39a6f3e0dc925c8ec4f77e0a65490e9   source=console
INFO[0002] /docs/icons/icon-144x144.png?v=b39a6f3e0dc925c8ec4f77e0a65490e9   source=console
INFO[0002] /docs/icons/icon-192x192.png?v=b39a6f3e0dc925c8ec4f77e0a65490e9   source=console
INFO[0002] /docs/icons/icon-256x256.png?v=b39a6f3e0dc925c8ec4f77e0a65490e9   source=console
INFO[0002] /docs/icons/icon-384x384.png?v=b39a6f3e0dc925c8ec4f77e0a65490e9   source=console
INFO[0002] /docs/icons/icon-512x512.png?v=b39a6f3e0dc925c8ec4f77e0a65490e9   source=console
INFO[0002] https://fonts.googleapis.com/css?family=Roboto+Mono:300,400  source=console
INFO[0002] /docs/sitemap/sitemap-index.xml       source=console

    data_received..................: 451 kB 212 kB/s
    data_sent......................: 1.8 kB 848 B/s
    http_req_blocked...............: avg=21.16ms  min=0s       med=21.16ms  max=42.33ms  p(90)=38.09ms  p(95)=40.21ms
    http_req_connecting............: avg=7.03ms   min=0s       med=7.03ms   max=14.06ms  p(90)=12.65ms  p(95)=13.36ms
    http_req_duration..............: avg=26.85ms  min=25.35ms med=26.85ms  max=28.35ms  p(90)=28.05ms  p(95)=28.2ms
      { expected_response:true }...: avg=26.85ms  min=25.35ms med=26.85ms  max=28.35ms  p(90)=28.05ms  p(95)=28.2ms
    http_req_failed................: 0.00% ✓ 0      ✗ 2
    http_req_receiving.............: avg=22.14ms  min=19.04ms med=22.14ms  max=25.23ms  p(90)=24.61ms  p(95)=24.92ms
    http_req_sending...............: avg=550.04µs min=534.4µs med=550.04µs max=565.69µs p(90)=562.57µs p(95)=564.13µs
```

Figure 12.26 – Executing a k6 test script parsing the "links" string

Grafana k6 is very configurable. We can define stages in which we can change the behavior of the probe during the check as well as many complex features that are completely outside the scope of this book. We suggest that you read the tool's documentation for a better understanding of its functionality and to customize it to your needs.

We will now jump into the instrumentation part. In the next section, we will learn how to integrate some tracing technologies into our application observability strategy.

Adding instrumentation to your application's code

When you code your applications, it should be easy to prepare appropriate monitoring endpoints and adequate monitoring tools to retrieve your application's metrics. Observability helps us understand applications without really having good knowledge of their internal code. In this section, we will explore some tools that provide traces, metrics, and logs for our applications, without actively knowing our applications' functionality. **Monitoring** and **logging** are part of the observation tasks but in a different context. We actively know where to retrieve the monitoring and logging information from, but sometimes, we need to go further – for example, when we run a third-party application or we don't have enough time to add monitoring endpoints to our application. It is important to prepare for monitoring your applications from the very beginning, even when you start to plan your application. The same applies to the logging part – you have to prepare your application to provide good logging information and not merely through the output of your processes as they are.

Distributing the same type of tracing, logging, and monitoring among your application's components will help you understand what is going on in your application and follow every request through the different steps taken in your application's processes. The traces, metrics, and logging data obtained are considered your application's telemetry.

OpenTelemetry has become a standard observability framework. It is open source and provides different tools and SDKs that help implement a telemetry solution to easily retrieve traces, logs, and metrics from your applications. This data can be integrated into Prometheus and other observability tools, such as Jaeger.

The main goal of the OpenTelemetry platform is to add observability to your applications without any code modification. Currently, the Java, Python, Go, and Ruby programming languages are supported. The simplest way of working with OpenTelemetry in Kubernetes is using the **OpenTelemetry Operator**. This Kubernetes operator will deploy the required OpenTelemetry components for us and create the associated CRDs that will allow us to configure the environment. This implementation will deploy the Collector's components, which will receive, process, filter, and export telemetry data to designed backends, and create the **OpenTelemetryCollector** and instrumentation resource definitions.

There are four different ways or modes of deploying an OpenTelemetry Collector:

- **Deployment Mode** allows us to control the Collector as if it were a simple application running in Kubernetes, and it is suitable for monitoring a simple application.

- **DaemonSet Mode** will run a collector replica as an agent, on each cluster node.

- **StatefulSet Mode**, as expected, will be suitable when you don't want to lose any tracing data. Multiple replicas can be executed and each one will have a dataset.

- **Sidecar Mode** will attach a collector container to the application's workloads, which is better if you create and remove applications often (perfect for a development environment). It also offers a more fine-grained configuration if you use different languages and want to choose specific collectors for each application component. We can manage which collector to use for a specific Pod with special annotations.

Let's run a quick demo environment from the OpenTelemetry community (`https://github.com/open-telemetry/opentelemetry-demo/`). This demo deploys a web store example application, Grafana, Jaeger, and the required OpenTelemetry components to obtain an application's metrics and traces.

> **Important note**
>
> **Jaeger** is a distributed tracing platform that maps and groups application flows and requests to help us understand workflow and performance issues, analyze our Services and their dependencies, and track down root causes when something goes wrong in our applications. You can find its documentation at the project's URL: `https://www.jaegertracing.io/`.

The full demo is installed via a Helm Chart (`open-telemetry/opentelemetry-demo`). We can deploy it on any namespace, but all the components will run together. The presented demo provides a very good overview of what we can include in our desktop environment in either Docker Desktop, Rancher Desktop, or Minikube (it may work on any other Kubernetes environment), although it doesn't follow some of the modern OpenTelemetry best practices for adding traces and managing collectors. This demo doesn't deploy the Kubernetes OpenTelemetry Operator, and the OpenTelemetry Collector is deployed as a simple deployment.

In the next section, we'll install the demo and review the application and tools that have been deployed.

Reviewing the OpenTelemetry demo

In this section, we will install and review some of the most important features of a ready-to-use demo prepared by the OpenTelemetry project. This demo will deploy a simple web store application and the tools required for managing and retrieving the tracing data from different components (you can find additional useful information at `https://opentelemetry.io/docs/demo`):

1. First, we will install the demo by following the simple steps described at `https://opentelemetry.io/docs/demo/kubernetes-deployment/`. We will add the OpenTelemetry project's Helm Chart repository and then issue `helm install` with the default values included in the demo package:

Figure 12.27 – OpenTelemetry demo application deployment

As you can see, different web UIs were deployed. We can quickly review the different Pods that were created:

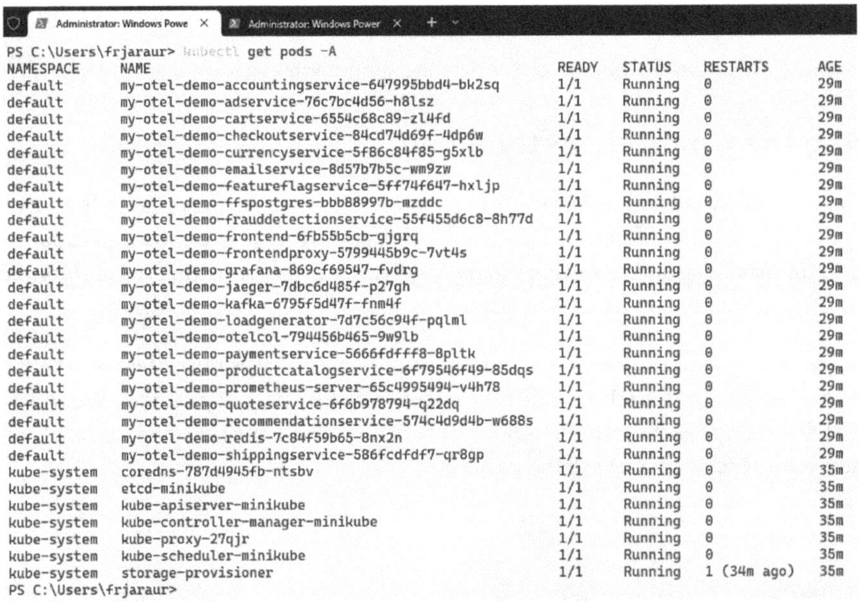

Figure 12.28 – OpenTelemetry demo application running Pods

In the preceding code snippet, we can see that OpenTelemetry Collector has been deployed using a deployment that will monitor all the applications instead of selected ones. Please read the OpenTelemetry documentation (`https://opentelemetry.io/docs`) and specific guides available for the associated Helm Charts (`https://github.com/open-telemetry/opentelemetry-operator` and `https://github.com/open-telemetry/opentelemetry-helm-charts/tree/main/charts/opentelemetry-operator`).

2. We will use port forwarding so that we can get quick and easy access to all the available demo web UIs. All will be reachable at once at localhost using different paths, thanks to a reverse proxy deployed with the demo Helm Chart. The web store application will be available at `http://localhost:8080`:

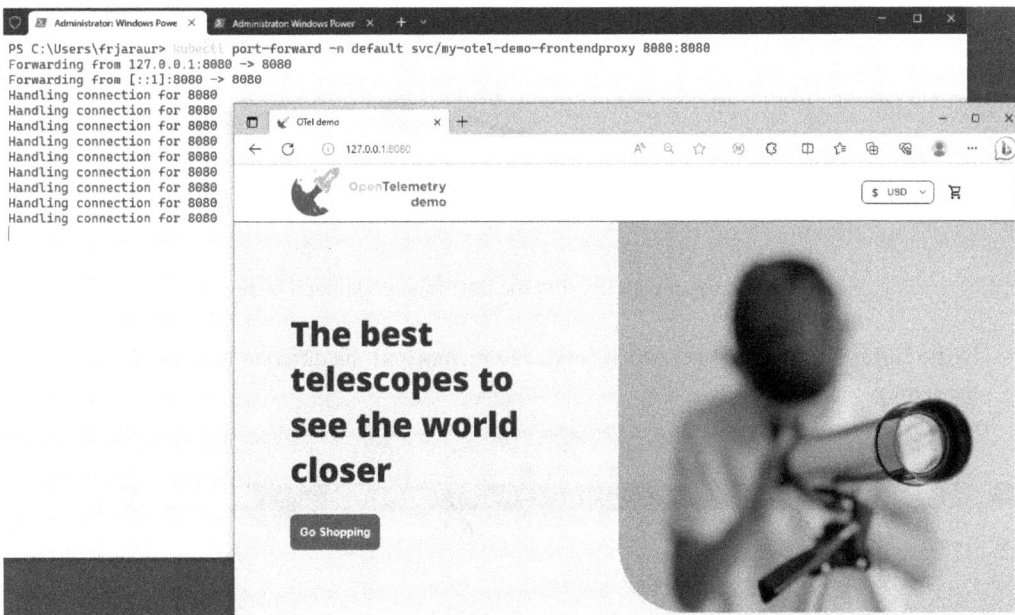

Figure 12.29 – Web store demo application accessible using the port-forward kubectl feature

The demo web store shows a catalog of telescopes, and it will simulate a shopping experience.

3. We can now access Jaeger at `http://localhost:8080/jaeguer/ui`:

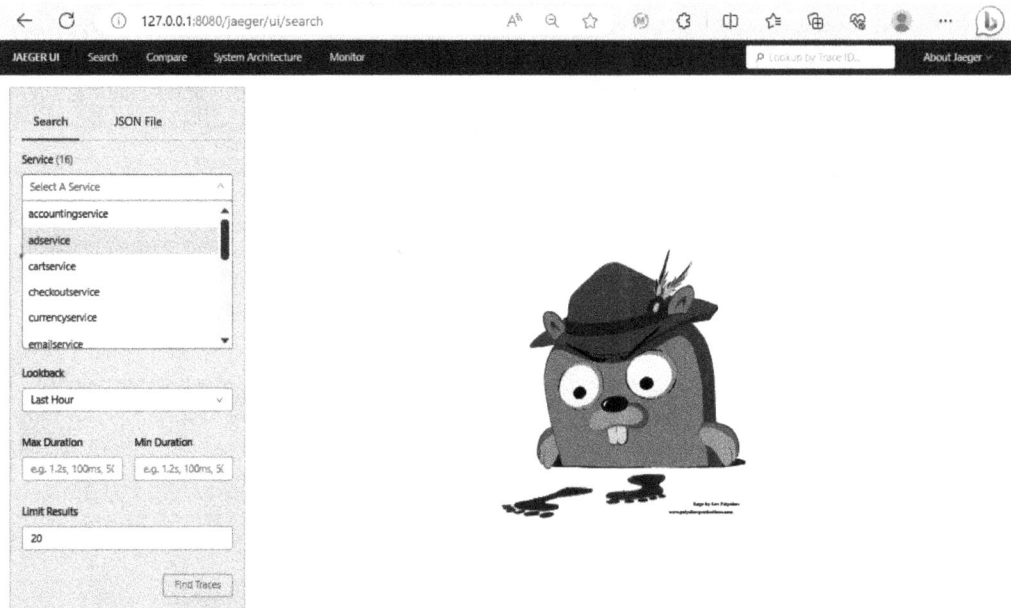

Figure 12.30 – The Jaeger UI using the port-forward kubectl feature

4.   In the Jaeger UI, we can select which Service to review, and the different traces will appear in the right panel:

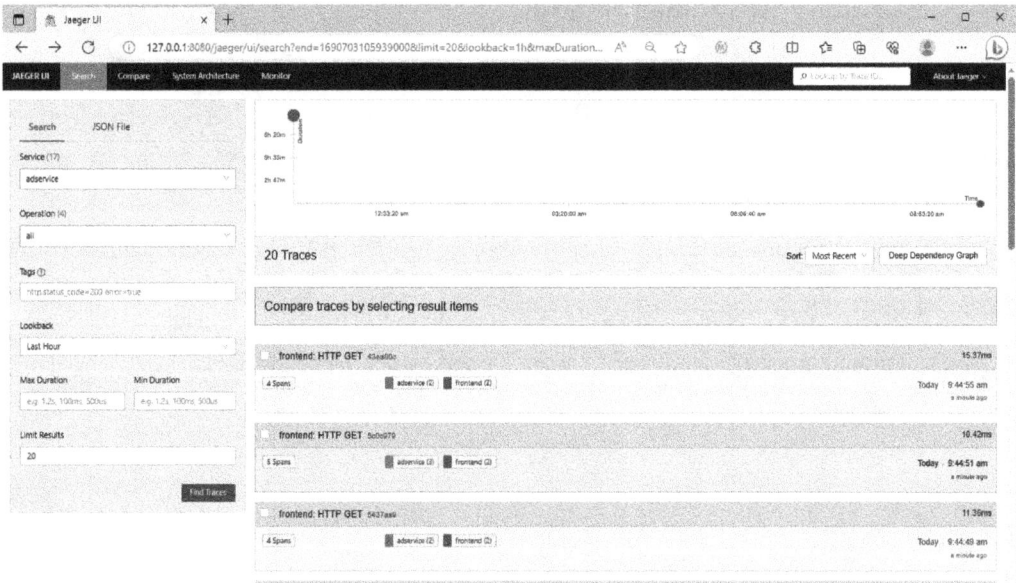

Figure 12.31 – The Jaeger UI showing traces for different application Services

5. An architecture overview is also available in the **System Architecture** tab so that you can verify the relations between the different application components:

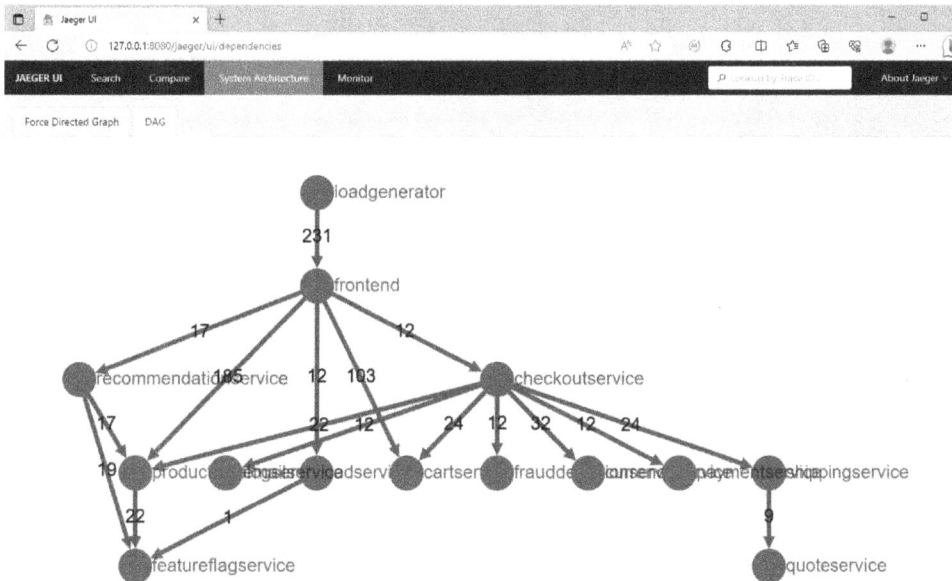

Figure 12.32 – The Jaeger UI showing the application components

Notice that there's a load generator workload, which is creating synthetic requests for us to be able to retrieve some statistics and traces from the demo environment.

6. The demo deployment also installs Grafana and Prometheus. The Grafana UI is available at `http://localhost:8080/grafana` and Prometheus and Jaeger data sources are configured for us:

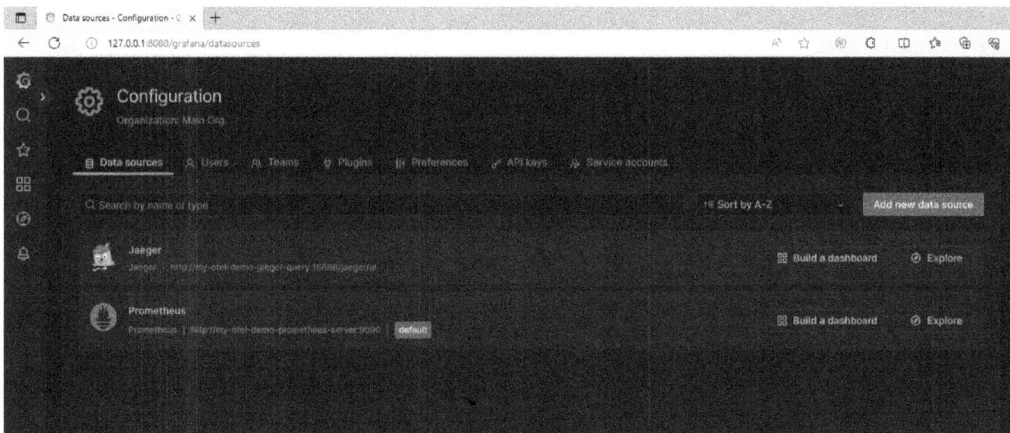

Figure 12.33 – The Grafana UI showing the configured data sources

Therefore, we can use Grafana to graph the data from both data sources and create an application dashboard. The people from OpenTelemetry have prepared some dashboards for us, showing the application's metrics and traces:

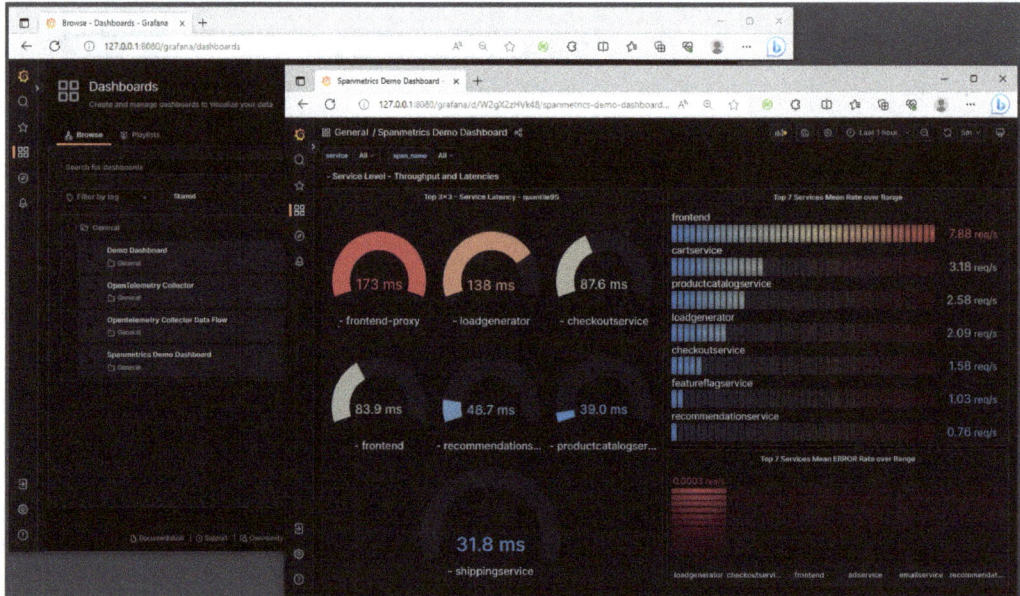

Figure 12.34 – Grafana's Dashboards page and the Spammetrics Demo
Dashboard showing current requests per Service

We could have included Grafana Loki in this scenario, added some of the logging entries, and, finally, created a custom dashboard with all the relevant metrics, traces, and log entries. The Grafana platform can even include certain databases as data sources, which will improve the overall overview of our application's health.

Labs

This section will show you how to implement some of the techniques that were covered in this chapter using the `simplestlab` three-tier application in Kubernetes. We will deploy a complete observability platform, including Grafana, Prometheus, and Loki, and prepare some exporters for our application's components.

The code for these labs is available in this book's GitHub repository at `https://github.com/ PacktPublishing/Containers-for-Developers-Handbook.git`. Ensure you have the latest revision available by simply executing `git clone https://github.com/ PacktPublishing/Containers-for-Developers-Handbook.git` to download all its content or `git pull` if you've already downloaded the repository before. All the manifests and steps required for running the labs are included in the `Containers-for-Developers-Handbook/Chapter12` directory. All the manifests required for the labs are included in the code repository. Detailed instructions for running the labs are included in the `Chapter12` directory, after you download the associated GitHub. Let's take a brief look at the steps that are to be performed:

1. First, we will create a Minikube Kubernetes environment on our desktop computer. We will deploy a simple cluster with one node and ingress and metrics-server plugins:

```
Chapter12$ minikube start --driver=hyperv /
--memory=6gb --cpus=2 --cni=calico /
--addons=ingress,metrics-server
```

2. Next, we will deploy the `simplestlab` application, which we used in previous chapters, on Kubernetes. The following steps will be executed inside the `Chapter12` folder:

```
Chapter12$ kubectl create ns simplestlab
Chapter12$ kubectl create -f .\simplestlab\ /
-n simplestlab
```

This will deploy our `simplestlab` application in the cluster.

3. Then, we will deploy a functional monitoring and logging environment on top of our Kubernetes desktop platform. To do this, we will first deploy the Kubernetes Prometheus stack. We prepared a custom `kube-prometheus-stack.values.yaml` values file with appropriate content for deploying a small environment with Grafana and Prometheus. We've provided the required Helm Charts inside the `Chapter12/charts` directory. In this case, we will use the `kube-prometheus-stack` subdirectory. To deploy the solution, we will use the following command:

```
Chapter12$ helm install kube-prometheus-stack /
--namespace monitoring --create-namespace /
--values .\kube-prometheus-stack.values.yaml .\charts\kube-
prometheus-stack\
```

This command deploys Grafana and Prometheus at the same time, with two data sources configured to integrate Prometheus and Loki (deployed in the next step) in Grafana. These data sources will be enabled inside the Grafana platform, and we will be able to use them later to create some easy dashboards.

4. When Grafana and Prometheus are available, we will change our `hosts` file (`/etc/hosts` or `c:\windows\system32\drivers\etc\hosts`) to include the Minikube IP address for `grafana.local.lab` and `simplestlab.local.lab`. We will now be able to access Grafana, published in our ingress controller, at `https://grafana.local.lab`.

Then, we will deploy Grafana Loki to retrieve the logs from all the applications and the Kubernetes platform. We will use a custom values file (`loki.values.yaml`) to deploy Loki that's included inside `Chapter12` directory. The file has already been prepared for you with the minimum requirements for deploying a functional Loki environment. We will use the following command:

```
helm install loki --namespace logging /
--create-namespace --values .\loki.values.yaml .\charts\loki\
```

5. Loki is available as a Grafana data source, but there isn't any data from that source yet. We will deploy Promtail to retrieve the logging data and push it to Loki. This application will be deployed using a chart located in the `Chapter12/charts/promtail` directory. We will leave the default values as they are because the communication with Loki will only be internal:

```
helm install promtail --namespace logging --create-namespace .\
charts\promtail\
```

6. Now, let's integrate monitoring and logging data from different sources to create a custom application dashboard in which you can review the logs from all your application's components and the resource usage of the application's processes. This will help you calculate the requirements for your application and the limits required to work properly. A detailed step-by-step example has been included to show you how to create a simple Grafana dashboard that integrates the `container_cpu_usage_seconds_total` and `container_memory_max_usage_bytes` metrics, which were retrieved using Prometheus, and the `simplestlab` application logs in Loki thanks to the Promtail component.

7. By the end of these labs, we will have modified our `simplestlab` application, adding some Prometheus exporters for monitoring different application components that will help us customize these components for the best performance. We will integrate a Postgres database exporter and the NGINX exporter for the `loadbalancer` component of `simplestlab`. Manifests for both integrations have been prepared for you in the `Chapter12/exporters` directory.

8. Prometheus must be configured to poll these new targets – the Postgres database and NGINX exporters. We will prepare a ServiceMonitor resource for each one to inform Prometheus to retrieve metrics from these new sources. The ServiceMonitor resources manifests are included in the `Chapter12/exporters` directory. Here is the one that's been prepared for you for the NGINX component:

```
apiVersion: monitoring.coreos.com/v1
kind: ServiceMonitor
metadata:
  labels:
    component: lb
    app: simplestlab
    release: kube-prometheus-stack
  name: lb
```

```
      namespace: simplestlab
  spec:
    endpoints:
    - path: /metrics
      port: exporter
      interval: 30s
    jobLabel: jobLabel
    selector:
      matchLabels:
        component: lb
        app: simplestlab
```

9. Detailed information about creating simple queries in Grafana to graph data using these new metrics is available in the `Readme.md` file for this chapter, in this book's GitHub repository.

The labs we've prepared for you in this chapter will give you an overview of how to implement a simple monitoring and logging solution for your applications and prepare some metrics to review the performance of some of your application's components.

Summary

In this chapter, we learned how to implement some tools and techniques for monitoring, logging, and tracing our application workloads in Kubernetes. We also took a quick look at the load testing task, with an overview of what you should expect from your probes. We talked about Grafana, Prometheus, and Loki, among other tools, but the principles we discussed in this chapter can be applied to any monitoring, logging, or tracing tool you use.

Monitoring how much of the hardware resources your application consumes and reading the logs of your application's components in a unified environment can help you understand your application's limits and requirements. If you test how it behaves under heavy load, it can help to improve your application's logic and predict the performance under unexpected circumstances. Adding traces to your code manually or using some of the automation mechanisms seen in this chapter will help you go further with your application's development by understanding their insights and integrations.

In the next chapter, we will make all the concepts seen so far fit in the application's life cycle management.

Part 4:
Improving Applications'
Development Workflow

In this part, we will review some well-known application life cycle management phases and best practices. We will cover how working with containers improves them, reviewing some continuous integration and continuous deployment logical patterns.

This part has the following chapter:

- *Chapter 13, Managing the Application Life Cycle*

13

Managing the Application Life Cycle

In this book, we've reviewed some modern architectures and the microservices concept, understanding how containers fit into this new application development logic and covering how to create applications using different containers to provide their differing functionalities. This concept really is a game changer: we can implement an application's components using different deployment strategies and scale processes up or down as needed. We used container registries for storing and managing the new artifacts and container images, which in turn are used for creating containers. Container runtimes allow us to run such components. We then introduced orchestration, which allows us to manage application availability and updates easily. Container orchestration requires new resources to solve different issues that arise from these new architectures. In this chapter, we will cover how all these pieces fit together in the management of your application life cycle. Then, we will learn how the automation of such actions allows us to provide a complete application **supply chain**, running **continuous integration/ continuous delivery (CI/CD)** on Kubernetes.

The following are the main topics covered in this chapter:

- Reviewing the application life cycle
- Shifting our application's security left
- Understanding CI patterns
- Automating continuous application deployment
- Orchestrating CI/CD with Kubernetes

Technical requirements

You can find the labs for this chapter at https://github.com/PacktPublishing/ Containers-for-Developers-Handbook/tree/main/Chapter13, where you will find some extended explanations, omitted in the chapter's content to make it easier to follow. The *Code In Action* video for this chapter can be found at https://packt.link/JdOIY.

Reviewing the application life cycle

When we talk about how applications are created and evolve, we have to consider all the creative and maintenance processes involved. The application life cycle includes the following stages:

1. **Planning** of a software solution

2. **Development** of the application's components

3. Different **testing** phases, including component integration and performance tests

4. **Deployment** of the solution

5. **Maintenance**

As we can see, a lot of people, processes, and tools are involved across the whole life cycle of an application. In this book, however, we will only cover those that can be resolved technically with the use of software containers. We can use the following schema to situate the aforementioned processes within a broader context:

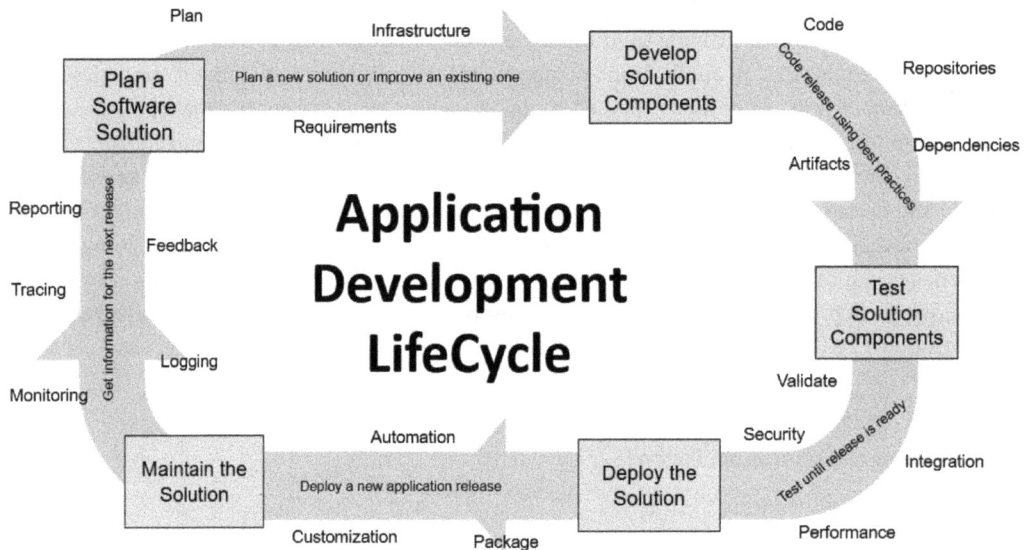

Figure 13.1 – Basic application life cycle schema

Let's think now about which of these phases can be implemented using software containers.

Planning a software solution

This phase covers the early stages of a software solution when an idea becomes a project. It includes the collection and analysis of the **requirements** of the users, customers, and other project stakeholders.

These requirements will always need validation to ensure the final characteristics of the developed solution. Depending on the size of the project, an exploration of alternatives currently available on the market and the viability of the solution may call a stop to the process. The success of the project is usually directly related to the effectiveness of the planning phase, in which different teams propose the architecture, infrastructure, software frameworks, and other resources that may be key for the resulting solution.

> **Important note**
>
> In this book, all the content presented is intended for working on either cloud environments or an on-premises data center infrastructure. You will be able to use your desktop computer for developing your application code and can use a variety of workflows to interact with different infrastructure platforms through the different project phases, as we will learn in this chapter.

Developing a good **timeline** for the project is always critical, and working with containers helps you improve delivery times, as they don't require dedicated or overly specific infrastructure. You can even start your project on one platform and then move to a new completely different one. Containers mitigate any friction and remove infrastructure vendor lock-in.

In this phase, you will also decide on the **architecture** for your application. Dividing your application into small, code-independent but cooperative services allows different groups of developers to work in parallel, which will always speed up project delivery. Working with microservices lets you as a developer focus on specific functionality and deliver your component following defined guidelines to ensure proper integrations. It is important to prepare the logic for scaling up or down any application's process if needed and to ensure components' **high availability** (**HA**) and resilience. This will add flexibility to your solution and increase overall availability for your users.

Developing the application's components

This stage involves writing the code for your application. When you are developing microservices applications, you can choose the most appropriate language for your code, but you must be aware of any issues in the dependencies you use and understand the risks that come with using certain components instead of others. Using open source libraries or frameworks always requires a good knowledge of the maintainer's activity and the maturity of their code.

In the microservices model, your applications serve their APIs, and resources and other components use them. If you plan to enable multiple instances, you must ensure that your application's logic allows this situation. To avoid infrastructure friction and provide maximum availability, ensure your application runs in different circumstances, manage its dependencies, and enable some circuit breakers. You will need to figure out how your processes behave when some components are down, how to reconnect in case some connection is lost and recovered, what will happen if you decide to execute your application's components in a cloud platform or on a different cluster, and so on.

Testing your application

Once your development is finished, a selection of test stages will be triggered. As this is an iterative process, you can deliver certain components of the application (or even the full solution), but it won't truly be finished until all the tests return positive results. We must always consider the following principles when preparing and running our tests:

- Tests must meet the expected requirements

- They should be executed by third-party groups, not involved in the design or development of the application to keep these tests independent

- Automation helps to reproduce tests under the same circumstances in different iterations

- Tests must be executed on either small components or a set of components running together

Let's see some of the testing types and how containers can integrate them:

- **Unit testing**: This type tests the *individual* components of an application. It is usually generated and executed in the *development phase* because developers need to know whether their code is working as expected. Depending on the complexity of the component's code and the returned objects of the requests, they may be included in the container probes. Components will not be considered healthy if the returned status isn't valid, although further pattern matching can be included in the validation of the returned data. If you are developing a component that works via an API, you should consider having a test request that always returns a valid value, or alternatively, you could use mock data. Unit tests will help you validate your code whenever changes have to be made to fix an issue, and they also make your code modular (microservices). Each component should include its own unit tests, and we can also include some code quality verification against defined standards.

- **Integration testing**: These tests validate how *different* components of your software solution work together. They help us to identify issues between components and fix the delivery and interaction of all the components. So, this type of test needs to be arranged between the developers of the different components and planned consistently. If our application's components run within containers, it would be very easy to prepare Docker Compose or some Kubernetes manifests to run all the required components together in our development environment – although these tests can also be automated on a remote CI/CD platform, as we will see later in this chapter in the *Orchestrating CI/CD within Kubernetes* section. If some components are key for your application's health, their endpoints or probes can be integrated into the monitoring platform to ensure everything works as expected.

- **Regression testing**: These tests validate that new changes made don't introduce new issues or break the overall project. Working with containers in these tests can significantly improve the overall process. We can go forward with new container image builds or roll backward using a previous image. If your code has changed significantly between releases, maybe having a completely different development platform as a result of moving to a new version of Python or

Java, this can be tricky, but using containers makes it smooth and simple. Regression tests help us solve any issues related to advancements or changes in our code (evolution of the solution) that can break the current application's behavior.

- **Smoke testing**: This stage is usually prepared in the early phases of integration testing to ensure nothing will *crash and burn* when an application's components start. These tests are used to solve dependency order. Running containers with Docker Compose allows us to change the order by changing the `depends_on` key, but it's recommended to solve any dependency order issues in your code because commonly used container orchestrators don't include such keys, requiring other mechanisms to manage dependencies. You can include additional `init` containers or sidecar containers that will check for the required components before other containers actually start.

- **Stress testing**: These tests validate your application's component under stress or heavy load. We learned in *Chapter 12* how to make tests using third-party tools. These tools can be deployed within containers and automated to create thousands of requests for our application's components. If we've already dealt with the monitoring of the application's components, we can get a good overview of the hardware requirements of our processes and use this to minimize resource usage within our container orchestrator clusters.

- **Performance testing**: Once you have integrated all your components and tested the requirements for each one, you can go further and verify different contexts for your application. You can test, for example, how your application behaves with multiple frontend components or work out how to distribute load between multiple databases. You can prepare both the application and the tests within containers, scale certain components up or down, and analyze the performance outcomes. This lets you distribute load automatically and add dynamism to your software solutions – but you do have to ensure that your code allows multiple instances at once of certain components. For example, you can have multiple instances of a distributed NoSQL database or multiple static frontends, but you can't run multiple database instances at once and write to the same data file. This also applies to your application's code. You can't simultaneously execute multiple instances of a process that write to a file if you don't block the file, so just one gets complete access to it. Another example is to allow requests from users on different instances without managing the response in a central database. You have to atomize the requests or integrate mechanisms to distribute them across the different instances.

- **Acceptance testing**: You should always define **user acceptance tests** (**UATs**) before delivering your solution because these will ensure that your code fits the requirements exposed at the beginning of the project. Multiple tests can be included in this stage (alpha, beta tests) depending on the complexity of your solution. New issues may arise in these tests, hence multiple iterations will probably be required. The automation of delivery and the simplicity inherited from working with software containers both help you to provide different testing environments to your users in a short period of time.

The testing phase is very important for a project because it helps you improve the quality and reliability of your software delivery, identify and fix problems before going to production, and increase the visibility of the project, improving stakeholders' confidence and user satisfaction. We can also reduce the maintenance costs of the solution because it was designed and tested with all the requirements in mind and validated multiple times, so errors that arise should have been ironed out before they impact production. On the other hand, testing is always time-consuming, but making different tests using containers will reduce both costs (as fewer environments are required for tests) and the time spent on each test (as we can deploy multiple releases at the time and test in parallel).

Deploying the solution

In this phase, we actually deploy our software solution in production. The solution often goes through multiple environments before this step is complete. For example, we can have a preproduction environment for validating certain releases and **Quality Assurance (QA)** environments where other more specific tests can be run. Using containers makes deployments in these testing stages simple – we just change our configuration; all the container images will be the same. Using containers as new **deployment artifacts** makes things easier. Let's quickly introduce some packaging solutions for containers:

- **Helm charts**: This package solution only works with Kubernetes. A Helm chart is just a packaged set of manifests that includes variables for modifying the deployment of an application and its components. Version-3-compatible Helm charts are the go-to now. A previous version of Helm that was deprecated some time ago used the Tiller privileged component for deploying manifests, which may affect cluster integrity and security. Newer releases simplify how applications are deployed without having to create any Helm-specific resources in your Kubernetes cluster. Helm charts are very popular, and software vendors provide their own supported chart repositories for installing their applications directly from the internet into your own Kubernetes clusters.

- **Kustomize**: This tool is growing in popularity and is now part of the Kubernetes ecosystem. Kustomize only works with Kubernetes and is also based on manifest templates that are refactored by users before being applied in a Kubernetes cluster. The kubectl command line includes Kustomize functionality, which makes it very usable out of the box without having to include new binaries in our environment.

- **Cloud-Native Application Bundle** (**CNAB**): CNAB goes a step further than Helm and Kustomize. It is designed to include the infrastructure and services required by our application to work. Multiple tools work together to provide both the infrastructure (with the Porter component providing integration of Helm, HashiCorp Terraform, and the cloud provider's API) and the application (managed by Duffle and Docker). This solution is not really in use today and many of its components have been deprecated, but it is worth mentioning as it can give you some ideas for fully packaging your software solutions (that is, the infrastructure and the application together).

- **Kubernetes operators**: Kubernetes operators are controllers that deploy and manage specific application deployments and have become very popular these days. An operator deploys its own specific controllers inside a Kubernetes cluster to manage application instances. Kubernetes operators are intended to self-manage all the tricky parts of your application's management and upgrades. You as a user just need to define certain required values for your instance, and the operator will handle installing the required components and dependencies and manage any upgrade during its lifetime. If you are planning to develop your application using a Kubernetes operator, make sure to include all the manifests of your application, dependencies, and the automation required for the application to come up. Third-party Kubernetes operators run as black boxes in your Kubernetes cluster and may not include all the functionality you expect for your applications to work; therefore, it may be worth reading the documentation before deploying a third-party Kubernetes operator.

Deploying your application using a microservices architecture allows you to integrate different components' releases. Depending on your software solution, you might use one full deployment or multiple small ones for each component of your application. Either way, the solution must provide all the functionality called for by your users and stakeholders in the project planning stage.

Maintaining the application

We might think that the deployment of the solution is the last phase, but it isn't. Once the application is in production, new functionalities may be required, new improvements to current functionality may be called for, and inevitably, new errors will appear. If your application is monitored, you can obtain feedback on the status of different components before actual errors appear. **Logging** also helps to identify problems, and tracing allows you to improve your code.

But in any case, the application's life cycle continues, and a new project may start adding new functionalities while issues are repaired for the current release. Monolithic architectures require multiple environments for such processes. Working on two releases at the same time will double the efforts for maintaining environments. Microservices architecture allows us to distribute the work according to the different components, and thus mitigate the need for having dedicated environments for building each component. And, more importantly, we can change one component at a time and focus on solving a specific issue, or have each application component managed by a different team with different release times. These teams develop their code using the programming language that best fits the functionality of their requirements, taking into account release times and proper integration within the application. However, note that each team also has to keep track of vulnerabilities and security issues in their implementation.

Throughout this book, we have learned some security practices that will make our applications safer when we work within containers (*Chapter 2* and *Chapter 3*) and with container orchestrators (*Chapter 6*, *Chapter 7*, and *Chapter 8*). **Shift-left security** goes beyond these recommendations and includes security from the very beginning of the project. We can consider shift-left security as a practice where we don't wait to address software security vulnerabilities until it's too late: when the application is already developed, built, and packaged. In the next section, we will learn how taking care of security from the very first phases of the application life cycle can significantly improve the overall security of the solution.

Shifting our application's security left

Shift-left security refers to the practice of starting security checks as early as possible in the development of our application. This doesn't mean we don't apply any security measures at other stages but that it will start improving security from the very beginning of the application life cycle. Shifting security left allows us to identify any vulnerabilities and other problems before it's too late and the application is already running in production. The benefits of shifting our security left include the following:

- It improves the delivery of software solutions because bugs are detected and fixed in early development stages

- It distributes application security into different stages, allowing different actions at each stage, starting from the code and ending in the infrastructure where the application will finally be deployed

- Different groups can implement different security policies and mechanisms, furthering the creation of a security culture in your organization

- It reduces overall development time and the costs of pushing back applications because of poor security

Now, let's understand some different methodologies for tackling the software development life cycle and how they impact security.

Software life cycle methodologies

Let's introduce some software life cycle methodologies here that will help us understand the importance of security when things begin to move faster in the stages of development:

- **Waterfall model:** In this model, stages must run *linearly*, hence a new stage begins when the previous one finishes. This model works very well when we don't expect to have many modifications from the planned requirements and our project tasks are well defined. However, this model lacks flexibility, which makes changes harder to implement, and issues usually remain hidden until the end of the project.

- **Agile model:** In this model, we *iterate* over stages to improve the final software solution. Flexibility and quick response times are key in this model. Iterations allow the introduction of new changes and the resolution of any issue found in the previous review. The main problem of this model is that it requires lots of collaboration between the groups or people involved in each stage, hence it may not work in big projects, but microservices architectures fit very well into this development model.

- **Spiral model:** This model can be considered a *mixture* of both the Waterfall and Agile models. The final software solution will be the result of different iterations that can be considered a complete software development cycle. In each iteration, we start from the very beginning, taking user requirements, designing a solution, developing the code, and testing, implementing, and maintaining the solution as is, before moving on to the next iteration. The Agile and spiral development models allow us to review and solve issues before the next iteration, which both accelerates the development process and makes the solution more secure.

Of these methods, Agile methodologies in particular have really changed how software is developed and delivered. Their adoption allows teams to go faster and swiftly adapt software solutions when users require new features. However, in such scenarios, the security team can be a bottleneck. These teams receive a software solution just before it goes into production, seeking to identify and resolve any vulnerabilities and security issues before malicious users find them in production. If we decouple our application into small pieces (that is, microservices), then the work required in the security review task is multiplied by the number of pieces, even if they are small. It gets even worse when we realize that most of the legacy tools used for reviewing security on monolith applications don't work on highly distributed and dynamic environments such as Kubernetes.

It is also the case that software containers and open source solutions have become so widely used in data centers and cloud platforms that we can find ourselves deploying third-party software solutions while barely even knowing their contents. Even software vendors provide open source products inside their own complex software solutions. Therefore, we cannot just keep using the same old security methodologies at the infrastructure and application levels.

Security at the application level

As discussed earlier, shifting the security of our applications left implies integrating security mechanisms and best practices as early as possible in our software development model. But this doesn't mean we leave security to the developers. We will prepare automated security validations in the testing phase and implement security policies in both the development environments and production clusters. This will ensure that everyone knows the security measures applied and how to implement them. The DevSecOps team prepares infrastructure and application rules and shares them with all the developer teams. Infrastructure rules include all policy enforcements in your execution environment, which is usually your Kubernetes cloud or on-premises platform. These policies may include, for example, the denial of any privileged container, the denial of Pods without limited resources, and the denial of access to hosts' filesystems. However, note that these rules are not part of the code, although they do affect the execution of your applications.

If we consider security from the application perspective, there are several techniques we can apply:

- **Software composition analysis (SCA)**: When we add open source libraries or other components to our code, we unconsciously add risk to our application. SCA tools help us identify these risks and in some cases mitigate them with patches and updates. While **static application security testing (SAST)** tools (which we will discuss next) are used to find vulnerabilities in the development cycle, within your code, SCA tools provide continuous vulnerability monitoring.

- **SAST**: These tests are used to find vulnerabilities in our code before it is actually compiled, hence they are run in the early stages of our development phase. The tools running these tests will search for well-known insecure patterns in our code and report them to us. Any hardcoded secret data and misconfigurations will be reported as issues in the analysis.

- **Dynamic application security testing (DAST)**: These tests are executed when the application is running, in the testing phase. They involve the execution of simulated attacks against our application's components. These tests can include code injection or malformed requests that may break your application at some point.

These three types of tests are very valuable in identifying vulnerabilities in our application before moving it to production, but SAST and SCA are the ones to focus on when talking about shifting security left. When automation is put in place, we can execute these tests continuously and use **integrated development environment (IDE)** plugins to help figure out problems before they are actually stored in our code. To start, we can use any good linter for our specific programming language. Let's discuss these next.

Introducing linters

A **linter** is a tool used to analyze our code looking for problems. Depending on the quality of the given linter, it can identify things from simple code improvements to more advanced issues. It is usual to use specific linters for different programming languages. You can check the extensions available in your favorite IDE.

Linters help us reduce the amount of code errors in the development stage, and improve our code style, construction consistency, and performance.

A simple code linter will do the following:

- Check for syntax errors
- Verify code standards
- Review *code smells* (well-known signs that something will go wrong in your code)
- Verify security checks
- Make your code look as if it were written by a single person

You should include linters in your code environment, but your specific choice will depend on the language you use. Good linters can be categorized based on the aspects they focus on, as outlined here:

- **Standardized coding**: Examples include SonarLint, Prettier, StandardJS, Brakeman, and StyleCop. Some languages such as .NET even include their own linter (Format).
- **Security**: GoSec, ESLint, or Bandit (Python module).

What's more, some linters can be used for both of these aspects when the appropriate configurations are used. You can check for additional code analysis tools at https://owasp.org/www-community/Source_Code_Analysis_Tools.

Let's see a quick example using a Dockerfile linter, **Hadolint** (https://github.com/hadolint/hadolint). We will simply check a valid Dockerfile that does not include the best practices we learned in *Chapter 1, Modern Infrastructure and Applications with Docker*. Let's see this in the following screenshot:

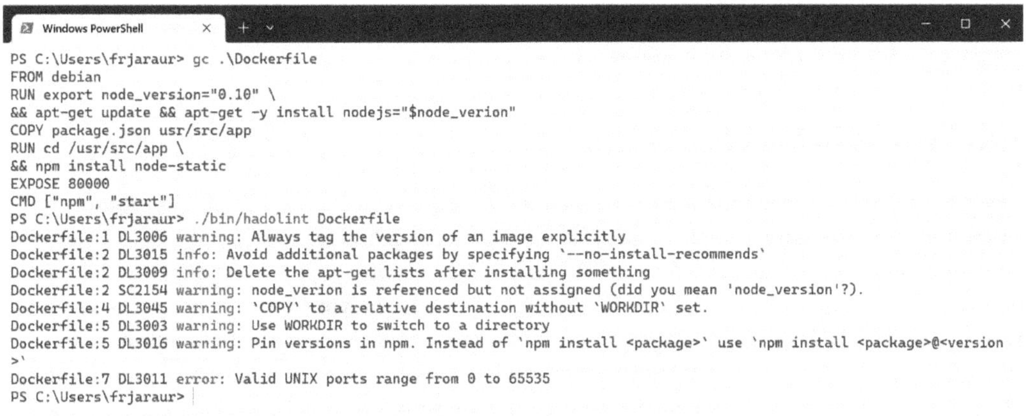

```
Windows PowerShell                                           ×    +   ∨                           —   □   ×

PS C:\Users\frjaraur> gc .\Dockerfile
FROM debian
RUN export node_version="0.10" \
&& apt-get update && apt-get -y install nodejs="$node_verion"
COPY package.json usr/src/app
RUN cd /usr/src/app \
&& npm install node-static
EXPOSE 80000
CMD ["npm", "start"]
PS C:\Users\frjaraur> ./bin/hadolint Dockerfile
Dockerfile:1 DL3006 warning: Always tag the version of an image explicitly
Dockerfile:2 DL3015 info: Avoid additional packages by specifying `--no-install-recommends`
Dockerfile:2 DL3009 info: Delete the apt-get lists after installing something
Dockerfile:2 SC2154 warning: node_verion is referenced but not assigned (did you mean 'node_version'?).
Dockerfile:4 DL3045 warning: `COPY` to a relative destination without `WORKDIR` set.
Dockerfile:5 DL3003 warning: Use WORKDIR to switch to a directory
Dockerfile:5 DL3016 warning: Pin versions in npm. Instead of `npm install <package>` use `npm install <package>@<version
>`
Dockerfile:7 DL3011 error: Valid UNIX ports range from 0 to 65535
PS C:\Users\frjaraur>
```

Figure 13.2 – Local Hadolint installation reviewing a simple Dockerfile

But the good thing here is that we can include this linter, or any other, inside container images and have a collection of linters ready to use for any language we might encounter. Let's see how this works within a container using docker run -i hadolint/hadolint hadolint -:

```
Windows PowerShell                                           ×    +   ∨                           —   □   ×

PS C:\Users\frjaraur> gc Dockerfile |docker run --rm -i hadolint/hadolint hadolint –
Unable to find image 'hadolint/hadolint:latest' locally
latest: Pulling from hadolint/hadolint
db4123164570: Pull complete
Digest: sha256:fff226bdf9ebcc08db47fb90ee144dd770120b35c2b1cbbb46e932a650cfe232
Status: Downloaded newer image for hadolint/hadolint:latest
-:1 DL3006 warning: Always tag the version of an image explicitly
-:2 DL3015 info: Avoid additional packages by specifying `--no-install-recommends`
-:2 DL3009 info: Delete the apt-get lists after installing something
-:2 SC2154 warning: node_verion is referenced but not assigned (did you mean 'node_version'?).
-:4 DL3045 warning: `COPY` to a relative destination without `WORKDIR` set.
-:5 DL3003 warning: Use WORKDIR to switch to a directory
-:5 DL3016 warning: Pin versions in npm. Instead of `npm install <package>` use `npm install <package>@<version>`
-:7 DL3011 error: Valid UNIX ports range from 0 to 65535
PS C:\Users\frjaraur>
```

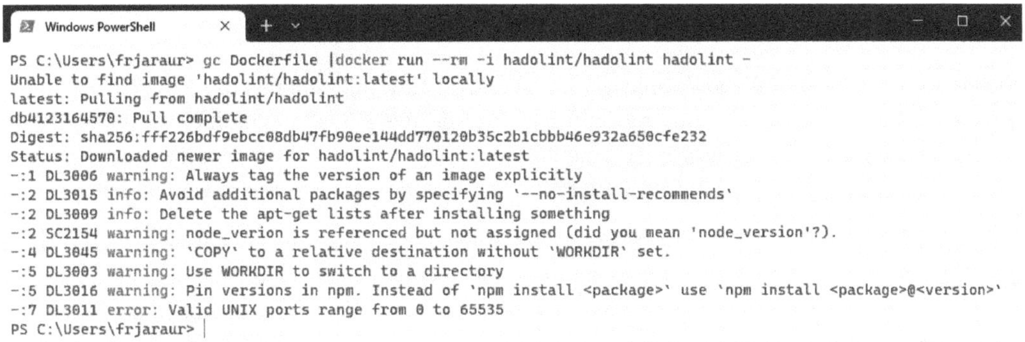

Figure 13.3 – Docker-based Hadolint execution reviewing a simple Dockerfile

> **Important note**
> There are tools such as **Conftest** (`https://www.conftest.dev/`) that can be integrated with different **Infrastructure as Code** (**IaC**) solutions and used to validate infrastructure scripts before they are deployed in our platform.

Linting tools can be executed automatically within our development processes to improve security. We will see this in action when we talk about CI/CD workflows.

In the next section, we will introduce simple methodologies and practices to learn how CI can help us manage the life cycle of our applications.

Understanding CI patterns

CI refers to the practice of automating the integration of code changes from multiple contributors (or even multiple projects) into a single project. These automated processes may happen once a day or several times per hour. We can consider CI as the part of the software supply chain where we build our application (or its components) and launch different tests before moving to production. The second part of this process is deploying the application or its components into production, although some intermediate environments can also be employed to test the quality of the solution or certification in special circumstances (for example, integrating our solution with a third-party solution release from a vendor).

In this section, we are going to review some of the most common patterns used for CI in the most intuitive logical order. Developers should always get the last version of their code to start developing a new feature or start over the creation of a new component, or new release with fixes. Therefore, we will start our development process by pulling the code from a **version control system** (**VCS**).

Versioning the code

A VCS is a tool that stores file and directory changes over time, allowing us to recover a specific version later. This tool is crucial from the developer's perspective as it allows multiple developers to work together and track the changes made to application code over time. Versioning the code and the artifacts created allows us to run specific integration tests and deploy a specific release of our code.

These tools usually run in *client-server* mode. Users interact using the relevant commands to push and pull the changes. The VCS stores and manages these changes. Once all the changes are synced (committed), you can proceed to build your applications' artifacts. This step may not be necessary if you are using an interpreted scripting language, although some bytecode artifacts may be created to speed up the application's execution. We can automate this process and trigger a compilation of our code in certain circumstances – for example, when we do a commit (synchronization of the code). As a result, we get a **binary artifact** with all its dependencies every time we simply commit our code. But we can go further and create different branches on our code repository to allow different users to interact with the code at the same time or solve different code functionalities. Once all required changes are made, we can consolidate these branches into a common one and build the artifact again. Depending on the final product, the issues found, and the functionalities required, this process can be complicated, but automation can be used to create a common workflow that is much easier to follow and reproduce.

When a project is developed by multiple developers or teams, certain types of management are required to avoid collisions between changes. VCSs offer mechanisms to resolve incompatibilities between different pulls when multiple developers change the same files at the same time. **Change management** is required to validate and merge the changes made by different users to the same parts of the application's code. These validation procedures significantly speed up the code workflow without us losing control of the code itself. Version control also allows us to go back and forward through the application's changes, and as such, is very useful even if you are the only developer of a project. But none of these mechanisms work if we don't define **standardized methodologies** and **code conventions** to minimize code integration tasks. It is common practice to ask developers to write down useful descriptions for their commits and follow the branch and release naming conventions standardized in your organization. For example, for releases, it is common to follow the MAJOR.MINOR.PATCH versioning syntax, where MAJOR indicates changes that may break compatibilities with previous releases, MINOR indicates that some functionality was added without breaking compatibility, and PATCH is used when some issues were solved without actually modifying any of the previous functionality. On the other hand, branch names can be used to reference any issues found and their solutions in the code.

At this early stage, we can add some **validation tests** using linters to ensure proper code syntax, code quality, and the presence of security features (such as valid external dependencies) and exclude any sensitive information that may have made its way into the code.

If we work with containers, our code should include at least one **Dockerfile** to allow us to create our container image artifact. Versioning of this file is also required, and thus it will be stored in our code repository (which is a VCS). Validation tests can be automated and executed to verify certain patterns such as the user executing the container's main process or exposed ports.

A CI pipeline, therefore, is a group of workflow processes intended to automate software application code validation, construction, and integration. Accordingly, let's quickly introduce the concept of DevOps here.

Introducing DevOps methodology

DevOps is a methodology that improves software engineering by integrating and automating some of the stages of software development, the operational tasks related to the operation and maintenance of the systems where the applications run, and the applications themselves. We should think of DevOps as a culture that goes *beyond* groups or teams in your organization; it applies to your entire organization with the goal of minimizing time and friction between the development, deployment, and maintenance stages.

The following list shows some of the key features of the DevOps methodology:

- Automate as many tasks as possible in the software life cycle

- Collaboration between different teams as part of this culture makes things work more effectively

- Continuous revision and feedback from tasks, automation, and code quality, all of which are key to improving the software development processes

- Monitoring and logging are part of the application life cycle, and they are important for improving its performance and finding code issues

Since DevOps covers a lot of tasks and disciplines, there are many tools available to help you with different tasks and stages. For example, for VCSs and code repositories, you can use very popular **cloud services** such as GitHub (acquired by Microsoft in 2018), Atlassian's Bitbucket, or GitLab, among others. If you are looking for **on-premise solutions**, you can use open source offerings such as Gitea, GitLab, or Azure DevOps Server. Choosing the right tool for your organization can be complicated because many tools offer multiple features. The following schema represents some of the more popular DevOps tools related to the application development stage, showing where they fit in best:

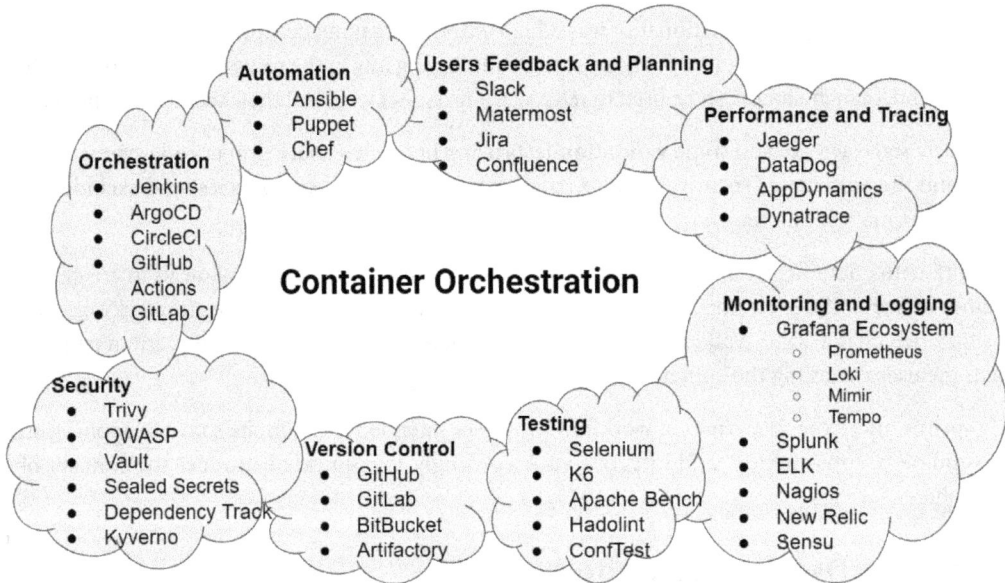

Figure 13.4 – Most popular DevOps tools

> **Important note**
>
> A new methodology was recently introduced that focuses heavily on the security of development, deployment, and maintenance processes, called **DevSecOps**. This methodology emphasizes an extension of security as part of the culture of the different teams involved in the process. This is why we reviewed shift-left security practices, which are an aspect of DevSecOps that lies closer to the development teams. A DevSecOps culture breaks the old mindset in which a singular team is given the security role and participates in the development process only at the end, validating the code just before the software is moved into production.

Once the code is synced and validated, we are able to build our software solution. Let's discuss this process next.

Building artifacts

Depending on the programming language and its dependencies, it may be tricky to prepare environments for different releases. For example, moving from one previous Node.js release to a newer one may require separate build environments, even if the language is interpreted and not compiled.

Imagine a situation where different code developers need to compile their software at the same time in the same environment. It would be complete chaos, and errors from different releases would appear. Automation allows us to package environments and build our software using the appropriate environment. But we can go further by using software containers because these environments need only to exist at runtime, specifically when required, and we can use software containers to build our software using the required builder environment. The resulting container images of the complete build process are stored in a container image registry right after the successful building and validation of the new artifact.

What is even more important is that we can prepare a full workflow in which all code is validated using our rules (code syntax, code quality, non-privileged execution, and so on), then the workflow triggers the build of the code, and finally, different tests (unity, integration, stress, performance, and so on) are triggered using the container images generated. We can forbid the execution of any application in production if it doesn't come from this standardized construction workflow. You as a developer are able to code on your laptop and test your application, but you must pass all the corporate validation checks on a shared environment or platform before actually deploying in production (or sometimes even earlier, in the quality or certification stages).

Testing your application's components

As mentioned before, automating different tests allows us to break the workflow whenever any test fails before moving on to the next step. To achieve this with containers, we can prepare some integrated processes using Docker Compose (for example) and validate how they work together. This can be done on your own desktop environment or using shared services, triggering the execution of the components by using defined tasks. These tasks also can be defined in Docker Compose format and be stored with your code. There are tools such as Jenkins that help us define these automated jobs and execute them on different systems. This tool is a very popular CI/CD orchestration tool created for managing build tasks on different systems that can be evolved to integrate the use of containers to simplify the overall workflow. Instead of having different nodes with separate releases for different languages or compilers, we can use software containers executed on a unique container runtime.

Monitoring the build processes and tests

To understand how changes can improve or have a negative impact on our applications, we need to continuously measure the performance and output of the different tests. We must always ensure we monitor the workflow processes because this will help us to improve the overall development process thanks to the iteration of the different tests. Popular CI orchestration tools always measure the build time, and we can retrieve the time spent during the execution of chained jobs, hence we will be able to trace how a certain change in our code (for example, the addition of new dependencies) impacts the build and modify the tests accordingly.

Sharing information about the development process

DevOps culture is all about communicating changes, exchanging feedback, and sharing information about any issues that arise to align all the teams involved in the process. Automation will avoid many misunderstandings; everything should be reproducible, hence the same results will be expected if we don't change anything. All changes must be traceable to allow us to quickly identify issues related to any given change and apply the appropriate patches. As we saw in *Figure 13.4*, there are many tools available to assist us in keeping our teams informed. One good practice is to implement automatic notifications sent by the different tools whenever a development task is executed (code changes, validated tests, and so on).

Excluding configurations from code

Although it might be obvious, we should keep any configuration or sensitive information for the application out of the code. It would be nice to include a set of default values and some documentation covering how to change them, but keep in mind that your application will pass through several phases and maybe different environments. In this book, we have looked at multiple mechanisms used to include sensitive information and configurations within containers (*Chapter 2*, *Chapter 4*, and *Chapter 5*). Never include certificates, even if they are just required for a simple step in which you download some artifact from a self-signed or corporate server. It is important to understand that sometimes, it is even necessary to use versioning for configurations. If you change the way you use a variable in your code, it may break a rollback to a previous release. In such cases, you may also need to store configurations in the versioning system. But keep in mind that the audience of this repository is probably different from the repository that stores your code. Automation helps us to keep track of the different code releases with the appropriate configurations and ensure that every task runs smoothly.

Now that we have reviewed the first part of the development process, where the application is coded, compiled, and validated, we can move on to the delivery stage.

Automating continuous application deployment

In this section, we are going to examine the second part of the software development process – the delivery of the product. Many organizations invest all their efforts into CI, leaving the determination of whether or not software should be executed in production to a manual decision.

A CD pipeline gets changes from the artifacts and code repositories, including required configurations, and deploys them into production in a fluent and continuous way. To achieve this, we need to somehow package all these artifacts and configurations in a reproducible and deployable state, aiming to keep the maximum stability and reliability in our systems. The following list shows some of the most notable benefits of using CD:

- We mitigate the risks of deploying new releases because automation ensures a quick rollback in case something goes wrong

- Automation may use blue–green and canary deployments, enabling new application releases while older processes are still serving

- Lower **time to market** (**TTM**) and reduced costs can be reliably expected due to the level of confidence generated by the application's life cycle

While CI automates the build and testing stages, CD on the other hand continues the process and goes a step further, automating the packaging, deployment, and testing throughout the rest of the life cycle.

While the benefits of CI are for developers, we might think that CD is more targeted at operations teams. However, in the DevOps culture, many stages are shared between the two groups. The major benefits of using CD extend even to the end users because applications are always kept updated and don't suffer outages between changes. Additionally, new functionalities can be added with less friction. Users can provide feedback using the defined channels (see the tools presented in *Figure 13.4*), and monitoring, logging, and tracing the software allows us to enrich this feedback, and then the cycle starts again to keep improving the application's code.

If we give some thought to how can we implement the different stages of CD automation, containers fit perfectly as we can package container images and the application's configurations for different environments and deploy the software solution. In case of errors, container runtimes provide **resilience**, and container orchestrators allow us to roll back to the previous release in seconds, informing us of the issues encountered during deployment. As mentioned before, blue–green and canary deployments allow us to progressively deploy a new release or test it with just a few users to avoid a massive outage if anything goes wrong.

> **Important note**
>
> Modern application life cycle models, such as **GitOps**, manage the deployment of software releases by defining a repository as the **source of truth (SOT)**. Any change within our applications or even the Kubernetes clusters themselves are managed as out-of-sync situations, requiring either manual intervention or automatic triggers to apply the appropriate changes and synchronize the situation with the required configuration (SOT). In such scenarios, we will just customize how the deployment packages will be executed on each environment by setting a required state for the application. Resource upgrades or rollbacks will be executed to synchronize the current status with the required one.

Monitoring the actual performance of the new deployment is key in situations where you are limiting access to a new release while most users are still using the old one. Should we go further with our new release, we must have a reliable **performance baseline** to fully understand how the changes are impacting our application's services. Although we may have passed all our performance tests successfully, deploying a new release may show different behaviors when accessed by real users. The better the tests in the testing stages, the lower the gap between the real user experience and the automated test, which lowers the risks of releasing a new version.

Logging is also important. We use logs to search for well-designed error patterns. The log standardization in your corporation can be used to easily implement common patterns for all your application's components and provide a single logging control plane for all processes at once, which will make it easy to find errors across multiple logs and verify how some requests affect different components at specific time frames.

Tracing in production is *not* recommended unless you have some dedicated instances of your project for that purpose or you are reviewing a critical error.

Retrieving **user feedback** is the final step of the complete application's life cycle. User feedback, alongside monitoring and logging (and eventually tracing) of the application components, feeds into the next iteration of the application's life cycle process to improve its overall performance and behavior, and the process starts over.

We examined some open source monitoring, logging, and tracing tools back in *Chapter 12*, *Gaining Application Insights*. To get users' feedback, any ticketing software will be fine, but the smoother it integrates into the full DevOps paradigm, the better. In *Figure 13.4*, we showed some of the most common and popular DevOps tools. All serious code repositories include an **issue tracking system** with which you can align users' comments and issues with actual code commits, solving these issues or adding requested functionalities.

As you can imagine, some of these tools can be deployed on and integrated into Kubernetes. The next section presents a sample DevOps environment in which we will use some of the tools presented in *Figure 13.4* to provide a full application life cycle management platform.

Orchestrating CI/CD with Kubernetes

This section will help us understand the full life cycle of an application prepared and managed within a Kubernetes cluster. Let's start by reviewing the CI part.

Understanding the CI component of the workflow

The CI part of our workflow is where we code, build, and test our solution.

At the beginning of the project, user requirements are collected. Subsequently, during the development stage, you can use your favorite code editor. Depending on the programming language you use, compilation may be necessary, which requires you to have installed compilers. Instead of that, you can use software containers to run the actual compilation steps. You will be able to use different releases of code compilers, with different environments and sets of tools at the same time without actually having to install any of them. Indeed, managing multiple releases of certain code environments on a single computer can be tricky. Building your application's code using containers will help you decide which container images would best fit your needs for each stage (building the application's artifacts, and running them for either testing or production).

Next, in the CI workflow, you build your binaries and prepare the Dockerfiles for your application's components. Multiple Dockerfiles can be created for a single component, specifying things such as the inclusion or omission of some debugging tools or flags that could be very useful during the testing stages.

Then, you build your container images. Production images must be clean and only include the binaries and libraries required for running your application's processes. You can build your code artifacts and container images for testing the application in your coding environment, although you may already have a shared environment for such tasks.

With containers, it becomes possible to locally test each application component (unit tests) or even the full application stack (integration tests) using Docker Compose. In such a case, you will need access to the other application components' container images and some mock configurations that will help you run a sample environment more easily. It's usual to include some mocked-up default values and perhaps some test connection strings, authentications, and tokens (which will be overwritten during execution with real values). Having *sample values* is key when you work in a team, and other developers may need to execute your artifacts and adjust their parameters to meet their needs.

You are likely to work on a specific branch of the code depending on the development stage you are in. Code branches are usually used to either fix issues or develop new functionalities and allow multiple developers to code in parallel on different resources at the same time. Once a given issue is solved and tested successfully, the code can be committed, pushed, and finally merged into the main code.

You may have your own code routine, but chances are it is quite similar to the one described here (the order of steps may vary, but ultimately the main code should contain your changes), and you probably apply similar steps for adding some new functionality or fixing an issue.

Pushing the new code to the code repository will trigger automation mechanisms that create appropriate artifacts (binaries, libraries, and container images) using tags and labels to help you track the changes associated and the issues or functionalities included. This allows you to either use your own built artifacts or those created by the automation system using your build rules and the Dockerfiles included in your code. It is recommended to use the artifacts created by the automated build environment because your DevOps team has most likely created a full supply chain, and this step is just the beginning of a longer process in which they will use these automatically created artifacts.

Code repositories will probably run on top of Kubernetes in your on-premise infrastructure, although you could use SaaS services instead. Depending on the integrations required for the different steps, it may be difficult to fully integrate cloud solutions with on-premises tools without taking on risks such as having certain data center credentials stored on your cloud platform (integration from cloud repositories to your on-premises Kubernetes clusters, for example). You should always ensure minimal required privileges for all your platform integrations, no matter whether they run on the cloud or your own data center.

Once your code is pushed to the code repository, different triggers can be configured to first validate the quality of your code, the maturity and security of the dependencies included in your project, and the security itself of your code and built binaries. For these tasks, the different tools presented in *Figure 13.4* can be used. For example, we can configure some container images with programming language linters and rules, and execute containers injecting our code for its validation. The process can be stopped whenever any test isn't passed or just inform us at the end of the check about some minor or major improvements we can make to our code. These tasks can be configured as jobs in our favorite CI/CD orchestration environment, probably also running on Kubernetes to leverage the availability of the cluster container runtimes. Some of the most popular CI/CD orchestrators are presented in *Figure 13.4*, but many advanced code repositories include task management functionality of their own, which simplifies the number of tools required for running our complete CI/CD workflows. For example, we can use GitLab for storing and versioning our code, storing and managing our artifacts (built artifacts and container images), and executing different CI/CD tasks. We will see platforms such as this in action in the *Labs* section with a full example.

As mentioned before, consecutive validation tasks (tests) can be triggered, and as a final step, we can build a container image ready for production. At this time, new tests can be executed for testing the integration of the new component release with other application components and validate the performance of the solution. Depending on the required integrations, this pipeline (that is, the definition of the different concatenated tasks to be executed) can be complex. It is usually recommended to group tasks and prepare the output of the different processes involved to provide easy-to-read reports. Most of the tools mentioned in the validation group of *Figure 13.4* provide summary reports that can be parsed to find any errors that should stop the workflow. The tasks associated with the pipeline can be executed within containers (isolated Pods on Kubernetes), and their logs should be available in the CI/CD orchestrators as these containers will be volatile.

Depending on the complexity of the application, it might be worthwhile to package the required components before the tests. You probably wouldn't execute simple manifests in your Kubernetes environments, and you would use Helm charts or Kustomize to create packages for either your full application or each component.

> **Important note**
>
> Some tools such as **Argo CD** can use Helm charts as templates for application deployments. Although we do not really deploy our application using a Helm chart, it will be used by the process to manage and manipulate the Kubernetes resources associated with your application. That's why it is always worthwhile preparing your applications as packages: it allows someone else to easily deploy your full application or some components therein without really knowing the contents back to front.

Before we continue, let's see some of the most important features of Helm and how to create a simple manifests package.

Using Helm to package our application's resource manifests

Helm is a tool that packages Kubernetes resource manifests using templated YAML files and automation scripts that allow us to completely configure and deploy applications using a simple command line and a configuration file.

Using Helm charts, we can replace all application resource manifests at once or only those that were changed, with a simple path to roll them back to a previous release at any time. Helm keeps track of all the changes made to a Helm instance and is capable of reverting those changes by applying a previously stored release version.

When we execute `helm create <NAME_OF_THE_CHART>`, Helm creates a directory structure that contains some example manifests and other files used to create a new Helm chart package:

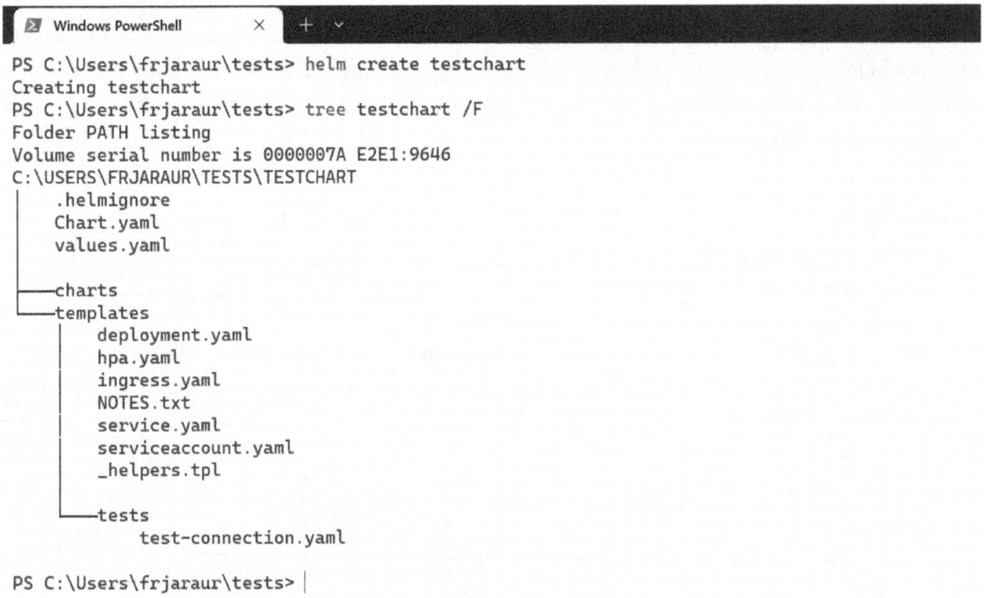

```
Windows PowerShell        ×     +  ⌄

PS C:\Users\frjaraur\tests> helm create testchart
Creating testchart
PS C:\Users\frjaraur\tests> tree testchart /F
Folder PATH listing
Volume serial number is 0000007A E2E1:9646
C:\USERS\FRJARAUR\TESTS\TESTCHART
    .helmignore
    Chart.yaml
    values.yaml

───charts
───templates
        deployment.yaml
        hpa.yaml
        ingress.yaml
        NOTES.txt
        service.yaml
        serviceaccount.yaml
        _helpers.tpl

    ───tests
            test-connection.yaml

PS C:\Users\frjaraur\tests>
```

Figure 13.5 – Helm chart file structure

In this code snippet, we used `helm create` to create a Helm chart tree structure. You may have noticed the existence of the `charts` directory. A Helm chart can contain other Helm charts as dependencies. This way, we can create an **umbrella chart** that includes all the Helm charts required to fully deploy an application with all its components. We could, for example, create a chart for the database component and other charts for the backend and frontend components. All these charts can be included inside an umbrella chart that defines the values required for deploying the full application. Each chart should contain a mocked values file, with example or real values that can be used to deploy it. We will write down and use a customized values file for deploying the application by overwriting the default values included on each chart. The `Chart.yaml` file describes the dependencies of your package and its version. You will find two versioning properties in your `Chart.yaml` file:

- The `version` key, which indicates the package release number
- The `appVersion` key, which is used to identify your application release

> **Important note**
>
> Helm charts can be uploaded to repositories for storage and to share with other users. This way, your applications can be deployed by anyone authorized to pull and execute the Helm chart containing the manifests. Many vendors and open source projects offer their Helm charts as a way to deploy their applications, and some community-driven repositories host thousands of charts ready to use in your projects. Two of the most popular repositories are *ArtifactHub* (`https://artifacthub.io`) and *Bitnami Application Stacks* (`https://bitnami.com/stacks/helm`).

The magic behind the management and composition of some key variables, such as the instance name, is included in the `_helpers.tpl` file, and the composed variables will be used in all the YAML manifest files included within the `templates` directory. We will include all the manifests required for our application or its components to work. All the PersistentVolumeClaims, Deployments, StatefulSets, DaemonSets, Secrets, and ConfigMaps should be included. Indeed, if our application requires specific permissions, we must also include ServiceAccounts and the appropriate Role and RoleBinding manifests. The `values.yaml` file included by default is used to validate the manifests that will be created with the `helm` command with a set of default values. This is another validation test that can be included in our pipeline just before the creation of the Helm chart package. If this `values.yaml` file implements all the required values (a mocked version), the pipeline process can continue and create the Helm chart package. The Helm chart's files should also be managed using a versioning system; hence, we will store them in our code repository. Whether or not to use a different repository depends on you as a developer, but it would be nice to manage different releases for the application's components and the Helm charts that deploy them, and it will be easier if we use different repositories.

In the *Labs* section, you will work through a full example using the `simplestlab` application. We prepared a Helm chart for each application's component and an umbrella chart that deploys the full application.

Let's summarize the steps described so far before continuing with the rest of the pipeline chain that describes the application's life cycle:

1. Write your code and push it to the code repository. Our code should include at least one Dockerfile for building the container image or images for the application's component. Although it is not required, it is recommended to maintain a separate code repository for storing your Helm chart files. This way, you can follow the same code workflow for both the application's code and the Helm chart's code, but isolating each repository allows us to manage a different release for the code and the Helm chart's package.

2. Code will be validated using the relevant linters to verify its quality, its compliance with your organization's coding rules, its dependencies, and its inner security (do not include sensitive information unless it is mocked).

3. Different artifacts will be created and stored in your repositories. When your code is built, the resulting artifacts (binaries and libraries) will be stored (in our example, in GitLab). Storing artifacts is important if they are shared between components, such as binaries and client libraries, for example. Container images are also stored in GitLab as it additionally provides image registry capabilities. You can use a different repository for each type of artifact, but GitLab is a good catch-all solution because it offers storage for code, artifacts, and container images.

4. When all the artifacts (the build and container images) are created, we can either automate the execution of the unit tests or pull the resulting release images (with fixes or new functionalities) and test them on our development computer, or even do both.

5. Integration tests may require packaging the application's components. If this is the case, validation of the Helm chart code will be triggered, and then the package will be created. Sometimes, we

just change the application's container image (that is, we change some code, which triggers a new artifact build and a new image is created) without actually changing the application's Helm charts. That's why it is always useful to keep track of Helm chart package template changes in a different repository from the application's code. You may need to upgrade your application's code without changing the templated deployment manifests. Here, we would just need the customized values for deploying a new container image and the `appVersion` key on your `Chart.yaml` file. This is a good practice because you will be able to track your package and application release at the same time.

6. Once the container images are created and stored correctly in the images registry, and the Helm chart packages are created, the application is ready to be deployed. Additional vulnerability tests can be triggered using the container images. Some tools such as AquaSec's Trivy use a **bill of materials** (**BOM**), which is a list of all the files included in all the container image layers, and search for known issues using both their own and internet-based vulnerability databases.

Let's continue now with the second part of the pipeline. As you can see, we usually refer to the complete CI/CD workflow because CI and CD are often concatenated automatically one after the other.

Adding CD to the workflow

Different integration and performance tests can be executed by using the container images directly using Docker Compose or Kubernetes manifests, or via the Helm chart packages, which provide a more customizable solution.

The CI/CD workflow continues with the tests, as follows:

1. Deployments for the different tests are triggered using the custom value files stored in the code repository. It is important to understand that we should never store sensitive data in clear text in our code repositories. Instead, use solutions such as HashiCorp's Vault or Bitnami's SealedSecrets to store sensitive data under encryption. Both solutions enable data decryption during the deployment stages.

2. The application's performance and workflow task metrics can be integrated into your favorite dashboard environment. Most of the tests in this stage provide helpful summaries of the validation tasks executed, with which we can get a good overview of the impact of newly added changes. Logs will highlight any errors from either the tasks or the application's processes. We should separate these into different dashboards because they will probably have different end users.

3. Once all the tests are passed, we are ready to deploy the new release in production. Whether or not to automatically trigger this process depends on how your organization manages the changes in production. If your applications are governed using a GitOps model, use your configurations repository as the SOT, and the CI/CD orchestrator will push the changes into the Kubernetes platform. The current state of the application's components may necessitate an upgrade to a new release or a rollback to a previous version to synchronize the desired state of the application. This model allows you to manage all your applications by changing their deployment configurations.

> **Important note**
>
> The **GitOps** model extends the use of repositories to improve the tracking of infrastructure and application changes by using custom values repositories as a **single SOT (SSOT)** to trigger the delivery process. We can include automation for requiring specific security configurations, solving application or infrastructure dependencies before they are deployed, or any other requirement for the applications to work. All changes made to code and the values used for deploying the applications are tracked, making updates and rollbacks easier than ever.

4. Automating the deployment of our applications requires access and authorization to our Kubernetes environment. We include the required credentials for a deployment user in our CI/CD platform. We can use Argo CD to implement a simple GitOps working model. This way, a simple change in the custom package parameters will trigger the deployment of a new release using updated manifests. As a result, the new application release will be delivered with the given fixes or new requested features.

5. The new release deployed will be kept in the maintenance stage until a new one is released to replace it. Monitoring the application and retrieving and analyzing feedback from the users will end this iteration. The process will start over, with the team planning the implementation of newly requested features and fixes to issues not yet solved in the latest release. The following schema represents the workflow presented in the preceding bullet points:

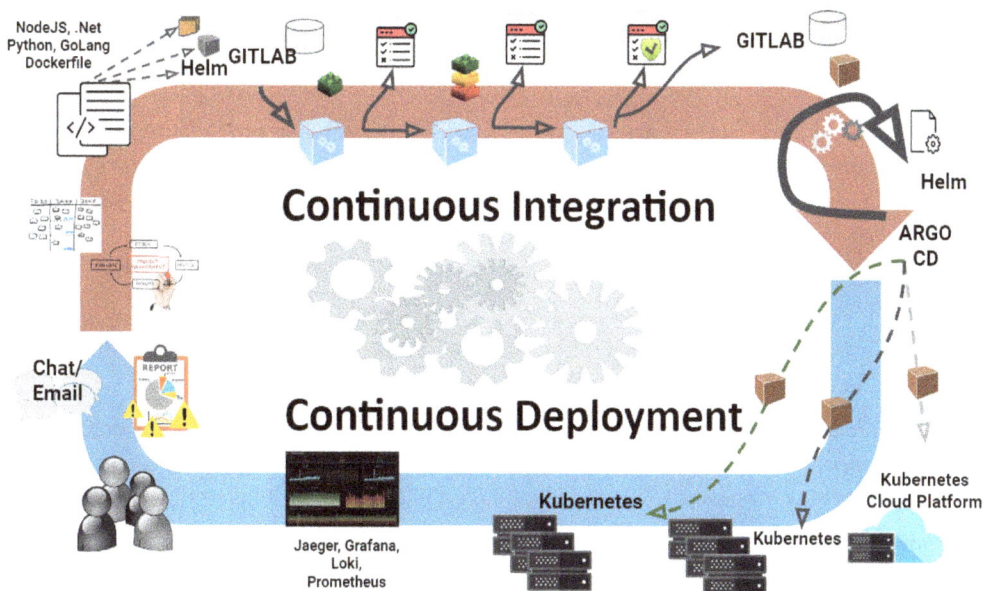

Figure 13.6 – Schema of the workflow followed to deliver a new application release

We will now review some of the aforementioned stages in the following *Labs* section, using a GitLab platform deployed on a Minikube desktop environment.

Labs

In this lab, we will reproduce a very simplified supply chain using automation and the GitOps deployment model by installing and configuring GitLab and Argo CD in a test environment for building, testing, and deploying the simplestlab application. You can use a fully working Kubernetes platform (on the cloud or on-premises) or a simplified Kubernetes desktop environment. The fully detailed steps of the process are explained in the GitHub repository of this book, in the Chapter13 folder, but here is the summary of the processes and some notable configurations you will find there:

1. First, we will prepare our environment with the tools required for the lab (Helm, kubectl, and the Argo CD CLI), and we will also use some environment variables for easier configuration of the Ingress resources and CA certificates for each application.

2. You will find complete Helm charts for the simplestlab application, alongside some value configurations for deploying the application. The specific values used in this file will depend on your environment, and we have provided an explanation to help. You can test and deploy the Helm charts using local configurations.

3. We will deploy and use GitLab to store all the application code, Helm charts, container images, and application configurations. Steps to create groups, subgroups, repositories, and required users are included.

4. The code and Helm charts folders included in the Chapter13 repository come with a .gitlab-ci.yml file that describes and prepares CI automation to validate our Dockerfile using **Hadolint** (a Docker linter) and finally build our image using **Kaniko** (a tool to build container images from a Dockerfile inside a container or Kubernetes cluster). This tool doesn't depend on the Docker container runtime and executes each command within a Dockerfile completely in user space, which is great for security. This way, we can build images inside any standard Kubernetes cluster.

5. We will use git commands, different branches, and tags to trigger the different automations included in the example pipeline for the code and the Helm charts.

6. The automation creates dev and release images using different container image tags. Development images will be added to the code repositories, but the release images will be considered ready for production and will be stored in a separate container images repository.

7. Helm charts are created using an umbrella structure; hence, the simplestlab chart deploys all the components at once. This chart includes dependencies for different applications' components, and these dependencies should be solved before it is deployed. We will see how this works with a local example and then automate the Helm chart creation.

8. Argo CD provides the CD part. While GitLab can be used to deploy directly on your Kubernetes cluster, Argo CD works by following the GitOps model. We will configure Argo CD to review

any change in the `values` repository, and it will deploy the application using the resources stored in GitLab (container images, Helm charts, and the file with the values required for deploying the application). We will give you a brief discussion of the steps included in this lab and recommend you follow the full description written in the `Chapter13/Readme.md` file.

We have prepared for you three main directories:

- `ArgoCD`: Contains the installation of the Argo CD component and the Application resource we will use to deploy our `simplestlab` application

- `GitLab`: Contains the installation of GitLab components

- `simplestlab`: This directory contains all the code, Helm charts, and values used for deploying a `simplestlab` application instance

We will need the following tools in our environment:

- **Minikube** (and Docker, if you follow my steps using a fixed IP for the lab)

- **Git**

- **Argo CD CLI**: To create and modify the integration of Argo CD inside our Kubernetes cluster

- **OpenSSL or Certutil on MS Windows**: To decode some `base64` strings in case you don't have `Base64`

- **Helm**: To deploy your Helm charts

- `kubectl`: To connect to our Kubernetes cluster

- `Base64`: For decoding some strings

Detailed steps for installing these tools are included in the code repository. We will start the lab by setting up a Minikube environment. We will use Linux and Docker for running this environment to be able to set up a fixed IP address. This will help you in case you decide to take your time for the lab and start and stop the Minikube environment without changing the setup. Follow these steps:

1. Start `minikube` using the following command line:

    ```
    Chapter13$ minikube start --driver=docker \
    --memory=8gb --cpus=4 \
    --addons=ingress,metrics-server --cni=calico \
    --insecure-registry="172.16.0.0/16" \
    --static-ip=172.31.255.254
    ```

2. We will now prepare the `simplestlab` application using a directory to avoid conflicts between different Git environments because you downloaded this repository from GitHub:

    ```
    Chapter13$ cp -R SimplestLab Simplestlab_WORKSPACE
    ```

3. Remove all the hidden `.git` folders in the `Simplestlab_WORKSPACE` folder and subfolders (if any) every time you start with the lab. We will use these folders to push some code changes inside `Code/simplestapp`, push Helm charts included in the `HelmCharts` directory, and push deployment values included inside the `Values` folder.

4. Install GitLab following the instructions included in the code repository. We have prepared a setup script to help you customize the values file for deploying GitLab using Helm. The chart is included under the `chapter13/GitLab` directory.

5. Once it is installed, we will review the secret created with the credentials and log in to the GitLab web UI, published at `https://gitlab.172.31.255.254.nip.io`.

> **Important note**
> We used the `nip.io` domain to simplify all the qualified domain names for your environment. You can read more about this simplified domain at `https://nip.io/`.

We include our GitLab environment inside the Minikube setup to allow Kubernetes to download images. Complete steps are described in the GitLab repository.

6. We will then install Argo CD using a setup script and Helm. The script will customize a values file for your environment, and we will use it to deploy Argo CD using the Helm chart included in the `Chapter13/ArgoCD` directory.

7. Detailed steps are provided in the code repository. Once installed, you will be able to access Argo CD at `https://argocd.172.31.255.254.nip.io`. You will use the admin user with the password obtained from the deployment secret, following the procedure described in the code repository.

8. We will then upload the code included in `SimplestLab/Code` directory to GitLab. But first, we will create a user (`coder` user) with developer privileges in GitLab. This user will be used to pull and push code only, without privileged access. Steps for creating this user and the different projects for managing the code, Helm charts, images, and the values for deploying the application are described in the code repository.

> **Important note**
> Different permissions will be declared for different projects in GitLab. We have simplified the environment, setting up some projects as `Public`. Follow the instructions detailed in the `Chapter13` repository.

9. Using this `coder` user, we will push the code for the `simplestapp` component, included in the `Chapter13/Simplestlab/Code/simplestapp` directory, to our GitLab instance.

10. Automation of a Docker image build is triggered thanks to the existence of the `.gitlab-ci.yml` file in our code repository. This file describes the automated process and steps for verifying and building a custom image using our code. We included three stages in the file:

- `test` (which basically validates our Dockerfile syntax)
- `security` (which reviews the content of the files to be included in the image before it is built)
- `build` (using Kaniko instead of Docker to improve security, avoiding the need to use the Kubernetes host's Docker or `containerd` engine)

The process is described in detail in the `Readme.md` file included in the code repository.

> **Important note**
> To avoid the need to add the GitLab environment SSL certificate to our client environment, we will configure Git to skip SSL verification (steps are included in the code repository).

11. This automation will use the following variables for executing the tasks defined in the `.gitlab-ci.yaml` file:

- PROJECTGROUP_USERNAME: coder
- PROJECTGROUP_PASSWORD: C0der000
- LABS_LOCAL_GITLAB_CERTIFICATE: Complete GitLab TLS certificate chain Base64-decoded value, obtained using the following command:

```
kubectl get secret -n gitlab -o yaml gitlab-wildcard-tls-chain
-o jsonpath='{.data.gitlab\.172\.31\.255\.254\.nip\.io\.crt}'
```

These variables should be included following the steps described in the repository in GitLab.

12. The GitLab automation file will trigger two types of image construction processes:

- **Dev images**: These images will be built and included in the `Code` project repository
- **Release images**: These images will be built and included in the `Images` project repository

We will create `dev` and `main` code branches, change some code, push it to GitLab, and switch between branches to see whether changes will trigger the build process or not. Once we are ready to build a release image, we will tag the commit with a release name, push it to GitLab, and verify how the automated pipeline will create the appropriate release image inside the image project in GitLab. Described steps for these tasks are included in the `Chapter13/Readme.md` file. Please follow them carefully, and review the pipeline results and files generated during the process in the different GitLab projects (`Code` and `Images`). Get familiar with the processes before continuing with the next step, in which we will push and build the Helm charts for deploying the different applications' components.

13. We will now manage the Helm charts' code files and their associated projects' repositories. We set up for you three Helm charts, one for each component (`simplestlab-db`, `simplestlab-app`, and `simplestlab-lb`), and one umbrella chart that will include the others as dependencies. Therefore, four project repositories must be created:

- `simplestlab`: This chart defines the umbrella Helm chart used to deploy all components at once and its Ingress resource. We didn't add any Ingress resource on any other component.

- `simplestlab-app`: Describes the application backend component Deployment resource deployment.

- `simplestlab-db`: Describes the database component StatefulSet deployment.

- `simplestlab-lb`: This describes the load balancer DaemonSet deployment.

This project should be `Public` in this demo because we will not declare any credentials in Argo CD. You will use credentials and `Private` repositories in your production and development platforms, but this will definitely require more configurations for this demo environment.

> **Important note**
>
> The `simplestlab` umbrella chart depends on `simplestlab-app`, `simplestlab-db`, and `simplestlab-lb` charts. Whatever change you make to any of these projects requires a Helm chart dependencies update on the `simplestlab` umbrella chart. While you use the prepared CI/CD environment, you will need to run the `simplestlab` umbrella chart project pipeline again to rebuild these dependencies. If you want to manually update them, you will use a Helm dependencies update in the `HelmCharts/simplestlab` directory. We prepared various scenarios in the `Chart.yaml` file in case you want to test it locally (review the `Chapter13/Simplestlab/Values/simplestlab/values.yaml` file comments).

Once the Helm charts' project repositories are created, we can push the Helm charts' code into their GitLab-associated repositories. The code for the charts is located in `Chapter13/Simplestlab/HelmCharts`. Push each component's code to the appropriate repository.

14. We have included in the charts' code the `.gitlab-ci.yaml` file for GitLab automation. This file describes three stages:

- `test` (which validates the Helm chart using its own linter)

- `dependencies` (which validates the chart dependencies if any are declared)

- `build` (which packages the code into a Helm chart `.tgz` file)

> **Important note**
>
> We need to include two new variables, DOCKERHUB_USERNAME (with your Docker Hub username) and DOCKERHUB_PASSWORD (with your Docker Hub password). These variables should be defined in the HelmChart/SimplestLab umbrella chart only. This repository is Public, and anyone will be able to read your password, but you are using your own demo environment. You can secure this password by making it Private, but you will need to prepare some username authentication (new user or even coder user here) and include it in the Argo CD OCI repository.

15. The GitLab automation file will trigger two types of package construction processes:

    • **Dev packages**: These packages will be built and included in each HelmChart project repository.

    • **Release packages**: These packages will be built, included, and pushed to Docker Hub. We will just use this procedure to build the umbrella simplestlab chart.

    We will create dev and main code branches and verify the build process when we push code to GitLab. Steps for making some changes and pushing them to GitLab are described in the Chapter13/Readme.md file. The simplestlab umbrella chart will be pushed to Docker Hub, and we will be ready to use it, but first, we will need to add the values.yaml file to the Values project repository.

16. We will create a Simplestlab/values/simplestlab repository to manage a simple values file that will be used to deploy the simplestlab application using the simplestlab umbrella Helm chart. The file contains different sections:

    • simplestlab-lb: Defines the values to overwrite when deploying the simplestlab-lb Helm chart, added as a dependency in the umbrella chart.

    • simplestlab-app: Defines the values to overwrite when deploying the simplestlab-app Helm chart, added as a dependency in the umbrella chart.

    • simplestlab-db: Defines the values to overwrite when deploying the simplestlab-db Helm chart, added as a dependency in the umbrella chart.

    • **Rest of the file**: The rest of the definitions included will manage the behavior of the umbrella chart. We included parameters for deploying the Ingress resource here.

    We have prepared this file for you with comments for two different labs using Argo CD:

    • The first test will deploy a configuration with the wrong database service name for the App component (__dbhost: db__). The correct data is commented: __dbhost: simplestlab-simplestlab-db__. Thus, when you create the Argo CD application for the first time, the application component and the load balancer components will fail. Until you change the correct mentioned value in the values YAML file, this will not fix the problem in the load balancer component.

- The second test will deploy a new configuration that will fix the load balancer component by deploying a completely new `nginx.conf` ConfigMap. To make this happen, uncomment the `nginxConfig` key in `simplestlab-lb`. Indentation is key; uncomment all the lines (you can leave the `###################################` line).

When an Application resource is created in Argo CD, the synchronization with the different reports starts, and every time you change either the Helm chart package or the values file, the misconfigurations will be reflected in the Argo CD environment.

17. Create a `simplestlab` values repository (`Project`) inside the `Values` project, and push the file from `Chapter13/Simplestlab/values/simplestlab` into this new repository.

18. We will now integrate our application into Argo CD. We will use the Argo CD CLI to manage the integration of our Kubernetes cluster with Argo CD. To connect Kubernetes with Argo CD, create a ServiceAccount resource with cluster privileges to manage applications cluster-wide. Detailed instructions for integrating our Minikube Kubernetes cluster are included in the `Chapter13` repository. Follow these instructions, and then log in to Argo CD to create the following repositories:

    - **Code repository type:** `https://gitlab.172.31.255.254.nip.io/simplestlab/values/simplestlab.git` This repository will be used to integrate our values YAML file, used to run the umbrella Helm chart and deploy the full application. This repository requires authentication; we will use `coder` as the username and `c0der000` as the password.

    - **OCI type** (`registry.172.31.255.254.nip.io/simplestlab/helmcharts/simplestlab.git`)This includes the `simplestlab-chart` package.

    - **OCI type** (`docker.io`): This includes the `simplestlab-chart` package uploaded at Docker Hub as a workaround for an issue in Argo CD with self-signed certificates (`https://github.com/argoproj/argo-cd/issues/12371`).

    Screenshots are provided in the instructions to guide you through the setup process.

19. Once the repositories are created in Argo CD, we can create an Argo CD Application resource. The Argo CD GUI does not allow us to use multiple repositories, hence we will not be able to use a code repository for the values file and another one for the Helm chart package artifact. In these circumstances, we need to prepare the Application resource using a YAML file. We included a YAML file for you in `Chapter13/ArgoCD/Applications`. The `minikube-simplestlab.yaml` file includes both the values file repository (`https://gitlab.172.31.255.254.nip.io/simplestlab/values/simplestlab.git`) and the Helm chart repository (`docker.io/frjaraur`). If you have followed all the steps, you can use your own Helm chart repository. Mine is public, and you will be able to use it at any time. The `Applications` manifest includes the sources for deploying an application and the destination environment – the Minikube lab environment in our case.

We will create this new resource using `kubectl`:

```
Chapter13$ kubectl create \
-f ArgoCD/Applications/minikube-simplestlab.yaml
```

20. As soon as the Argo CD application is set and the repositories are available, Argo CD deploys the application for us. Review the Argo CD environment and verify the synchronization between the different repositories. Screenshots of the different environment views are included in the `Chapter13` repository.

> **Important note**
>
> We have included in the Argo CD Application resource the `simplestlab` namespace. This namespace should be created before the application is actually deployed.

21. Next, we change the database host lab. The first thing you will notice is that the application's App component does not work. This is due to the fact that the connection string is wrong (check the comments included in the `Chapter13/Simplestlab/Values/simplestlab/values.yaml` file). Change the `dbhost` key to `simplestlab-simplestlab-db` and verify the changes in Argo CD.

22. Verify the new name, automatically created by the Helm chart template (these names could have been fixed, but this is a common error and we can see how to solve it in this example):

```
Chapter13$ kubectl get svc -n simplestlab
NAME                         TYPE        CLUSTER-IP
EXTERNAL-IP    PORT(S)       AGE
simplestlab-simplestlab-app  ClusterIP   10.96.6.93      <none>
        3000/TCP    2d14h
simplestlab-simplestlab-db   ClusterIP   10.98.159.97    <none>
        5432/TCP    2d14h
simplestlab-simplestlab-lb   ClusterIP   10.103.79.186   <none>
        8080/TCP    2d14h
```

23. We now know the new name for the database server, and we can change the `dbhost` value in the `values.yaml` file:

```
envVariables:
    dbhost: simplestlab-simplestlab-db
```

24. Commit the new changes and push the file to our repository in GitLab using Git.

The changes will be shown on Argo CD in a few seconds. We haven't configured auto-sync, hence we will see a misconfiguration of the values (out of sync). Current values in the cluster are different from those expected by the configuration. We will just proceed to sync the application (screenshots are included in the repository). This will create a new Secret resource. We will delete the App component Pods, and the new changes will be applied to this component.

25. Once the first problem is solved, you will find a new error because the Loadbalancer component isn't able to reach the App component. So, next, we need to fix the Loadbalancer component. In this case, we will change the __nginx.conf__ file required by Nginx ____Lb____. It is included as a ConfigMap resource and managed by the ____nginxConfig____ key in the values file. We need to change the name of the application backend service (____App____ component). By default, it uses ____app____, as you can see in the default values file included in the ____simplest-lb____ Helm chart (SimplestLab/HelmCharts/simplestlab/values.yaml).

We first verify the name of the App component service:

```
Chapter13$ kubectl get svc -n simplestlab
NAME                               TYPE        CLUSTER-IP
EXTERNAL-IP    PORT(S)       AGE
simplestlab-simplestlab-app        ClusterIP   10.96.6.93      <none>
    3000/TCP    2d14h
simplestlab-simplestlab-db         ClusterIP   10.98.159.97    <none>
    5432/TCP    2d14h
simplestlab-simplestlab-lb         ClusterIP   10.103.79.186   <none>
    8080/TCP    2d14h
```

26. Next, we again review the Chapter13/Simplestlab/Values/simplestlab/values.yaml file. This time, you will need to uncomment the nginxConfig key. Please be very careful with the indentation as it may break the integration with Argo CD. If the application isn't synced, verify the values fail because it may contain some unexpected characters.

We uncomment the ____nginxConfig____ key value prepared for you. After uncommenting the value, you should have something like this:

```
# Second Test Update -- Uncomment this section
nginxConfig: |
  user  nginx;
  worker_processes  auto;
  error_log  /tmp/nginx/error.log warn;
  pid          /tmp/nginx/nginx.pid;
  events {
    worker_connections  1024;
  }
  http {
    server {
      listen 8080;
      location /healthz {
        add_header Content-Type text/plain;
        return 200 'OK';
      }
      location / {
```

```
            proxy_pass http://simplestlab-simplestlab-app:3000;
          }
        }
      }
```

27. We commit and push the new changes. Argo CD will show the changes in a few seconds, and
    we will sync the resources and delete the Lb Pod, associated with the DaemonSet, to fix the
    NGINX configuration issue. After the synchronization and removal of the Pod, the new Pod
    works fine, and Argo CD will show the application as healthy and synced.

We've now reached the end of this long and complex lab, but we divided it into different stages to
make it easier to follow. You can make changes to either your configurations, code, or Helm charts
and trigger pipelines or GitOps integration to manage your application status and behavior. We
can't explain in a single lab all the configurations we have done to make all the workflow work; we
gave you some tips that will help, and you can deep dive by yourself, exploring the already prepared
configuration and script steps.

It would be useful to follow the lab by including the NetworkPolicy resources created in *Chapter 11*
and the NGINX and Postgres Prometheus exporters prepared in *Chapter 12*. After the completion of
this lab, you will understand how the different automations work and will be ready to create your own
using any other popular DevOps tool because the basic concepts are the same, no matter whether you
use a cloud solution or deploy your DevOps tools in your own data center.

Summary

In this chapter, we described the life cycle of an application using software containers. We used most
of the content learned in this book so far to prepare a CI/CD workflow, while we quickly reviewed the
different stages involved in the creation of an application based on containers. We also presented some
of the most popular applications used by DevOps teams to implement and automate the complete
supply chain of an application and learned how to use them in the *Labs* section. This final lab showed
you the different stages involved in the life cycle of an application. We coded our application, prepared
our container images to use as our application's artifacts, and prepared Helm charts, which we used
to deploy the application in Kubernetes. Finally, we triggered the execution of the application in
the Kubernetes cluster using Argo CD to deliver the application after its configuration was done.
All changes will be tracked, and the automation and orchestration functionalities help us to deliver
changes quickly and reliably. You are now ready to employ the content of this book to create your own
supply chain or use one already created using other common DevOps tools. Best of luck preparing
and delivering your applications using software containers!

Index

‹packt›

Other Books You May Enjoy

If you enjoyed this book, you may be interested in these other books by Packt:

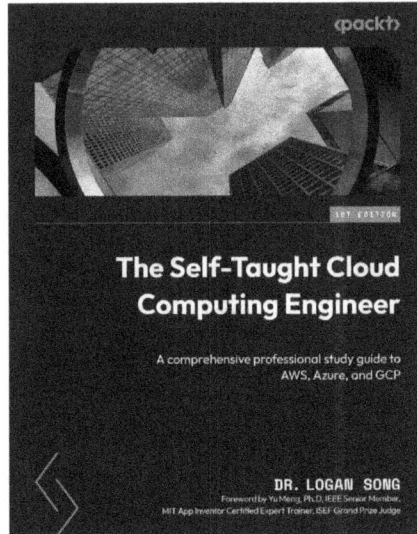

The Self-Taught Cloud Computing Engineer

Dr. Logan Song

ISBN: 978-1-80512-370-5

- Develop the core skills needed to work with cloud computing platforms such as AWS, Azure, and GCP
- Gain proficiency in compute, storage, and networking services across multi-cloud and hybrid-cloud environments
- Integrate cloud databases, big data, and machine learning services in multi-cloud environments
- Design and develop data pipelines, encompassing data ingestion, storage, processing, and visualization in the clouds
- Implement machine learning pipelines in multi-cloud environment
- Secure cloud infrastructure ecosystems with advanced cloud security services

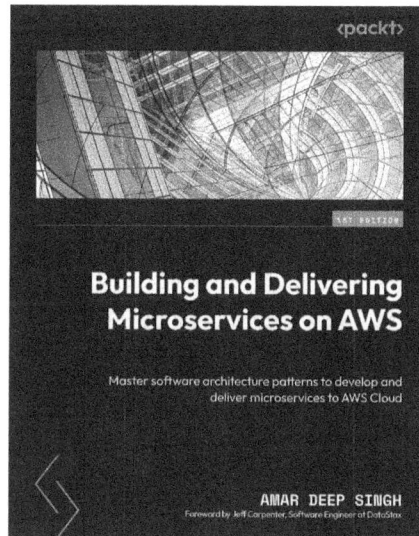

Building and Delivering Microservices on AWS

Amar Deep Singh

ISBN: 9781803238203

- Understand the basics of architecture patterns and microservice development

- Get to grips with the continuous integration and continuous delivery of microservices

- Delve into automated infrastructure provisioning with CloudFormation and Terraform

- Explore CodeCommit, CodeBuild, CodeDeploy, and CodePipeline services

- Get familiarized with automated code reviews and profiling using CodeGuru

- Grasp AWS Lambda function basics and automated deployment using CodePipeline

- Understand Docker basics and automated deployment to ECS and EKS

- Explore the CodePipeline integration with Jenkins Pipeline and on premises deployment

Packt is searching for authors like you

If you're interested in becoming an author for Packt, please visit `authors.packtpub.com` and apply today. We have worked with thousands of developers and tech professionals, just like you, to help them share their insight with the global tech community. You can make a general application, apply for a specific hot topic that we are recruiting an author for, or submit your own idea.

Share Your Thoughts

Now you've finished *Containers for Developers Handbook*, we'd love to hear your thoughts! Scan the QR code below to go straight to the Amazon review page for this book and share your feedback or leave a review on the site that you purchased it from.

`https://packt.link/r/1805127985`

Your review is important to us and the tech community and will help us make sure we're delivering excellent quality content.

Download a free PDF copy of this book

Thanks for purchasing this book!

Do you like to read on the go but are unable to carry your print books everywhere? Is your eBook purchase not compatible with the device of your choice?

Don't worry, now with every Packt book you get a DRM-free PDF version of that book at no cost.

Read anywhere, any place, on any device. Search, copy, and paste code from your favorite technical books directly into your application.

The perks don't stop there, you can get exclusive access to discounts, newsletters, and great free content in your inbox daily

Follow these simple steps to get the benefits:

1. Scan the QR code or visit the link below

https://packt.link/free-ebook/978-1-80512-798-7

2. Submit your proof of purchase

3. That's it! We'll send your free PDF and other benefits to your email directly

www.ingramcontent.com/pod-product-compliance
Lightning Source LLC
Chambersburg PA
CBHW081221220326
41598CB00037B/6851